6·19·80

THE GUINNESS BOOK OF
ASTRONOMY
FACTS & FEATS

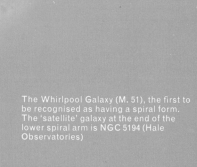

The Whirlpool Galaxy (M. 51), the first to be recognised as having a spiral form. The 'satellite' galaxy at the end of the lower spiral arm is NGC 5194 (Hale Observatories)

THE **GUINNESS** BOOK OF
ASTRONOMY
FACTS & FEATS

Patrick Moore

GUINNESS SUPERLATIVES LIMITED
2 CECIL COURT, LONDON ROAD, ENFIELD, MIDDLESEX

© Patrick Moore
and Guinness Superlatives Ltd 1979

Published in Great Britain by
Guinness Superlatives Ltd, 2 Cecil Court,
London Road, Enfield, Middlesex

ISBN 0 900424 76 1

Guinness is a registered trade mark of
Guinness Superlatives Ltd

Every endeavour has been made to trace and make
acknowledgement to the copyright owners of
photographs used in this book.

Set in Monotype Imprint and Grotesque

Design and layout by Bernard Crossland Associates

Printed and bound in Great Britain by
Hazell Watson & Viney Ltd., Aylesbury, Bucks.

CONTENTS

PREFACE 6

METRIC CONVERSION 6

GLOSSARY 2099458 7

THE SOLAR SYSTEM 16

The Sun 16

The Moon 32

Mercury 64

Venus 70

Earth 74

Mars 75

The Minor Planets 88

Jupiter 98

Saturn 104

Uranus 109

Neptune 113

Pluto 115

Comets 118

Meteors 126

Meteorites 128

Glows and Atmospheric Effects 132

THE STARS 135

Classification and Evolution 135

Double Stars 144

Variable Stars and Novæ 148

Star Clusters and Nebulæ 157

Galaxies 160

The Constellations 169

THE STAR CATALOGUE 174

TELESCOPES AND OBSERVATORIES 261

THE HISTORY OF ASTRONOMY 267

ASTRONOMERS 275

PREFACE

Compiling this book has been no easy task. Data had to be collected from many sources, and translation alone proved to be very much of a problem – particularly from Russian! I can only hope that no errors are incorporated, and that some readers will find the book useful.

I must express my sincere thanks to the publishers, and in particular to Norris McWhirter and to Alex Reid, without whose help and encouragement the book would never have been completed.

I am also most grateful to Lawrence Clarke for the line drawings, and to Barney D'Abbs and Roger Prout for invaluable help in proof-reading, and to Derek French for assistance with the photographs.

PATRICK MOORE

Selsey, Sussex.

METRIC CONVERSION

In this book I have followed the current practice of giving lengths in metric units rather than the familiar imperial ones. To help in avoiding confusion, the following table may be found useful.

centimetres		inches	kilometres		miles
2·54	1	0·39	1·61	1	0·62
5·08	2	0·79	3·22	2	1·24
7·62	3	1·18	4·83	3	1·86
10·16	4	1·58	6·44	4	2·49
12·70	5	1·97	8·05	5	3·11
15·24	6	2·36	9·66	6	3·73
17·78	7	2·76	11·27	7	4·35
20·32	8	3·15	12·88	8	4·97
22·86	9	3·54	14·48	9	5·59
25·40	10	3·94	16·09	10	6·21
50·80	20	7·87	32·19	20	12·43
76·20	30	11·81	48·28	30	18·64
101·6	40	15·75	64·37	40	24·86
127·0	50	19·69	80·47	50	31·07
152·4	60	23·62	96·56	60	37·28
177·8	70	27·56	112·7	70	43·50
203·2	80	31·50	128·7	80	49·71
228·6	90	35·43	144·8	90	55·92
254·0	100	39·37	160·9	100	62·14

GLOSSARY

Aberration of starlight. As light does not move infinitely fast, but at a rate of practically 300 000 km/s., and as the Earth is moving round the Sun at an average velocity of 28 km/s., the stars appear to be shifted slightly from their true positions. The best analogy is to picture a man walking along in a rainstorm, holding an umbrella. If he wants to keep himself dry, he will have to slant the umbrella forward; similarly, starlight seems to reach us 'from an angle'. Aberration may affect a star's position by up to 20·5 seconds of arc.

Absolute magnitude. The **apparent magnitude** that a star would have if it could be observed from a standard distance of 10 **parsecs** (32·6 light-years).

Achromatic object-glass. An **object-glass** which has been corrected so as to eliminate **chromatic aberration** or false colour as much as possible.

Aerolite. A **meteorite** whose main composition is stony.

Albedo. The reflecting power of a planet or other non-luminous body. The Moon is a poor reflector; its albedo is a mere 7% on average.

Altazimuth mounting for a telescope. A mounting on which the telescope may swing freely in any direction.

Altitude. The angular distance of a celestial body above the horizon.

Ångström unit. One hundred-millionth part of a centimetre.

Aphelion. The furthest distance of a planet from the Sun in its orbit or of a satellite from its primary planet.

Apogee. The furthest point of the Moon from the Earth in its orbit.

Apparent magnitude. The apparent brightness of a celestial body. The lower the magnitude, the brighter the object: thus the Sun is approximately —27, the Pole Star +2, and the faintest stars detectable by modern techniques around +26.

Areography. The official name for 'the geography of Mars'.

Asteroids. One of the names for the minor planet swarm.

Astrograph. An astronomical telescope designed specially for astronomical photography.

Astrolabe. An ancient instrument used to measure the altitudes of celestial bodies.

Astronomical unit. The mean distance between the Earth and the Sun. It is equal to 149 598 500 km.

Aurora. Auroræ are 'polar lights'; Aurora Borealis (northern) and Aurora Australis (southern). They occur in the Earth's upper atmosphere, and are caused by charged particles emitted by the Sun.

Azimuth. The bearing of an object in the sky, measured from north (0°) through east, south and west.

Baily's beads. Brilliant points seen along the edge of the Moon just before and just after a total solar eclipse. They are caused by the sunlight shining through valleys at the Moon's limb.

Barycentre. The centre of gravity of the Earth-Moon system. Because the Earth is 81 times as massive as the Moon, the barycentre lies well inside the Earth's globe.

Binary star. A stellar system made up of two stars, genuinely associated, and moving round their common centre of gravity. The revolution periods range from millions of years for very widely-separated visual pairs down to less than half an hour for pairs in which the components are almost in contact with each other. With very close pairs, the components cannot be seen separately, but may be detected by spectroscopic methods.

Black hole. A region round a very small, very massive collapsed star from which not even light can escape.

BL Lacertæ objects. Variable objects which are powerful emitters of infra-red radiation, and appear to be very luminous and remote. Their nature is uncertain; they may be associated with quasars.

Bode's law. A mathematical relationship linking the distances of the planets from the Sun. It may or may not be genuinely significant. Strictly speaking it should be called Titius' Law, since it was discovered by J. D. Titius some years before J. E. Bode popularized it in 1772.

Bolide. A brilliant exploding meteor.

Bolometer. An instrument used to measure small quantities of heat radiation.

Carbon stars. Red stars of spectral types R and N with unusually carbon-rich atmospheres.

Cassegrain reflector. A reflecting telescope in which the secondary mirror is convex; the light is passed back through a hole in the main mirror. Its main advantage is that it is more compact than the Newtonian reflector.

Celestial sphere. An imaginary sphere surrounding the Earth, whose centre is the same as that of the Earth's globe.

Cepheid. A short-period **variable star**, very regular in behaviour; the name comes from the prototype star, Delta Cephei. Cepheids are astronomically important because there is a definite law linking their variation periods with their real luminosities, so that their distances may be obtained by sheer observation.

Chromatic aberration. A defect in all lenses, due to the fact that light

Conjunction of Venus and Jupiter

is a mixture of all wavelengths – and these wavelengths are refracted unequally, so that false colour is produced round a bright object such as a star. The fault may be reduced by making the lens a compound arrangement, using different kinds of glasses.

Chromosphere. That part of the Sun's atmosphere which lies above the bright surface or photosphere.

Circumpolar star. A star which never sets. For instance, Ursa Major (the Great Bear) is circumpolar as seen from England; Crux Australis (the Southern Cross) is circumpolar as seen from New Zealand.

Cluster variables. An obsolete name for the stars now known as RR Lyræ variables.

Cœlostat. An optical instrument making use of two mirrors, one of which is fixed, while the other is movable and is mounted parallel to the Earth's axis; as the Earth rotates, the light from the star (or other object being observed) is caught by the rotatable mirror and is reflected in a fixed direction on to

the second mirror. The result is that the eyepiece of the instrument need not move at all.

Collapsar. The end product of a very massive star, which has collapsed and has surrounded itself with a **black hole.**

Colour index. The difference between a star's visual magnitude and its photographic magnitude. The redder the star, the greater the positive value of the colour index; bluish stars have negative colour indices. For stars of type Ao, colour index is zero.

Colures. Great circles on the celestial sphere.

Conjunction. (1) A planet is said to be in conjunction with a star, or with another planet, when the two bodies are apparently close together in the sky. (2) For the inferior planets, Mercury and Venus, inferior conjunction occurs when the planet is approximately between the Earth and the Sun; superior conjunction, when the planet is on the far side of the Sun and the three bodies are again lined up. Planets beyond the Earth's orbit can never

come to inferior conjunction, for obvious reasons!

Corona. The outermost part of the Sun's atmosphere, made up of very tenuous gas. It is visible with the naked eye only during a total solar eclipse.

Coronagraph. A device used for studying the inner **corona** at times of non-eclipse.

Cosmic rays. High-velocity particles reaching the Earth from outer space. The heavier cosmic-ray particles are broken up when they enter the upper atmosphere.

Cosmogony. The study of the origin and evolution of the universe.

Cosmology. The study of the universe considered as a whole.

Counterglow. The English name for the sky-glow more generally called by its German name of the **Gegenschein.**

Culmination. The maximum altitude of a celestial body above the horizon.

Dawes' limit. The practical limit for the resolving power of a telescope; it is $4\cdot56/d$, where d is the aperture of the telescope in inches.

Day, sidereal. The interval between successive meridian passages, or **culminations**, of the same star: 23h 56m 4s·091.

Day, solar. The mean interval between successive meridian passages of the Sun: 24h 3m 56s·555. It is longer than the sidereal day because the Sun seems to move eastward against the stars at an average rate of approximately one degree per day.

Declination. The angular distance of a celestial body north or south of the celestial equator. It corresponds to latitude on the Earth.

Dewcap. An open tube fitted to the upper end of a refracting telescope. Its rôle is to prevent condensation upon the object-glass.

Dichotomy. The exact half-phase of the Moon or an **inferior planet**.

Diffraction grating. A device used for splitting up light; it consists of a polished metallic surface upon which thousands of parallel lines are ruled. It may be regarded as an alternative to the prism.

Direct motion. Movement of revolution or rotation in the same sense as that of the Earth.

Doppler effect. The apparent change in wavelength of the light from a luminous body which is in motion relative to the observer. With an approaching object, the wavelength is apparently shortened, and the spectral lines are shifted to the blue end of the spectral band; with a receding body there is a red shift, since the wavelength is apparently lengthened.

Double star. A star made up of two components – either genuinely associated (binary systems) or merely lined up by chance (optical pairs).

Driving clock. A mechanism for driving a telescope round at a rate which compensates for the axial rotation of the Earth, so that the object under observation remains fixed in the field of view.

Dwarf novæ. A term sometimes applied to the U Geminorum (or SS Cygni) variable stars.

Earthshine. The faint luminosity of the night side of the Moon, frequently seen when the Moon is in its crescent phase. It is due to light reflected on to the Moon from the Earth.

Eclipse, lunar. The passage of the Moon through the shadow cast by the Earth. Lunar eclipses may be either total or partial. At some eclipses, totality may last for approximately 1¾ hours, though most are shorter.

Eclipse, solar. The blotting-out of the Sun by the Moon, so that the Moon is then directly between the Earth and the Sun. Total eclipses can last for over 7 minutes under exceptionally favourable circumstances. In a partial eclipse, the Sun is incompletely covered. In an annular eclipse, exact alignment occurs when the Moon is in the far part of its orbit, and so appears smaller than the Sun; a ring of sunlight is left showing round the dark body of the Moon. Strictly speaking, a solar 'eclipse' is the **occultation** of the Sun by the Moon.

Eclipsing variable (or Eclipsing Binary). A **binary star** in which one component is regularly **occulted** by the other, so that the total light which we receive from the system is reduced. The prototype eclipsing variable is Algol (Beta Persei).

Ecliptic. The apparent yearly path of the Sun among the stars. It is more accurately defined as the projection of the Earth's orbit on to the celestial sphere.

Electron. Part of an atom; a fundamental particle carrying a negative electric charge.

Electron density. The number of 'free' (unattached) electrons in unit volume of space.

Elongation. The angular distance of a planet from the Sun, or of a satellite from its primary planet.

Ephemeris. A table showing the predicted positions of a celestial body such as a comet, asteroid or planet.

Epoch. A date chosen for reference purposes in quoting astronomical data.

Equator, celestial. The projection of the Earth's equator on to the **celestial sphere**.

Equatorial mounting for a telescope. A mounting in which the telescope is set up on an axis which is parallel with the axis of the Earth. This means that one movement (east to west) will suffice to keep an object in the field of view.

Equinox. The equinoxes are the two points at which the **ecliptic** cuts the **celestial equator**. The vernal equinox or First Point of Aries now lies in the constellation of Pisces; the Sun crosses it about 21 March each year. The autumnal equinox is known as the First Point of Libra; the Sun reaches it about 22 September yearly.

Escape velocity. The minimum velocity which an object must have in order to escape from the surface of a planet, or other celestial body, without being given any extra impetus.

Evection. An inequality in the Moon's motion, due to slight changes in the shape of the lunar orbit.

Exosphere. The outermost part of the Earth's atmosphere.

Extinction. The apparent reduction in brightness of a star or planet when low down in the sky, so that more of its light is absorbed by the Earth's atmosphere. With a star 1° above the horizon, extinction amounts to 3 magnitudes.

Eyepiece (or Ocular). The lens, or combination of lenses, at the eye-end of a telescope. It is responsible for magnifying the image of the object under study. With a positive eyepiece (for instance, a Ramsden, Orthoscopic or Monocentric) the image plane lies between the eyepiece and the object-glass (or main mirror); with a negative eyepiece (such as a Huyghenian or Tolles) the image-plane lies inside the eyepiece. A Barlow lens is concave, and is mounted in a short tube which may be placed between the eyepiece and the object-glass (or mirror). It increases the effective focal length of the telescope, thereby providing increased magnification.

Faculæ. Bright, temporary patches on the surface of the sun.

Filar micrometer. A device used for measuring very small angular distances as seen in the eyepiece of a telescope.

Finder. A small, wide-field telescope attached to a larger one, used for sighting purposes.

Fireball. A very brilliant **meteor**.

Flares, solar. Brilliant eruptions in the outer part of the Sun's atmosphere. Normally they can be detected only by spectroscopic means (or the equivalent), though a few have been seen in integrated light. They are made up of hydrogen, and emit charged particles which may later reach the Earth, producing magnetic storms and displays of auroræ. Flares are generally, though not always, associated with sunspot groups.

Flare stars. Faint Red Dwarf stars which show sudden, short-lived increases in brilliancy, due possibly to intense flares above their surfaces.

Flash spectrum. The sudden change-over from dark to bright lines in the Sun's spectrum, just before the onset of totality in a **solar eclipse.** The phenomenon is due to the fact that at this time the Moon has covered up the bright surface of the Sun, so that the chromosphere is shining 'on its own'.

Flocculi. Patches of the Sun's surface, observable with spectroscopic equipment. They are of two main kinds; bright (calcium) and dark (hydrogen).

Fraunhofer lines. The dark absorption lines in the spectrum of the Sun.

Forbidden lines. Lines in the spectrum of a celestial body which do not appear under normal conditions, but may be seen in bodies where conditions are exceptional.

Galaxies. Systems made up of stars, nebulæ, and interstellar matter. Many, though by no means all, are spiral in form.

Galaxy, the. The system of which our Sun is a member. It contains approximately 100 000 million stars, and is a rather loose spiral.

Gamma-rays. Radiation of extremely short wavelength.

Gauss. Unit of measurement of a magnetic field. (The Earth's field, at the surface, is on average about 0·3 to 0·6 gauss.)

Gegenschein. A faint sky-glow, opposite to the Sun and very difficult to observe. It seems to be due to thinly-spread interplanetary material.

Geodesy. The study of the shape, size, mass and other characteristics of the Earth.

Gibbous phase. The phase of the Moon or planet when between half and full.

Globules. Small dark patches inside gaseous nebulæ. They may be embryo stars.

Gnomon. In a sundial, the gnomon is a pointer whose function is to cast the Sun's shadow on to the dial. The gnomon always points to the celestial pole.

Great circle. A circle on the surface of a sphere whose plane passes through the centre of that sphere.

Green Flash. Sudden, brief green light seen as the last segment of the Sun disappears below the horizon. It is purely an effect of the Earth's atmosphere. Venus has also been known to show a Green Flash.

Gregorian reflector. A telescope in which the secondary mirror is concave, and placed beyond the focus of the main mirror. The image obtained is erect. Few Gregorian telescopes are in use nowadays.

H.I and H.II regions. Clouds of hydrogen in the Galaxy. In H.I regions the hydrogen is neutral; in H.II regions the hydrogen is ionized, and the presence of hot stars will make the cloud shine as a nebula.

Halo, galactic. The spherical-shaped cloud of stars round the main part of the Galaxy.

Heliacal rising. The rising of a star or planet at the same time as the Sun, though the term is generally used to denote the time when the object is first detectable in the dawn sky.

Herschelian reflector. An obsolete type of telescope in which the main mirror is tilted, thus removing the need for a secondary mirror.

Hertzsprung-Russell diagram (usually known as the H-R Diagram). A diagram in which stars are plotted according to their spectral types and their **absolute magnitudes.**

Horizon. The great circle on the celestial sphere which is everywhere 90 degrees from the observer's zenith.

Hour angle (of a celestial object). The time which has elapsed since the object crossed the meridian. If RA = right ascension of the object and LST = the local sidereal time, then Hour Angle = LST — RA.

Hour circle. A great circle on the **celestial sphere,** passing through both celestial poles. The zero hour circle coincides with the observer's meridian.

Hubble's constant. The rate of increase in the recession of a galaxy

with increased distance from the Earth.

Inferior planets. Mercury and Venus, whose distances from the Sun are less than that of the Earth.

Infra-red radiation. Radiation with wavelength longer than that of visible light (approximately 7500 Ångströms).

Interferometer, stellar. An instrument for measuring star diameters. The principle is based upon light-interference.

Ion. An atom which has lost or gained one or more of its planetary electrons, and so has respectively a positive or negative charge.

Ionosphere. The region of the Earth's atmosphere lying above the stratosphere.

Irradiation. The effect which makes very brilliant bodies appear larger than they really are.

Julian day. A count of the days, starting from 12 noon on 1 January, 4713. BC Thus 1 January 1977 was Julian Day 2 443 145. (The name 'Julian' has nothing to do with Julius Cæsar! The system was invented in 1582 by the mathematician Scaliger, who named it in honour of his father, Julius Scaliger.)

Kepler's laws of planetary motion. These were laid down by Johannes Kepler, from 1609 to 1618. They are: (1) The planets move in elliptical orbits, with the Sun occupying one focus. (2) The radius vector, or imaginary line joining the centre of the planet to the centre of the Sun, sweeps out equal areas in equal times. (3) With a planet, the square of the sidereal period is proportional to the cube of the mean distance from the Sun.

Kiloparsec. One thousand **parsecs** (3260 light-years).

Latitude, celestial. The angular distance of a celestial body from the nearest point on the **ecliptic.**

Libration. The apparent 'tilting' of the Moon as seen from Earth. There are three librations: latitudinal, longitudinal and diurnal. The overall effect is that at various times an observer on Earth can see a total of 59 per cent of the total surface of the Moon, though, naturally, no more than 50 per cent at any one moment!

Light-year. The distance travelled by light in one year: 9·4607 million million kilometres.

Local group. A group of more

M31 ANDROMEDAE
25cm. aperture Wray Anastigmat f/4
40 mins. exposure on H.P.3 Plate.
D.G.Daniels

Local Group: the Great Spiral in Andromeda is the largest member of the Local Group and is the most distant object (over 2 million light-years) that is clearly visible with the naked eye (D. G. Daniels)

than two dozen galaxies, one member of which is our own **Galaxy**. The largest member of the Local Group is the Andromeda Spiral, M.31.

Longitude, celestial. The angular distance of a celestial body from the **vernal equinox**, measured in degrees eastward along the ecliptic.

Lunation. The interval between successive new moons: 29d 12h 44m. (Also known as the Synodic Month.)

Magnetosphere. The region of the magnetic field of a planet or other body. In the Solar System, only the Earth, Jupiter and Mercury are known to have detectable magnetospheres, though as yet our information about the other giant planets is incomplete.

Main Sequence. A band along an **H-R Diagram**, including most normal stars except for the giants.

Maksutov telescope. An astronomical telescope involving both mirrors and lenses.

Mass. The quantity of matter that a body contains. It is not the same as 'weight'.

Mean sun. An imaginary sun travelling eastward along the celestial equator, at a speed equal to the average rate of the real Sun along the **ecliptic**.

Megaparsec. One million **parsecs**.

Meridian, celestial. The great circle on the **celestial sphere** which passes through the **zenith** and both celestial poles.

Meteor. A small particle, friable in nature and usually smaller than a sand grain, moving round the Sun, and visible only when it enters the upper atmosphere and is destroyed by friction. Meteors may be re-

garded as cometary débris.

Meteorite. A larger object, which may fall to the ground without being destroyed in the upper atmosphere. A meteorite is fundamentally different from a **meteor**. Meteorites are not associated with comets, but may be closely related to asteroids.

Micrometeorite. A very small particle of interplanetary material, too small to cause a luminous effect when it enters the Earth's upper atmosphere.

Micrometer. A measuring device, used together with a telescope to measure very small angular distances – such as the separations between the components of double stars.

Micron. One-thousandth of a millimetre. The usual symbol is μ.

Month. (1) Anomalistic: the interval between successive **perigee** passages of the Moon (27·55 days). (2) Sidereal: the revolution period of the Moon with reference to the stars (27·32 days). (3) Synodical: the interval between successive new moons (29·53 days). (4) nodical or Draconitic: the interval between successive passages of the Moon through one of its nodes (27·21 days). (5) Tropical: the time taken for the Moon to return to the same celestial longitude (about 7 seconds shorter than the sidereal month).

Nadir. The point on the celestial sphere directly below the observer.

Nebula. A cloud of gas and dust in space. Galaxies were once known as 'spiral nebulæ' or 'extragalactic nebulæ'.

Neutrino. A fundamental particle with no mass and no electric charge.

Neutron. A fundamental particle with no electric charge, but a mass practically equal to that of a **proton**.

Neutron star. The remnant of a very massive star which has exploded as a **supernova**. Neutron stars send out rapidly-varying radio emissions, and are therefore called 'pulsars'. Only two (the Crab and Vela pulsars) have as yet been identified with optical objects.

Newtonian reflector. A reflecting telescope in which the light is collected by a main mirror, reflected on to a smaller flat mirror set at an angle of 45°, and thence to the side of the tube.

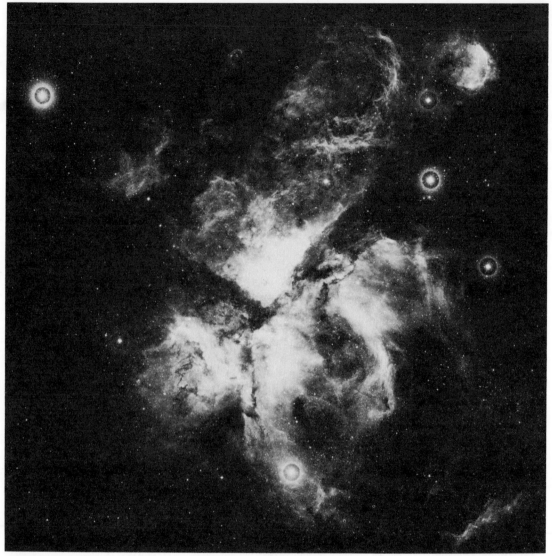

Nebula: NGC 3372, which surrounds η Carinæ the intensely luminous irregular variable star (Cerro-Tololo Observatory)

Nodes. The points at which the orbit of the Moon, a planet or a comet cuts the plane of the **ecliptic**; south to north (Ascending Node) or north to south (Descending Node).

Nova. A star which suddenly flared up to many times its normal brilliancy, remaining bright for a relatively short time before fading back to obscurity.

Nutation. A slow, slight 'nodding' of the Earth's axis, due to the gravitational pull of the Moon on the Earth's equatorial bulge.

Object-glass (or Objective). The main lens of a refracting telescope.

Objective prism. A small prism placed in front of the **object-glass** of a telescope. It produces small-scale spectra of the stars in the field of view.

Obliquity of the ecliptic. The angle between the **ecliptic** and the celestial equator: 23° 26′ 54″.

Occultation. The covering-up of one celestial body by another.

Opposition. The position of a planet when exactly opposite to the Sun in the sky; the Sun, the Earth and the planet are then approximately lined up.

Orbit. The path of a celestial object.

Orrery. A model showing the Sun and the planets, capable of being moved mechanically so that the planets move round the Sun at their correct relative speeds.

Parallax, trigonometrical. The apparent shift of an object when observed from two different directions.

Parsec. The distance at which a star would have a parallax of one second of arc: 3·26 **light-years**, 206 265 **astronomical units,** or 30·857 million million kilometres.

Penumbra. (1) The area of partial

shadow to either side of the main cone of shadow cast by the Earth. (2) The lighter part of a sunspot.

Perihelion. The position in orbit of a planet when closest to the Sun; a comet when closest to the Sun, or a satellite when closest to its primary.

Perigee. The position of the Moon in its orbit when closest to the Earth.

Perturbations. The disturbances in the orbit of a celestial body produced by the gravitational effects of other bodies.

Phases. The apparent changes in shape of the Moon and the inferior planets from new to full. Mars may show a **gibbous phase**, but with the other planets there are no appreciable phases as seen from Earth.

Photoelectric cell. An electronic device; light falling on the cell produces an electric current, the strength of which depends upon the intensity of the light.

Photoelectric photometer. A **photoelectric cell** used together with a telescope for measuring the magnitudes of celestial bodies.

Photometer. An instrument used to measure the intensity of light from any particular source.

Photometry. The measurement of the intensity of light.

Photon. The smallest 'unit' of light.

Photosphere. The bright surface of the Sun.

Planetary nebula. A small, dense, hot star surrounded by a shell of gas. The name is ill-chosen, since planetary nebulæ are neither planets nor nebulæ!

Planetoid. An **asteroid** or minor planet.

Poles, celestial. The north and south points of the celestial sphere.

Populations, stellar. Two main types of star regions: I (in which the brightest stars are hot and white), and II (in which the brightest stars are old Red Giants).

Position angle. The apparent direction of one object with reference to another, measured from the north point of the main object through east, south and west.

Precession. The apparent slow movement of the celestial **poles.** This also means a shift of the celestial equator, and hence of the equinoxes; the vernal **equinox** moves by 50″ of arc yearly, and has

moved out of Aries into Pisces. Precession is due to the pull of the Moon on the Earth's equatorial bulge.

Prime meridian. The meridian on the Earth's surface which passes through the Airy Transit Circle at Greenwich Observatory. It is taken as longitude 0°.

Prominences. Masses of glowing gas rising from the surface of the Sun. They are made up chiefly of hydrogen.

Proper motion, stellar. The individual movement of a star on the celestial sphere.

Proton. A fundamental particle with a positive electric charge. The nucleus of the hydrogen atom is made up of a single proton.

Quadrant. An ancient astronomical instrument used for measuring the apparent positions of celestial bodies.

Quadrature. The position of the Moon or a planet when at right-angles to the Sun as seen from the Earth.

Quantum. The amount of energy possessed by one photon of light.

Quasar. A very remote, super-luminous object. The nature of quasars is still uncertain; they may be special kinds of galaxies, but are much more powerful than ordinary galaxies even though they seem to be so much smaller. They are strong radio emitters.

Radial Velocity. The movement of a celestial body toward or away from the observer; positive if receding, negative if approaching.

Radiant. The point in the sky from which the meteors of any particular shower seem to radiate.

Regression of the nodes. The nodes of the Moon's orbit move slowly westward, making one complete revolution in 18·6 years. This regression is caused by the gravitational pull of the Sun.

Retardation. The difference in the time of moonrise between one night and the next.

Retrograde motion. Orbital or rotational movement in the sense opposite to that of the Earth's motion.

Reversing layer. The gaseous layer above the Sun's **photosphere.**

Right ascension. The angular distance of a celestial body from the vernal equinox, measured westward. It is usually given in hours, minutes

and seconds of time, so that the right ascension is the time-difference between the **culmination** of the vernal **equinox** and the culmination of the body.

Roche limit. The distance from the centre of a planet within which a second body would be broken up by the planet's gravitational pull. Note, however, that this would be the case only for a body which had no appreciable gravitational cohesion.

Saros. The period after which the Earth, Moon and Sun return to almost the same relative positions: 18 years 11·3 days. The saros may be used in eclipse prediction, since it is usual for an eclipse to be followed by a similar eclipse exactly one saros later.

Schmidt camera (or Schmidt telescope). An instrument which collects its light by means of a spherical mirror; a correcting plate is placed at the top of the tube. It is a purely photographic instrument.

Schwarzschild radius. The radius that a body must have if its **escape velocity** is to be equal to the velocity of light.

Scintillation. Twinkling of a star; it is due to the Earth's atmosphere. Planets may also show scintillation when low in the sky.

Secular acceleration of the Moon. The apparent speeding-up of the Moon in its orbit as measured over a long period of time, caused by the gradual slowing of the Earth's rotation (by 0·000 000 02 second per day).

Selenography. The study of the surface of the Moon.

Sextant. An instrument used for measuring the altitude of a celestial object.

Seyfert galaxies. Galaxies with relatively small, bright nuclei and weak spiral arms. Some of them are strong radio emitters.

Sidereal period. The revolution period of a planet round the Sun, or of a satellite round its primary planet.

Sidereal time. The local time reckoned according to the apparent rotation of the **celestial sphere.** When the vernal **equinox** crosses the observer's **meridian,** the sidereal time is 0 hours.

Solar wind. A flow of atomic

particles streaming out constantly from the Sun in all directions.

Solstices. The times when the Sun is at its maximum **declination** of approximately 23½ degrees; around 22 June (summer solstice, with the Sun in the northern hemisphere of the sky) and 22 December (winter solstice, Sun in the southern hemisphere).

Specific gravity. The density of any substance, taking that of water as 1. For instance, the Earth's specific gravity is 5·5, so that the Earth 'weighs' 5·5 times as much as an equal volume of water would do.

Spectroheliograph. An instrument used for photographing the Sun in the light of one particular wavelength only. The visual equivalent of the spectroheliograph is the spectrohelioscope.

Spectroscopic binary. A binary system whose components are too close together to be seen individually, but which can be studied by means of spectroscopic analysis.

Speculum. The main mirror of a reflecting telescope.

Spherical aberration. Blurring of a telescopic image; it is due to the fact that the lens (or mirror) does not bring the light-rays falling on its edge and on its centre to exactly the same focal point.

Superior planets. All the planets lying beyond the orbit of the Earth in the Solar System (that is to say, all the principal planets apart from Mercury and Venus).

Supernova. A very massive star which suffers a cataclysmic outburst, ending its career as a patch of expanding gas together with a neutron star remnant.

Synodic period. The interval between successive **oppositions** of a **superior planet.**

Syzygy. The position of the Moon in its orbit when new or full.

Tektites. Small, glassy objects found in a few localized parts of the Earth. Nobody is yet certain whether or not they come from the sky!

Terminator. The boundary between the day- and night-hemispheres of the Moon or a planet.

Thermocouple. An instrument used for measuring very small amounts of heat.

Transit. (1) The passage of a celestial body across the observer's meridian. (2) The projection of Mercury or Venus against the face of the Sun.

Transit instrument. A telescope mounted so that it can move only in **declination**; it is kept pointing to the meridian, and is used for timing the passages of stars across the meridian. Transit instruments were once the basis of all practical timekeeping. The Airy transit instrument at Greenwich is accepted as the zero for all longitudes on the Earth.

Troposphere. The lowest part of the Earth's atmosphere; its top lies at an average height of about 11 km. Above it lies the stratosphere; and above the stratosphere come the ionosphere and the exosphere.

Twilight. The state of illumination when the Sun is below the horizon by less than 18 degrees.

Umbra. (1) The main cone of shadow cast by the Earth. (2) The darkest part of a sunspot.

Van Allen zones. Zones of charged particles around the Earth. There are two main zones; the outer (made up chiefly of **electrons**) and the inner (made up chiefly of **protons**).

Variable stars. Stars which change in brilliancy over short periods. They are of various types.

Variation. An inequality in the Moon's motion, due to the fact that the pull of the Sun on the Moon is not constant for all positions in the lunar orbit.

White dwarf. A very small, very dense star which has used up its nuclear energy, and is in a very late stage of its evolution.

Widmanstätten patterns. If an iron **meteorite** is cut, polished and then etched with acid, characteristic figures of the iron crystals appear; these are the Widmanstätten patterns.

Wolf-Rayet stars. Very hot, greenish-white stars which are surrounded by expanding gaseous envelopes. Their spectra show bright (emission) lines.

Year. (1) Sidereal: the period taken for the Earth to complete one journey round the Sun (365·26 days). (2) Tropical: the interval between successive passages of the Sun across the vernal equinox (365·24 days). (3) Anomalistic: the interval between successive perihelion passages of the Earth (365·26 days; slightly less than 5 minutes longer than the sidereal year, because the position of the perihelion point moves along the Earth's orbit by about 11 seconds of arc every year). (4) Calendar: the mean length of the year according to the Gregorian calendar (365·24 days, or 365d 5h 49m 12s).

Zenith. The observer's overhead point (altitude 90°).

Zenith distance. The angular distance of a celestial object from the **Zenith.**

Zodiac. A belt stretching round the sky, 8° to either side of the **ecliptic,** in which the Sun, Moon and principal planets are to be found at any time. (Pluto is the only planet which can leave the Zodiac, though many asteroids do so.)

Zodiacal light. A cone of light rising from the horizon and stretching along the **ecliptic**; visible only when the Sun is a little way below the horizon. It is due to thinly-spread interplanetary material near the main plane of the Solar System.

THE SOLAR SYSTEM

THE SUN

DATA

Mean distance from the Earth:
149 597 900 km (=1 astronomical unit,
a.u.)
Maximum distance from the Earth:
152 100 000 km
Minimum distance from the Earth:
147 100 000 km
Mean parallax: 8″·794
Distance from centre of Galaxy:
32 000 light-years
**Period of revolution round centre of
Galaxy:** about 225 000 000 years (=1
'cosmic year')
Velocity round centre of Galaxy:
2150 km/s
Velocity toward solar apex: 19·5 km/s
Apparent diameter: max 32′35″,
mean 32′01″, min 31′31″
Equatorial diameter: 1 392 000 km
Density (water = 1): 1·409
Mass (Earth = 1): 332 946
Mass: 2 × 10²⁷ tonnes (99 per cent of the
mass of the entire Solar System)
Volume (Earth = 1): 1 303 600
Surface gravity (Earth = 1): 27·90
Escape velocity: 617·5 km/s
Mean apparent magnitude: −26·8
(= 600 000 times as brilliant as the full
moon)
Absolute magnitude: +4·83
Spectrum: G2
Surface temperature: 6000°C
Core temperature: 14 000 000°C*
Rotation period,
sidereal: mean 25·380 days
synodic: mean 27·275 days
**Time taken for light from the Sun to
reach the Earth:** mean 499·012 sec
= 8·3 minutes

*Some authorities prefer a value of
15 000 000°C.

The Sun is by far the nearest star; it is 270 000 times closer than α Centauri. It is therefore the only star which may be studied in detail. It is a normal Main Sequence star.

The first known estimate of the distance of the Sun was made by Aristarchus of Samos, about 270 BC. His value, derived from observations of the angle between the Sun and the exact half-moon, was approximately 4 800 000 km. Ptolemy (circa AD 150) increased this to 8 000 000 km, but about AD 1543 Copernicus reverted to only 3 200 000 km. Kepler, in 1618, gave a value of 22 500 000 km.

The first reasonably accurate estimate of the mean distance of the Sun was made by G. D. Cassini in 1672. He gave a value of 138 370 000 km. Successive determinations have been as follows:

Year	Authority	Method	Distance, km
1672	Cassini	Parallax of Mars	138 370 000
1770	Euler	Transit of Venus, 1769	151 225 000
1771	Lalande	Transit of Venus, 1769	154 198 000
1814	Delambre	Transit of Venus, 1769	153 841 000
1823	Encke	Transits of Venus, 1761 and 1769	153 303 000
1862	Foucault	Velocity of light	147 459 000
1867	Newcomb	Parallax of Mars	148 626 000
1872	Le Verrier	Masses of the planets	148 459 000
1875	Galle	Parallax of asteroid Flora	148 290 000
1877	Airy	Transit of Venus, 1874	150 152 000
1878	Stone	Transit of Venus, 1874	148 125 000
1881	Puiseux	Transit of Venus, 1874	146 475 000
1931	Spencer Jones	Parallax of Eros	149 645 000
1976	various	Radar to Venus	149 597 000

The first suggestion of measuring the Sun's distance (astronomical unit) by using transits of Venus was made by J. Gregory in 1663, and was extended by Edmond Halley in 1678. The method was sound in theory, but was affected by the 'Black Drop' – the apparent effect of Venus drawing a strip of blackness after it has passed on to the Sun's disk, thus making exact timings difficult. (Captain Cook's famous voyage, during which he discovered Australia, was made in order to take the astronomer Green to a suitable position from which to observe the transit of 1769.) Parallax measurements of the planets and asteroids were more accurate, but Spencer Jones' value as derived from the close approach of Eros in 1931 was too high, and was later revised by Rabe to 149 493 000 km. The modern method – radar – was introduced in the early 1960s in the United States. The present accepted value for the astronomical unit is accurate to a tiny fraction of one per cent.

Surface of the Sun photographed from the US Office of Naval Research Stratoscope balloon at 24 688 m above the Earth's surface on 25 September 1957

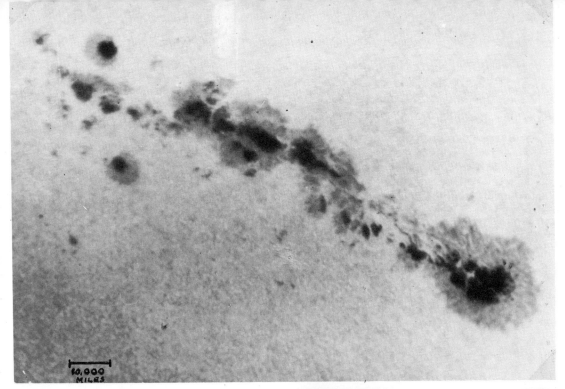

Gigantic sunspot group photographed
on 28 February 1967 (W. M. Baxter)

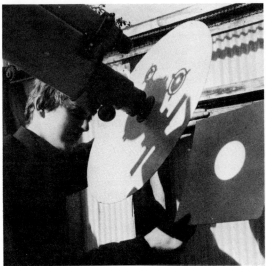

The only safe way to observe the Sun—
never look at it directly (Patrick Moore)

The first comments upon the Sun's rotation were made by Galileo, following his observations of sunspots from 1610. He gave a value of rather less than one month.

The discovery of the Sun's differential rotation, i.e. that the Sun does not rotate as a solid body would do, the equatorial rotation being shorter than the polar – was made by Richard Carrington in 1863. In order to help in identifying specific rotations of the Sun, Carrington had introduced a numbering system, beginning with Rotation No 1 on 9 November 1853. Rotation No 1500 began on 19 October 1965. **The first observations of Doppler shifts at opposite limbs due to the solar rotation** were made by H. C. Vogel in 1871.

Synodic rotation periods for features at various heliographic latitudes as are follows:

Latitude	Average synodic rotation period, days	Latitude	Average synodic rotation period, days
0	24·6	50	29·2
10	24·9	60	30·9
20	25·2	70	32·4
30	25·8	80	33·7
40	27·5	90	34·0

The first serious attempt to measure the solar constant was made by Sir John Herschel in 1837–8, using an actinometer (basically a bowl of water; the estimate was made by seeing the rate at which the bowl was heated). He gave a value which is about half the actual figure. The solar constant may be defined as the amount of energy in the form of solar radiation which is normally received on unit area at the top of the Earth's atmosphere; it is roughly equal to the

amount of energy reaching ground level on a
clear day. The modern value is 1·95 calories per
square cm per minute.

The first photograph of the Sun (a Daguerreotype) seems to have been taken by Lerebours, in
France, in 1842. However, the first good Daguerreotype was taken by Fizeau and Foucault, also in
France, on 2 April 1845, at the request of F
Arago. In 1854 J. B. Reade used a dry collodion
plate to show mottling on the Sun. **The first
systematic series of solar photographs** was
taken at Kew (London) from 1858 to 1872, using
equipment designed by the British astronomer
Warren de la Rue. Nowadays the Sun is photographed daily at many observatories throughout
the world.

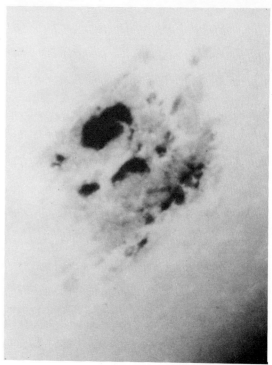

Sunspot on 25 October 1972 large enough to be seen without
optical aid (R. Lane, Shiremoor Solar Observatory)

SUNSPOTS

The discovery of sunspots was made in 1610–11.
(Naked-eye spots had been previously recorded,
but had not been explained; one given in a Chinese
record of 28 BC is described as 'a black vapour as
large as a coin'). The first observer to publish
telescopic observations of them was J. Fabricius,
from Holland, in 1611, and though his drawings
are undated he probably saw the spots toward the
end of 1610. C. Scheiner at Ingolstädt recorded
spots in March 1611, with his pupil C. B. Cysat.
Scheiner wrote a tract, which came to the notice of
Galileo, who claimed to have been observing
sunspots since November 1610. No doubt all
these observers recorded spots telescopically at
about the same time (the period was close to solar
maximum). However, interpretations differed.
Galileo's explanation was basically correct;
Scheiner regarded the spots as dark bodies moving
round the Sun at a distance close to the solar
surface; Cassini, later, regarded them as mountains protruding through the bright surface!

**The first to describe the projection method
of observing sunspots** may have been Galileo's
pupil B. Castelli. Galileo himself certainly used
the method, and said (correctly) that it was 'the
method that any sensible person will use'. (This
seems to dispose of the legend that he ruined his

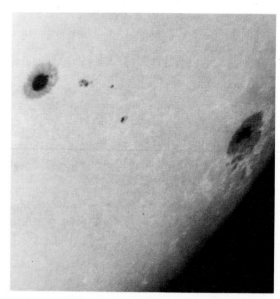

Sunspots on the east limb on 30 August 1970 at 08h 55m UT
(R. Lane, Shiremoor Solar Observatory)

eyesight by looking directly at the Sun through
a telescope.)

The Wilson effect was announced by A.
Wilson, of Glasgow, in 1774. He observed that

Day-by-day progress of the great
sunspot group of 1947 (Mount Wilson
and Palomar Observatories)

The Sun. Sunspots   19

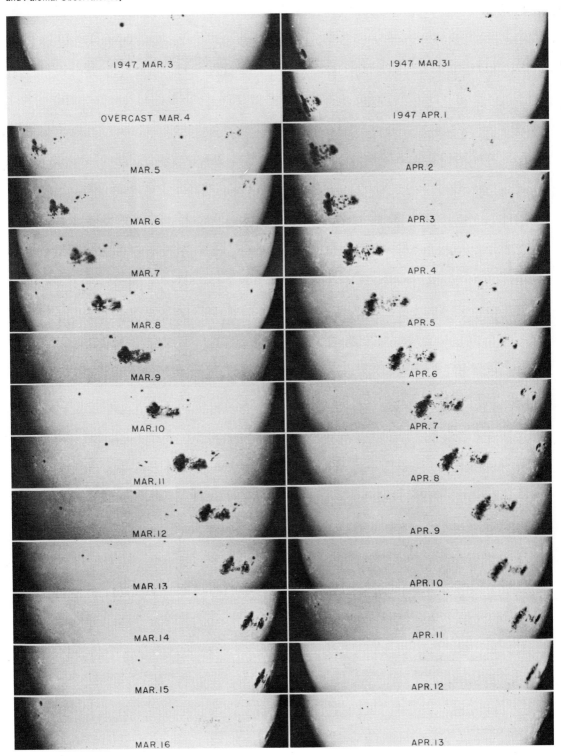

1947 MAR.3

OVERCAST MAR.4

MAR.5

MAR.6

MAR.7

MAR.8

MAR.9

MAR.10

MAR.11

MAR.12

MAR.13

MAR.14

MAR.15

MAR.16

1947 MAR.31

1947 APR.1

APR.2

APR.3

APR.4

APR.5

APR.6

APR.7

APR.8

APR.9

APR.10

APR.11

APR.12

APR.13

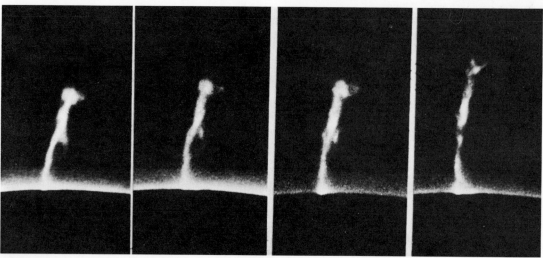

Development of a solar prominence on the west limb during 12 minutes on 12 June 1970 (R. Lane, Shiremoor Solar Observatory)

with a regular spot, the penumbra to the limbward side seemed to become broadened, as against the opposite side, as the spot neared the edge of the disk. From this, Wilson deduced that the spots must be hollows. The original observations were made in 1769.

The largest spot-group on record was that of April 1947; it covered an area of 18 130 000 000 km², reaching its maximum on 8 April. To be visible with the naked eye, a spot-group must cover 500 millionths of the visible hemisphere. (One millionth of the hemisphere is equal to 3 000 000 km².)

The longest-lived spot group lasted for 200 days, between June and December 1943. Very small spots (pores) may have lifetimes of less than an hour.

The first suggestion of a solar cycle seems to have come from the Danish astronomer Horrebow in 1775–6, but his work was not published until 1859, by which time the cycle had been definitely identified.

The 11-year solar cycle was discovered by H. Schwabe, a Dessau pharmacist, who began observing the Sun regularly in 1826 – mainly to see whether he could observe the transit of an intra-Mercurian planet. In 1851 his findings were popularized by Humboldt. A connection between solar activity and terrestrial magnetic phenomena was found by E. Sabine in 1852, and in 1870 E. Loomis, at Yale, established the link between the solar cycle and the frequency of auroræ.

The mean value of the length of the solar cycle since 1715 is 11·04 years. **The longest interval between successive maxima** has been 17·1 years (1788 to 1805); **the shortest interval** has been 7·3 years (1829·9 to 1837). Since 1715, when reasonably accurate records began, the **most energetic maximum** has been that of 1957·9. The **least energetic maximum** was that of 1816.

The 'Maunder Minimum' was discovered, from examination of old records, by the British astronomer E. W. Maunder in 1890. (It had been noted independently by F. G. W. Spörer.) He found that between 1645 and 1715 there were virtually no spots at all, so that the solar cycle was suspended; more recent research indicates that the corona may also have been virtually absent. It may be significant that there was freak weather in England during the 1680s, when the Thames froze regularly and 'frost fairs' were held on it. Auroræ, too were lacking; Halley noted that he saw his first aurora only in 1716, after forty years of watching. There may have been an earlier spotless period of the same type from 1400 to 1510, though the records are very incomplete. Further evidence of these prolonged minima comes from tree-ring studies, such as that carried out by F. Vercelli, who examined a tree which had lived between about 275 BC to AD 1914. Tree-rings are affected by events on the Sun; the solar cycle effects are well marked, and it is clear that conditions during the Maunder Minimum were decidedly abnormal.

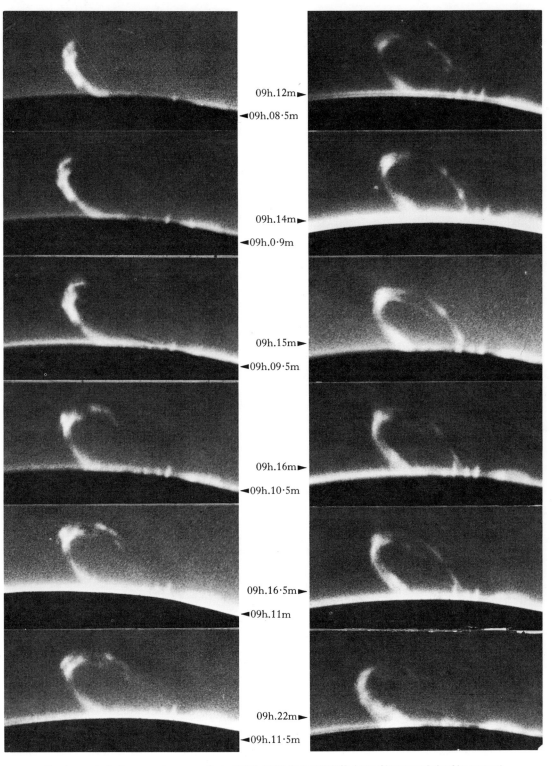

09h.12m ▶
◀ 09h.08·5m

09h.14m ▶
◀ 09h.0·9m

09h.15m ▶
◀ 09h.09·5m

09h.16m ▶
◀ 09h.10·5m

09h.16·5m ▶
◀ 09h.11m

09h.22m ▶
◀ 09h.11·5m

Development of a loop prominence on the east limb on 29 August 1970 (R. Lane, Shiremoor Solar Observatory)

Maxima

1718·2	1805·2	1894·1
1727·5	1816·4	1907.0
1738·7	1829·9	1917·6
1750·5	1837·2	1928·4
1761·5	1848·1	1937·4
1769·7	1860·1	1947·5
1778·4	1870·6	1957·9
1788·1	1883·9	1968·9

Minima

1723·5	1810·6	1901·7
1734·0	1823·3	1913·6
1745·0	1833·9	1923·6
1755·2	1843·5	1933·8
1766·5	1856·0	1944·2
1775·5	1867·2	1954·3
1784·7	1878·9	1964·7
1798·3	1889·6	±1975

There is strong evidence for a longer cycle superimposed on the 11-year one.

The law relating to the latitudes of sunspots (Spörer's Law) was discovered by the German amateur F. G. W. Spörer in 1861. At the start of a new cycle after minimum, the first spots appear at latitudes between 30 and 45 degrees north or south. As the cycle progresses, spots appear closer to the equator, until at maximum the average latitude of the groups is only about 15 degrees north or south. After maximum the spots become less common, but the approach to the equator continues, reaching only about 7 degrees north or south. The spots of the old cycle then die out (before reaching the equator), but even before they have completely disappeared the first spots of the new cycle are seen at the higher latitudes.

The first 'butterfly diagram', showing the effects of Spörer's Law, was drawn by Maunder in 1904.

The Wolf or Zürich sunspot number for any given day, indicating the state of the Sun at that time, was worked out by R. Wolf of Zürich in 1852. The formula is: $R = k\,(10\,g + f)$, where R is the Wolf number, g is the number of groups seen, f is the total number of individual spots seen, and k is a constant depending on the equipment and site of the observer. (k is not usually far from unity.) The Wolf number may range from 0 for a clear disk up to over 200. A spot less than about 2500 km in diameter is officially known as a **pore**.

The magnetic fields associated with sunspots were discovered by G. E. Hale, in the United States, in 1908. This resulted from the Zeeman effect (discovered in 1896 by the Dutch physicist P. Zeeman), according to which the spectral lines of a light source are split into two components if the source is associated with a magnetic field. Hale also found that in a spot-group the leader and the follower are of opposite polarity – and that the conditions are the same over a complete hemisphere of the Sun, though reversed in the opposite hemisphere. At the end of each cycle the whole situation is reversed, and there are grounds for supposing that the true cycle is 22 years in length rather than 11. Thus in the present cycle (beginning in 1975) the leader of a northern-hemisphere group is a north-polarity spot, the follower a south-polarity; in the southern hemisphere of the Sun, it is the leader which has south polarity. The magnetic fields of spots are very powerful, and may exceed 4000 gauss. **The strongest field observed** was with a group seen in 1967; the field was 5000 gauss, according to Steshenlio.

The discovery of faculæ was made by C. Scheiner, probably about 1611. Faculæ (Latin, 'torches') may be regarded as luminous clouds lying above the brilliant surface; they are composed largely of hydrogen, and are best seen near the limb, where the photosphere is less bright than at the centre of the disk (in fact, the limb has only two thirds of the brilliance of the centre, because at the centre we are looking more directly down into the hotter material). Faculæ may last for over two months, though the average lifetime is about 15 days. Also, faculæ often appear in areas where a spot-group is about to appear, and persist after the group has disappeared.

The first flare was observed by R. Carrington on 1 September 1859, in white light. A flare is a sudden, short-lived outbreak in the chromosphere in the region of a spot; large amounts of energy are released, and effects upon the Earth include magnetic storms, radio disturbance due to interference with the ionosphere, and bright aur#ae. Flares are graded into classes of importance, from 3+ (largest area) down to 1—. Very few are visible in integrated light; most are seen with spectroscopic equipment or the equivalent. A violent flare may eject material at up to 1500 km/s.

The modern sunspot theory was proposed by H. Babcock in 1961. It may be assumed that the solar magnetic lines of force run from one magnetic pole to the other below the bright surface. The differential rotation means that the lines are distorted and drawn out into loops. Over a period of years, the lines are coiled right round the Sun and are bunched near the poles, creating knots. Eventually a loop of magnetic energy erupts through the surface, steadying and cooling it, and producing the two spots characteristic of a group – with, predictably, opposite polarities. After about 11 years, the knots have become so complex that they break. The Sun 'snaps back' to its original state, but 'overshoots', so that the polarities in the two hemispheres are reversed.

Every spot-group has its own characteristics, but in general an 'average' two-spot group begins as two tiny pores at the limit of visibility. The pores develop into proper spots, growing and separating in longitude. Within two weeks the group has reached its maximum length, with a fairly regular leading spot together with a less regular follower of opposite polarity – and, of course, various minor spots and clusters. The darkest part of a spot – the umbra – still has a temperature of about 4500 °C, while the surrounding penumbra is at about 5000 °C; this means that a spot is by no means black, and if it could be seen shining on its own the surface brilliancy would be greater than that of an arc-lamp. After the group has reached its peak a slower decline sets in. The leader is generally the last survivor. Roughly 75 per cent of groups fit into this general pattern, but of the remainder some do not conform, and there are also frequent single spots.

Even in non-spot zones, the solar surface is not calm. The brilliant surface or photosphere has a granular structure; each granule is about 1000 km in diameter (1″·3) and has a life of about 8 minutes. It is estimated that the whole surface includes about 4 000 000 granules at any one time; they represent upcurrents, and the general situation has been compared with 'boiling' of a liquid, though the photosphere is of course entirely gaseous. The granular structure is easy to observe, though the first really good pictures of it were obtained from a balloon (*Stratoscope II*). Rising from the surface are **spicules**, not visible in integrated light; the average size of a spicule is 1000 km, and a typical height is 7000 km. The average lifetime of a spicule is no more than 5 minutes.

The first solar spectrum was obtained by Isaac Newton in 1666, but he never took his investigations much further, though he did of course prove the complex nature of sunlight.

The discovery of dark lines in the solar spectrum was made in England by W. H. Sollaston, in 1802. However, Wollaston merely took the lines for the boundaries between different colours of the rainbow spectrum.

The first systematic studies of the dark lines were carried out in Germany by J. von Fraunhofer, from 1814. Fraunhofer realised that the lines were permanent; he recorded 574 of them, and mapped 324.

The first explanation of the dark 'Fraunhofer lines' was given by G. Kirchhoff in 1859 (initially working with R. Bunsen). Kirchhoff found that the photosphere yields a rainbow or continuous spectrum; the overlying gases produce a line spectrum, but since these lines are seen against the rainbow background they appear dark instead of bright. Since their positions and intensities are not affected, each line may be tracked down to a particular element or group of elements. In 1861-2 Kirchhoff produced **the first detailed map of the solar spectrum.** (His eyesight was affected, and the work was actually finished by his assistant, K. Hofmann.) In 1869 Å. Angström, the Swedish physicist, studied the solar spectrum by using a grating instead of a prism, and in 1889 Rowland produced a detailed photographic map of the spectrum.

The most prominent Fraunhofer lines in the visible spectrum are:

Letter	Wavelength, Ångströms	Identi-fication	Letter	Wavelength, Ångströms	Identi-fication
A	7593	O_2			
a	7183	H_2O			
B	6867	O_2			

(These three are telluric lines—due to the Earth's intervening atmosphere.)

Letter	Wavelength, Ångströms	Identi-fication	Letter	Wavelength, Ångströms	Identi-fication
C (Hα)	6563	H	b_4	5167	Mg
D_1 { {	5896		F (Hβ)	4861	H
D_2 } {	5890	Na	f (Hγ)	4340	H
E {	5270	Ca, Fe	G	4308	Fe, Ti
{	5269	Fe	g	4227	Ca
b_1	5183	Mg	h(Hδ)	4102	H
b_2	5173	Mg	H {	3967	Ca^{II}
b_3	5169	Fe	K {	3933	

(One Ångström is equal to one hundred-millionth part of a centimetre; it is named in honour of Anders Ångström. The diameter of a human hair is roughly 500 000 Å.)

By now many of the chemical elements have been identified in the Sun. The list of elements which have and have not been identified is as follows:

THE CHEMICAL ELEMENTS, AND THEIR OCCURRENCE IN THE SUN

The following is a list of elements 1 to 103
* = detected in the Sun.
R = included in H. A. Rowland's list published in 1891.

Atomic No.		Name	Atomic Weight	Occurrence in the Sun	
1	H	Hydrogen	1·008	*	R
2	He	Helium	4·003	*	
3	Li	Lithium	6·939	* (in sunspots)	
4	Be	Beryllium	9·013	*	R
5	B	Boron	10·812	* (in compound)	
6	C	Carbon	12·012	*	R
7	N	Nitrogen	14·007	*	
8	O	Oxygen	16·000	*	
9	F	Fluorine	18·999	* (in compound)	
10	Ne	Neon	20·184	*	
11	Na	Sodium	22·991	*	R
12	Mg	Magnesium	24·313	*	R
13	Al	Aluminium	26·982	*	R
14	Si	Silicon	28·090	*	R
15	P	Phosphorus	30·975	*	
16	S	Sulphur	32·066	*	
17	Cl	Chlorine	35·434		
18	A	Argon	39·949	* (in corona)	
19	K	Potassium	39·103	*	R
20	Ca	Calcium	40·080	*	R
21	Sc	Scandium	44·958	*	R
22	Ti	Titanium	47·900	*	R
23	V	Vanadium	50·944	*	R
24	Cr	Chromium	52·00	*	R
25	Mn	Manganese	52·94	*	R
26	Fe	Iron	55·85	*	R
27	Co	Cobalt	58·94	*	R
28	Ni	Nickel	58·71	*	R
29	Cu	Copper	63·55	*	R
30	Zn	Zinc	65·37	*	R
31	Ga	Gallium	69·72	*	
32	Ge	Germanium	72·60	*	R
33	As	Arsenic	74·92		
34	Se	Selenium	78·96		
35	Br	Bromine	79·91		
36	Kr	Krypton	83·80		
37	Rb	Rubidium	85·48	* (in spots)	
38	Sr	Strontium	87·63	*	R
39	Y	Yttrium	88·91	*	R
40	Zr	Zirconium	91·22	*	R
41	Nb	Niobium	92·91	*	R
42	Mo	Molybdenum	95·95	*	R
43	Tc	Technetium	99		
44	Ru	Ruthenium	101·07	*	
45	Rh	Rhodium	102·91	*	
46	Pd	Palladium	106·5	*	R
47	Ag	Silver	107·87	*	R
48	Cd	Cadmium	112·41	*	R
49	In	Indium	114·82	* (in spots)	
50	Sn	Tin	118·70	*	R
51	Sb	Antimony	121·76	*	
52	Te	Tellurium	127·61		
53	I	Iodine	126·91		
54	Xe	Xenon	131·30		
55	Cs	Cæsium	132·91		
56	Ba	Barium	137·35	*	R
57	La	Lanthanum	138·92	*	R
58	Ce	Cerium	140·13	*	R
59	Pr	Praseodymium	140·91	*	
60	Nd	Neodymium	144·25	*	R
61	Pm	Promethium	147		
62	Sm	Samarium	150·36	*	
63	Eu	Europium	151·96	*	
64	Gd	Gadolinium	157·25	*	
65	Tb	Terbium	158·93	*	
66	Dy	Dysprosium	162·50	*	
67	Ho	Holmium	164·94		
68	Er	Erbium	167·27	*	R
69	Tm	Thulium	168·94	*	
70	Yb	Ytterbium	173·04	*	
71	Lu	Lutecium	174·98	*	
72	Hf	Hafnium	178·50	*	
73	Ta	Tantalum	180·96	*	
74	W	Tungsten	183·86	*	
75	Re	Rhenium	186·3		
76	Os	Osmium	190·2	*	
77	Ir	Iridium	192·2	*	
78	Pt	Platinum	195·1	*	
79	Au	Gold	197·0	*	
80	Hg	Mercury	200·6		
81	Tl	Thallium	204·4		
82	Pb	Lead	207·2	*	R
83	Bi	Bismuth	209·0		
84	Po	Polonium	210		
85	At	Astatine	211		
86	Rn	Radon	222		
87	Fr	Francium	223		
88	Ra	Radium	226		
89	Ac	Actinium	227		
90	Th	Thorium	232	*	
91	Pa	Protoactinium	231		
92	U	Uranium	238		

The remaining elements are 'transuranic' and radioactive, and have not been detected in the Sun. They are:

Atomic No.		Name	Atomic Weight	Occurrence in the Sun
93	Np	Neptunium	237	
94	Pu	Plutonium	239	
95	Am	Americium	241	
96	Cm	Curium	242	
97	Bk	Berkelium	243	
98	Cf	Californium	244	
99	Es	Einsteinium	253	
100	Fm	Fermium	254	
101	Md	Mendelevium	254	
102	No	Nobelium	254	
103	Lw	Lawrencium	257	

For elements 43, 61, 85-89, 91, 93-103 the mass number is that of the most stable isotope.

The fact that the remaining elements have not been identified in the Sun does not necessarily indicate that they are completely absent. They may be present, though no doubt in very small amounts.

So far as relative mass is concerned, the most abundant element by far is hydrogen (71 per cent). It is followed by helium (27 per cent). All the others combined make up only 2 per cent.

Helium was identified in the Sun (by Sir Norman Lockyer, in 1868) before being found on Earth. Lockyer named it after the Greek ἥλιος, the Sun. It was detected on Earth in 1894, by Sir William Ramsay, as a gas occluded in cleveite. For a time it was believed that the corona contained another element unknown on Earth, and it was even given a name – Coronium – but the lines, described initially by Harkness and Young at the

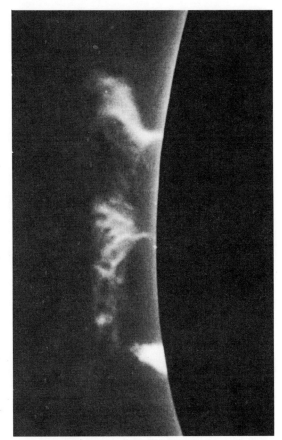

Solar prominence on the west limb on 28 June 1972 at 11h 33m UT (R. Lane, Shiremoor Solar Observatory)

eclipse of 1869, proved to be due to elements already known. The brilliant photosphere is surprisingly thin – only about 300 km thick. **The limb darkening was first explained** by K. Schwarzschild in 1906; as already noted, it is due to the fact that when we look at the centre of the disk we are seeing into deeper and hotter layers. Old ideas about the Sun sound strange today. Sir William Herschel believed that below the bright surface there was a cool layer, which could well be inhabited; up to the time of his death, in 1822, he maintained this view. As recently as 1952 a German lawyer, Godfried Büren, stated that the Sun had a vegetation-covered inner globe, and offered a prize of 25 000 marks to anyone who could prove him wrong. The leading German astronomical society took up the challenge, and won a court case. (Whether the prize was actually paid does not seem to be on record!)

The spectroheliograph, enabling the Sun to be photographed in the light of one element only, was invented by G. E. Hale in 1892. The visual equivalent, the **spectrohelioscope,** was invented in 1923, also by Hale. In 1933 B. Lyot, of France, developed the **Lyot filter,** which is less versatile but more convenient, and also allows the Sun to be studied in the light of one element only.

The meteoritic theory of solar energy was discussed by J. R. Mayer in 1848. Mayer found that a globe of hot gas the size of the Sun would cool down in 5000 years or so if there were no other energy source, while a Sun made of coal, and burning furiously enough to produce as much heat as the Sun actually does, would last for only 4600 years. He therefore assumed that the energy was produced by meteorites striking the solar surface.

The contraction theory was proposed in 1834 by H. von Helmholtz. He calculated that if the Sun contracted by 60 m per year, the energy produced would suffice for 15 000 000 years. This theory was supported later by the great British physicist Lord Kelvin. However, it had to be abandoned when astronomers concluded that the Sun is certainly at least as old as the Earth (about 4700 million years) and probably older.

The nuclear transformation theory was worked out by H. Bethe in 1938, during a train journey from Washington to Cornell University. Hydrogen is being converted to helium, so that energy is released and mass is lost: the decrease in mass amounts to 4 000 000 tonnes per second. Bethe assumed that carbon and nitrogen were used as catalysts, but C. Critchfield, also in America, subsequently showed that in solar-type stars the proton-proton reaction is dominant. Eventually the Sun will enter the red giant stage, with a probable diameter of about 300 000 000 km, before collapsing into a white dwarf. Fortunately, no dramatic changes in the Sun are likely for at least 5000 million years in the future, though slight variations may occur, and some authorities maintain that it is these minor changes which have produced the Ice Ages which have affected the Earth now and then throughout its history.

The solar wind is a radial outflow of charged particles from the Sun, 'blowing' continuously. The particles are made up of protons and electrons, and steam past the Earth at about 600 km/s. The solar wind affects the Earth's magnetosphere, and during periods of great activity on the Sun, when the wind is enhanced, the particles overload the Van Allen belts surrounding the Earth; they then enter the upper atmosphere, producing auroræ. The solar wind is also mainly responsible for directing the tails of comets away from the Sun.

Cosmic rays from the Sun were detected by Forbush in 1942. In 1954 Forbush established that cosmic-ray intensity decreases when solar activity increases (Forbush effect).

Solar neutrinos are presenting astronomers with a definite problem. Neutrinos are particles with no mass and no electric charge, so that they are extremely difficult to detect. Theoretical considerations indicated that the Sun should emit quantities of them, and efforts to identify them were made by R. Davis, of the Brookhaven National Laboratory in the United States, in an 'underground observatory' in South Dakota, at the bottom of a mine-shaft. The 'telescope' consisted of a tank of 454 600 litres of cleaning fluid (tetrachloroethylene). Only neutrinos could penetrate so far below the Earth, and on the rare occasions when a chlorine atom in the cleaning fluid happened to be struck by a neutrino, argon would be produced; the amount of argon found after a set period would therefore provide a key to the number of solar neutrinos. So far (1978) far fewer neutrinos than expected have been found, so that either the theory is inaccurate or else there is something wrong with the experiment!

An even deeper 'observatory' has just been completed in the USSR.

The chromosphere of the Sun lies above the photosphere. The temperature increases to 8000 °C at an altitude of 1500 km, and then increases rapidly until the chromosphere merges with the corona. The dark lines in the solar spectrum are produced in the lower chromosphere, which is known as the **reversing layer.** It is 8000 to 16 000 km deep.

The corona, visible with the naked eye only during a total eclipse, lies above the chromosphere. The mean temperature is nearly 2 000 000 °C, but the density is so low (less than one million millionth of the density of the Earth's air at sea-level) that there is little 'heat'. Indeed, the corona sends out only one millionth as much light as the photosphere. It has no definite boundary, but merely thins out until the density is no higher than that of the interplanetary medium.

Prominences were first described in detail by the Swedish observer Vassenius at the total eclipse of 1733, though he believed that they belonged to the Moon rather than the Sun. (They may have been seen earlier – by Stannyan in 1706, from Berne.) It was only after the eclipse of 1842 that astronomers became certain that they were solar rather than lunar. They were formerly termed 'red flames', but are in fact regions of hot hydrogen gas, reddish in colour. **Quiescent** prominences may persist for months; **eruptive** prominences show violent motions, and may reach heights of at least 2 000 000 km.

Prominences are visible with the naked eye only during a total eclipse. However, following the eclipse of 19 August 1868, J. Janssen (France) and Sir Norman Lockyer (England) developed the method of observing them spectroscopically without an eclipse. By observing at hydrogen wavelengths, prominences may be seen against the bright disk of the Sun as dark filaments, sometimes termed flocculi. (Bright flocculi are due to calcium.)

The first coronagraph was built by B. Lyot in 1930, and tested at the Pic du Midi Observatory (altitude 2870 m). It depends upon producing an 'artificial eclipse' instrumentally, and with it Lyot studied the inner corona and its spectrum.

The discovery of radio emission from the Sun was due to J. S. Hey and his team in 1942 (27–8 February). Initially, the effect was thought to be due to German jamming of the radar transmitters! Various types of emission are now known, and there are 'bursts', some of which are associated with flares. In June every year the 'radio sun' occults the Crab Nebula, itself a radio source, and the phenomenon yields valuable information about the corona. The first observations of this kind were carried out in June 1952.

The first radar contact with the Sun was made in 1959, by Eshleman and his colleagues at the Stanford Research Institute in the United States.

Sunrise is defined as the moment when the Sun's upper limb appears above the horizon. As refraction reduces the apparent zenith distance of the Sun to 34 minutes at the horizon, and as the Sun's semi-diameter is about 16′, the moment of sunrise is defined as being the instant when the Sun has a zenith distance of 90 degrees 50 minutes. **Sunset** is the moment when the upper limb of the Sun disappears below the horizon; the zenith distance is again 90 degrees 50 minutes. Because of refraction effects, it has sometimes been possible to see the Sun and the Full Moon visible simultaneously above opposite horizons.

The apex of the Sun's way (that is to say, the point in the sky toward which the Sun is at present moving) is in Hercules, at R.A. 18h, declination +34°. The **antapex** is in Columba (R.A. 6h, declination —34°.)

ECLIPSES OF THE SUN
An eclipse of the Sun occurs when the Moon passes in front of the Sun; strictly speaking, the phenomenon is an occultation of the Sun by the Moon. Solar eclipses may be total (when the whole of the photosphere is hidden), partial, or annular (when the Moon's apparent diameter is smaller than that of the Sun, so that a ring of light is left showing round the lunar disk; Latin **annulus,** a ring.)

The greatest number of eclipses possible in one year is 7; thus in 1935 there were 5 solar and 2 lunar eclipses, and in 1982 there will be 4 solar and 3 lunar.

The least number of eclipses possible in one year is 2, both of which must be solar, as in 1969.

The longest possible duration of totality for a solar eclipse is 7m 31s. This has never been actually observed, but at the eclipse of 20 June 1955 totality over the Philippine Islands lasted for 7m 8s.

The shortest possible duration of totality may be a fraction of a second. This will happen at the eclipse of 3 October 1986, which will be annular along most of the central track, but will be total for about a tenth of a second over a restricted area in the North Atlantic Ocean.

The longest possible duration of the annular phase of an eclipse is 12m 24s.

The longest totality ever observed was during the eclipse of 30 June 1973. A Concorde aircraft, specially equipped for the purpose, flew underneath the Moon's shadow and kept pace with it, so that the scientists on board (including the British astronomer John Beckman) saw a totality lasting for 72 minutes! They were carrying out observations at millimetre wavelengths, and at their height of 55 000 feet were above most of the water vapour in our atmosphere, which normally hampers such observations. They were also able to see definite changes in the corona and prominences over the full period. The Moon's shadow moves over the Earth at over 3000 km/h.

Total eclipse of the Sun in 1898

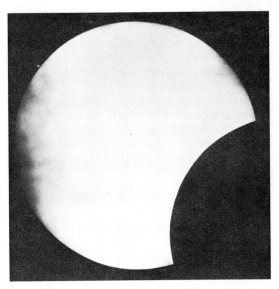

Partial eclipse of the Sun on 29 April 1976 at 10h 35m UT (K. Kennedy)

The widest track for totality across the Earth's surface is 272 km. In most total eclipses, of course, the width is considerably less than this;

and a partial eclipse is seen to either side of the track of totality.

The first eclipse predictions were made by studies of the Saros period. This is the period after which the Sun, Moon and node arrive back at almost the same relative positions; it amounts to 6585·321 solar days, or approximately 18 years 11 days. Therefore, an eclipse tends to be followed by another eclipse in the same Saros series 18 years 11 days later, though conditions are not identical, and the Saros is at best a reasonable guide. **The first known prediction** was made by the Greek philosopher Thales, who forecast the eclipse of 25 May 585 BC. This occurred near sunset in the Mediterranean area, and is said to have put an end to a battle between the forces of King Alyattes of the Lydians and King Cyaxares of the Medes; the combatants were so alarmed by the sudden darkness that thay concluded a hasty peace. One Saros series lasts for 1150 years; it includes 64 eclipses, of which 43 or 44 are total, while the rest are partial eclipses seen from the polar zones of the Earth.

The first recorded solar eclipse seems to have been that of 2136 BC (22 October), seen in China during the reign of the Emperor Chung K'ang. The Chinese believed that eclipses were due to an attack on the Sun by a hungry dragon, and they endeavoured to scare the dragon away by making as much noise as possible. (It always worked!) There is a story – probably apocryphal – that on this occasion the two Court Astronomers, Hi and Ho, were executed for their negligence in failing to predict the eclipse. The next observed eclipse which may be dated with any certainty is that of 780 BC.

The first solar eclipse recorded in Britain was that of 15 February 538, described in the Anglo-Saxon Chronicle; it occurred four years after the death of Cerdic, first king of the West Saxons. The Sun was two-thirds eclipsed in London. English total eclipses over the past thousand years have been those of 1140, 1715, 1724 and 1927. The eclipse of 3 May 1715 was observed by Halley from London and by Flamsteed from Greenwich; that of 22 May 1724 was observed only by Dr. Stukeley from Haradon Hill, near Salisbury, because of generally cloudy conditions. Totality was, however, seen from France. The eclipse of 29 June 1927 was total over part of

Total eclipse of the Sun on 30 June 1973 at 10h 37m UT (James Shepherd)

north England; the conditions were generally poor, and the best photographs were obtained from Giggleswick – totality lasted for less than half a minute. The eclipse of 30 June 1954 was total over parts of Norway and Sweden; the extreme southern edge of the track just brushed the Shetland Isles. The next English total eclipse will be that of 11 August 1999, when the track will cross Cornwall; it will be followed by the total eclipses of 14 June 2051 (total in London), 23 September 2090 and 7 October 2135. No total eclipse was seen from London between 875 and 1715, though those of 968 and 1140 were nearly total there. The 1927 eclipse was 95 per cent total in London.

There were total eclipses over parts of Scotland in 1133, and on 17 June 1433, when the length of totality reached 4·5 minutes at Inverness and caused what became known as 'Black Hour'. The eclipses of 1598 and 1652 were also total over parts of Scotland, and that of 1699 nearly so.

The maximum theoretical length for a British total eclipse is 5·5 minutes. That of 15 June 885 lasted for almost 5 minutes, and the same will be true for the Scottish total eclipse of 20 July 2381.

The first total eclipse recorded in the United States was that of 24 June 1778, when the track passed from Lower California to New England. **The first official American total eclipse expedition** was that of 21 October 1790, when a party went to Penobscot, Maine; it was led by S. Williams of Harvard, and was given 'free passage' by the British forces, but unfortunately a mistake in the calculations meant that the party remained outside the track of totality! **The first American eclipse expedition to Europe** was that of 28 July 1851, when G. P. Bond led a team to Scandinavia.

The only emperor to have died of fright because of an eclipse was Louis of Bavaria, in 840 (his three sons then proceeded to indulge in a ruinous war over the succession).

The only astronomer to have escaped from a besieged city in a balloon to study a total eclipse was Jules Janssen. The eclipse was that of 22 December 1870, and Janssen flew out from Paris, which was surrounded by the German forces. He made his way safely to Oran, but clouds prevented him from making any observations.

The first mention of the corona may have been due to Plutarch, who lived from about AD 46 to 120. Plutarch's book 'On the Face in the Orb of the Moon' contains a reference to 'a certain splendour' round the eclipsed Sun which could well have been the corona. The corona was definitely recorded from Corfu during the eclipse of 22 December 968. The astronomer Clavius saw it at the eclipse of 9 April 1567, but regarded it as merely the uncovered edge of the Sun; Kepler showed that this could not be so, and attributed it to a lunar atmosphere. After observing the eclipse of 16 June 1806 from Kindehook, New York, the Spanish astronomer Don José Joaquin de Ferrer pointed out that if the corona were due to a lunar atmosphere, then the height of this atmosphere would have to be 50 times greater than that of the Earth, which was clearly unreasonable. However, it was only after careful studies of the eclipses of 1842 and 1851 that the corona and the prominences were shown unmistakably to belong to the Sun rather than to the Moon.

There is some evidence that during eclipses which occurred during the 'Maunder Minimum' (1645–1715) the corona was absent, though it is impossible to be sure. Certainly the shape of the corona at spot-maximum is more symmetrical than at spot-minimum, when there are 'polar streamers'. This was first recognized after studies of the eclipses of 1871 and 1872.

The first observation of Baily's Beads seems to have been made by Halley during the eclipse of 1715. They were also seen by Maclaurin, from Edinburgh, at the annular eclipse of 1 March 1737. They were described in detail by the English astronomer Francis Baily (after whom they are named) in 1836 during the annular eclipse of 15 May. The brilliant bead-like effect is caused by the Sun's rays shining through valleys on the lunar limb immediately before and immediately after totality. They were **first photographed** at the eclipse of 7 August 1869 by C. F. Hines and members of the Philadelphia Photographic Corps, observing from Ottumwa, Iowa.

The first observation of the shadow-bands was made by H. Goldschmidt in 1820. These are wavy lines seen across terrestrial features before and after totality; they are, of course, produced in the Earth's atmosphere.

The first attempt to photograph a total eclipse was made by the Austrian astronomer Majocci on 8 July 1842. He failed to record totality, though he did succeed in photographing the partial phase.

The first successful photograph of a total eclipse, showing the corona and prominences, was taken by Berkowski on 28 July 1851, using the 6·25 in Königsberg heliometer and giving an exposure time of 24 s.

The first photograph of the flash spectrum was taken by the American astronomer C. Young on 22 December 1870. (The flash spectrum is the sudden, brief change in the Fraunhofer lines from dark to bright, when the Moon blots out the photosphere in the background, and the chromosphere is left shining 'on its own'). The flash spectrum was first observed during an annular eclipse by Pogson in 1872.

The first attempt to show a total eclipse on television from several stations along the track was made by the BBC at the eclipse of 15

February 1961. The track passed from France through Italy and Jugoslavia, and thence into Russia. The attempt was successful; totality was shown from St Michel in France (commentator, Dr Hugh Butler); from Florence in Italy (C. A. Ronan); and from the top of Mount Jastrebac in Jugoslavia (myself). This must also have been the most peculiar way in which a television commentator has spoken to the technical crew. I talked French to a Belgian astronomer, who relayed it in German to the senior Jugoslav, who passed it on to his companions in Serbo-Croat. This was no doubt why, at one stage, we showed pictures of mountain oxen chewing the cud rather than the eclipsed Sun!

SOLAR ECLIPSES, 1923-1977

T = total, P = partial, A = annular

Date		Type	Area
1923	March 16	A	S. Africa
1923	Sept. 10	T	California, Mexico
1924	March 5	P	S. Africa
1924	July 31	P	Antarctic
1924	Aug. 29	P	Iceland, N. Russia, Japan
1925	Jan. 24	T	North-eastern USA
1925	July 20/1	A	New Zealand, Australia
1926	Jan. 14	T	E. Africa, Indian Ocean, Borneo
1926	July 9/10	A	Pacific
1927	Jan. 3	A	New Zealand, S. America
1927	June 29	T	England, Scandinavia
1927	Dec. 24	P	Polar zone
1928	May 19	T	S. Atlantic
1928	June 17	P	N. Siberia
1928	Nov. 12	P	England to India
1929	May 9	T	Indian Ocean, Philippines
1929	Nov. 1	A	Newfoundland. C. Africa. Indian Ocean
1930	April 28	T	Pacific
1930	Oct. 21/2	T	S. Pacific to S. America
1931	April 17/18	P	Arctic
1931	Sept. 12	P	Alaska, N. Pacific
1931	Oct. 11	P	S. America, S. Pacific, Antarctic
1932	March 7	A	Antarctic
1932	Aug. 31	T	USA
1933	Feb. 24	A	S. America, C. Africa
1933	Aug. 21	A	Iran, India, N. Australia
1934	Feb. 13/14	T	Pacific
1934	Aug. 10	A	S. Africa
1935	Jan. 5	P	No land surface
1935	Feb. 3	P	N. America
1935	June 30	P	Britain
1935	July 30	P	No land surface
1935	Dec. 25	A	New Zealand, south S. America
1936	June 19	T	Greece, Turkey, Siberia, Japan
1936	Dec. 13/14	A	Australia, New Zealand
1937	June 8	T	Pacific, Chile
1937	Dec. 2/3	A	Pacific
1938	May 29	T	S. Atlantic
1938	Nov. 21/2	P	E. Asia, Pacific coast of N. America
1939	April 19	A	Alaska, Arctic
1939	Oct. 12	T	Antarctic
1940	April 7	A	USA, Pacific
1940	Oct. 1	T	Brazil, S. Atlantic, S. Africa
1941	March 27	A	S. Pacific, S. America
1941	Sept. 21	T	China, Pacific
1942	March 16/17	P	S. Pacific, Antarctic
1942	Aug. 12	P	Invisible in Britain
1942	Sept. 10	P	Britain
1943	Feb. 4/5	T	Japan, Alaska
1943	Aug. 1	A	Pacific
1944	Jan. 25	T	Brazil, Atlantic, Sudan
1944	July 20	A	India, New Guinea
1945	Jan. 14	A	Australia, New Zealand
1945	July 9	T	Canada, Greenland, N. Europe
1946	Jan. 3	P	Invisible in Britain
1946	May 30	P	S. Pacific
1946	June 29	P	Arctic
1946	Nov. 23	P	N. America
1947	May 20	T	Pacific, Equatorial Africa, Kenya
1947	Nov. 12	A	Pacific
1948	May 8/9	A	E. Asia
1948	Nov. 1	T	Kenya, Pacific
1949	April 28	P	Britain
1949	Oct. 21	P	New Zealand, Australia
1950	March 18	A	S. Atlantic
1950	Sept. 12	A	Siberia, N. Pacific
1951	March 7	A	Pacific
1951	Sept. 1	A	Eastern USA, C. and S. Africa
1952	Feb. 25	T	Africa, Arabia, Russia
1952	Aug. 20	A	S. America
1953	Feb. 13/14	P	E. Asia
1953	July 11	P	Arctic
1953	Aug. 9	P	Pacific
1954	Jan. 5	A	Antarctic
1954	June 30	T	Iceland, Norway, Sweden, Russia, India
1954	Dec. 25	A	S. Africa, S. Indian Ocean
1955	June 20	T	S. Asia, Pacific, Philippines
1955	Dec. 14	A	Sudan, Indian Ocean, China
1956	June 8	T	S. Pacific
1956	Dec. 2	P	Europe, Asia
1957	April 29/30	A	Arctic
1957	Oct. 23	T	Antarctica
1958	April 19	A	Indian Ocean, Pacific
1958	Oct. 12	T	Pacific
1959	April 8	A	S. Indian Ocean, Pacific
1959	Oct. 2	T	N. Atlantic, N. Africa
1960	March 27	P	Australia, Antarctica
1960	Sept. 20/1	P	N. America, E. Siberia
1961	Feb. 15	T	France, Italy, Greece, Yugoslavia, Russia
1961	Aug. 11	A	S. Atlantic, Antarctica
1962	Feb. 4/5	T	Pacific
1962	July 31	A	S. America, C. Africa
1963	Jan. 25	A	Pacific, S. Africa
1963	July 20	T	Japan, north N. America, Pacific
1964	Jan. 14	P	Tasmania, Antarctica
1964	July 9	P	N. Canada, Arctic
1964	Dec. 3/4	P	N.E. Asia, Alaska, Pacific
1965	May 30	T	Pacific, New Zealand, Peruvian coast
1965	Nov. 23	A	Russia, Tibet, E. Indies
1966	May 20	A	Greece, Russia
1966	Nov. 12	T	S. America, Atlantic
1967	May 9	P	N. America, Iceland, Scandinavia
1967	Nov. 2	T	S. Atlantic
1968	March 28/9	P	Pacific, Antarctica
1968	Sept. 22	T	Arctic, Mongolia, Siberia
1969	March 18	A	Indian Ocean, Pacific
1969	Sept. 11	A	Peru, Bolivia
1970	March 7	T	Mexico, USA, Canada
1970	Aug. 3/Sept. 1	T	East Indies, Pacific
1971	Feb. 25	P	Europe, N.W. Africa
1971	July 22	P	Alaska, Arctic
1971	Aug. 20	P	Australasia, S. Pacific
1972	Jan. 16	A	Antarctica
1972	July 10	T	Alaska, Canada
1973	Jan. 4	A	Pacific, S. Atlantic
1973	June 30	T	Atlantic, N. Africa, Kenya, Indian Ocean
1973	Dec. 24	A	Brazil, Atlantic, N. Africa
1974	June 20	T	Indian Ocean
1974	Dec. 13	P	N. and C. America
1975	May 11	P	Europe, N. Asia, Arctic
1975	Nov. 3	P	Antarctic
1976	April 29	A	N.W. Africa, Turkey, China
1976	Oct. 23	T	Tanzania, Indian Ocean, Australia

SOLAR ECLIPSES 1977-1999

Date	Area	Type	Notes
1977 April 18	Atlantic, S.W. Africa, Indian Ocean	Annular	
1977 October 12	Pacific, Peru, Brazil	Total	Max. length 2m 37s
1978 April 7	Antarctic	Partial	79% eclipsed
1978 October 2	Arctic	Partial	69% eclipsed
1979 February 26	Pacific. U.S.A., Canada, Greenland	Total	Max. length 2m 48s
1979 August 22	Pacific, Antarctica	Annular	
1980 February 16	Atlantic, Kenya, India, China	Total	Max. length 4m 8s
1980 August 10	S. Pacific, Brazil	Annular	
1981 February 4	Pacific, S. Australia and New Zealand	Annular	
1981 July 31	Russia, N. Pacific	Total	Max. length 2m 3s
1982 January 25	Antarctic	Partial	57% eclipsed
1982 June 21	Antarctic	Partial	62% eclipsed
1982 July 20	Arctic	Partial	46% eclipsed
1982 December 15	Arctic	Partial	74% eclipsed
1983 June 11	Indian Ocean, E. Indies, Pacific	Total	Max. length 5m 11s
1983 December 4	Atlantic, Equatorial Africa	Annular	
1984 May 30	Pacific, Mexico, U.S.A., Atlantic, N. Africa	Annular	
1984 November 22/3	E. Indies, S. Pacific	Total	Max. length 1m 59s
1985 May 19	Arctic	Partial	84% eclipsed
1985 November 12	S. Pacific, Antarctica	Total	Max. length 1m 55s
1986 April 9	Antarctic	Partial	82% eclipsed
1986 October 3	N. Atlantic	Total	Max. length 0m 1s Annular along most of track
1987 March 29	Argentina, Atlantic, Central Africa, Indian Ocean	Total	Max. length 56s, in Atlantic. Annular along most of track
1987 September 23	Russia, China, Pacific	Annular	
1988 March 18	Indian Ocean, E. Indies, Pacific	Total	Max. length 3m 46s
1989 March 7	Arctic	Partial	83% eclipsed
1989 August 31	Antarctic	Partial	63% eclipsed
1990 January 26	Antarctic	Annular	
1990 July 22	Finland, Russia, Pacific	Total	Max. length 2m 33s
1991 January 15/16	Australia, N. Zealand, Pacific	Annular	
1991 July 11	Pacific, Mexico, Brazil	Total	Max. length 6m 54s
1992 January 4/5	Central Pacific	Annular	
1992 June 30	S. Atlantic	Total	Max. length 5m 20s
1992 December 24	Arctic	Partial	84% eclipsed
1993 May 21	Arctic	Partial	74% eclipsed
1993 November 13	Antarctic	Partial	93% eclipsed
1994 May 10	Pacific, Mexico, U.S.A., Canada	Annular	
1994 November 3	Peru, Brazil, S. Atlantic	Total	Max. length 4m 23s
1995 April 29	S. Pacific, Peru, S. Atlantic	Annular	
1995 October 24	Iran, India, E. Indies, Pacific	Total	Max. length 2m 5s
1996 April 17	Antarctic	Partial	88% eclipsed
1996 October 12	Arctic	Partial	76% eclipsed
1997 March 9	Russia, Arctic	Total	Max. length 2m 50s
1997 September 2	Antarctic	Partial	90% eclipsed
1998 February 26	Pacific, Atlantic	Total	Max. length 3m 56s
1998 August 22	Indian Ocean, E. Indies, Pacific	Annular	
1999 February 16	Indian Ocean, Australia, Pacific	Annular	
1999 August 11	Atlantic, England (Cornwall), France, Turkey, India	Total	Max. length 2m 23s

Progress of an annular eclipse on 29 April 1976 (Patrick Moore)

THE MOON

South-east of the near side of the Moon

The Moon is officially ranked as the Earth's satellite. Relative to its primary, it is however extremely large and massive, and it might well be better to regard the Earth-Moon system as a double planet.

The first suggestion that the Moon is mountainous was made by the Greek philosopher Democritus (460–370 BC). Earlier, Xenophanes (*c.* 450 BC) had supposed that there were many suns and moons according to the regions, divisions and zones of the Earth!

The first map of the Moon was drawn by W. Gilbert, probably around 1600 (Gilbert died in 1603) though not published until 1651. This was compiled with the naked eye, as telescopes had not then come into use.

The first telescopic map of the Moon was drawn by Thomas Harriot in July 1609. It shows a number of identifiable features, and was much more accurate than the drawings made by Galileo from 1610.

The first British telescopic observations of the Moon following those of Harriot seem to have been made by Sir William Lower, from Wales, from about 1611. Lower's drawings have not survived, but he compared the appearance of the Moon with a tart that his cook had made!

The first serious attempts to measure the heights of lunar mountains were made by Galileo, from 1610. He measured the heights of the Lunar Apennines, and though he over-estimated the altitudes his results were of the right order. Much better results were obtained by J. H. Schröter, from 1778.

The first explanation of the faint visibility of the 'night' side of the Moon, when the Moon is in the crescent stage, was given by Leonardo da Vinci (1452–1519). Leonardo correctly stated that the phenonenon is due to light reflected on to the Moon from the Earth.

The first systems of lunar nomenclature were introduced in 1645 by Langrenus, and in 1647 by Hevelius; but these have not survived. (For instance, the crater now called Plato was called by Hevelius 'the Greater Black Lake'.) The modern system was introduced in 1651 by G. Riccioli, who named the features in honour of scientists – plus a few others. He was not impartial; for instance he allotted a major formation to himself and another to his pupil Grimaldi, but he did not believe in the 'Copernican' theory that the Earth moves round the Sun, and so he 'flung Copernicus into the Ocean of Storms'. Riccioli's principle has been followed since, though clearly all the major craters were 'used up' and later distinguished scientists have had to be given formations of lesser importance.

DATA

Distance from Earth, centre to centre: mean: 384 400 km; **closest (perigee):** 356 410 km; **furthest (apogee)** 406 697 km

Distance from Earth, surface to surface: mean: 376 284 km; **closest (perigee):** 348 294 km; **furthest (apogee):** 398 581 km

Revolution period: 27·321 661 days

Axial rotation period: 27·321 661 days

Synodic period: 29d 12h 44m 2·9s

Mean orbital velocity: 3680 km/h

Axial inclination of equator, referred to ecliptic: 1°32′

Orbital inclination: 5°09′

Orbital eccentricity: 0·0549

Diameter: 3475·6 km

Apparent diameter seen from Earth: max 33′31″, mean 31′5″, min 29′22″

Reciprocal mass, Earth = 1 : 81·3

Density, water = 1 : 3·342

Mass, Earth = 1 : 0·0123

Volume, Earth = 1 : 0·0203

Escape velocity: 2·38 km/s

Surface gravity, Earth = 1 : 0·1653

Albedo: 0·07

Mean magnitude, at full: −12·7

Earthshine, photographed on 13 February 1975. The bright crescent is illuminated directly by the Sun, the remainder is seen in light reflected from the Earth (P. W. Foley)

The first system of lunar co-ordinates was introduced by T. Mayer in 1775. The first really accurate measures with a heliometer were made by F. W. Bessel in 1839.

The first detailed statement that the Moon has a synchronous rotation was made by G. D. Cassini in 1693, though Galileo may also have realized it. The Moon's rotation is equal to its revolution period (27·3 Earth-days), and so the same face is turned Earthward all the time, though the eccentricity of the lunar orbit means that there are 'libration zones' which are brought alternately into and out of view. From the Earth, 59 per cent. of the Moon's surface can be studied at one time or another. The remaining 41 per cent. is inaccessible. There is no mystery about the behaviour; tidal forces over the ages have been responsible. Other planetary satellites also have synchronous rotation with respect to their prominences.

The first detailed descriptions of lunar rills were given by J. H. Schröter, from 1778, though rills had been observed earlier by Huygens.

The first really good map of the Moon was published in 1837–8 by W. Beer and J. H. Mädler, of Berlin. Though they used a small telescope (Beer's 3·75 in refractor) their map was amazingly good, and remained the best for several decades. They also published a book, *Der Mond*, the first detailed description of the lunar surface.

The first detailed English book about the Moon was published in 1876 by E. Neison (real name, Nevill). Neison included a map, which was basically a revision of that by Beer and Mädler.

The largest lunar map compiled by a visual observer was that by the Welsh selenographer H. P. Wilkins. It was 300 in in diameter. The final revision, to one-third this scale, appeared in 1959.

The first photograph of the Moon (a Daguerreotype) was taken on 23 March 1840 by J. W. Draper, using a 5 in reflector. The image was 1 in across, and the exposure time was 20 minutes. The first good photographic atlas was published in 1899 by the French astronomers Loewy and Puiseux.

The first report of a 'lunar city' was made in 1822 by the German astronomer F. von P. Gruithuisen, who described a structure with 'dark gigantic ramparts' – which, however, prove to be low and quite haphazard ridges! In 1790 Sir William Herschel had stated his belief in an inhabited Moon, and in 1835 there came the famous 'lunar hoax', when a paper, the New York *Sun*, published some quite imaginary reports of discoveries made by Sir John Herschel at the Cape of Good Hope. The reports were written by a reporter, R. A. Locke, and included descriptions of bat-men and quartz mountains. The first article appeared on 25 August, and the *Sun* admitted the hoax on 16 September.

The first serious suggestion of change on the Moon was made in 1866 by J. Schmidt, from Athens. Schmidt believed that a crater, Linné on the Mare Serenitatis, had been transformed into a pit surrounded by a white patch. The controversy continued for many years, but it is not now believed than any change occurred there.

The first reports of active volcanoes on the Moon were made by Sir William Herschel, in 1783 (May 4) and again in 1787 (April 19 and 20).

However, it does not seem that Herschel saw anything more significant than bright areas shining by earthlight. In modern times, lunar events or Transient Lunar Phenomena (TLP) have been reported on many occasions. The first photographic confirmation was obtained on 3 November 1958 by N. A. Kozyrev, at the Crimean Astrophysical Observatory, of an event in Alphonsus. On 30 October 1963 an event in the Aristarchus area was observed at the Lowell observatory by J. Greenacre and J. Barr. The first comprehensive catalogue of TLP reports was published by NASA, in 1967, by B. Middlehurst and Patrick Moore; Moore published a supplement in 1971. The total number of events reported in these catalogues is 713: though there is no doubt that many of them are due to observational error or misinterpretation, it is generally agreed that some are genuine, and that the Moon is not totally inert.

The most famous 'blue moon' of recent times was that of 26 September 1950. The phenomenon was due to dust in the Earth's atmosphere, raised by vast forest fires in Canada. There was also a blue moon on 27 August 1883, caused by material sent up from the Krakatoa outburst; and green moons were seen in Sweden in 1884 – at Kalmar on 14 February, for 3 minutes, and at Stockholm on 17 January, also for 3 minutes.

The temperatures on the lunar surface were first measured with reasonable accuracy by the fourth Earl of Rosse, from Birr Castle in Ireland. Lord Rosse's papers, from 1869, indicated that at noon the surface temperature approached 100 °C, though later Langley erroneously concluded that the temperature never rose above 0 °C. It is known that the equatorial noon temperature is about 117 °C, falling to -163 °C at minimum.

The first radar echoes from the Moon were obtained in 1946 by Z. Bay, from Hungary. The most recent determination of the Moon's distance, by a laser measure (J. Faller, Lick) gives 353 911·218, $\pm$ 45 m. At the Moon, the laser beam has a diameter of 4 km. The telescope used was the 1·4 m reflector at Catalina (Arizona).

The largest regular 'sea' is the Mare Imbrium, diameter 1300 km. It is 7 km deeper in its centre than at its periphery, reckoning from the mean reference sphere for the Moon. Of the other regular maria, Orientale has a diameter of 965 km; Serenitatis, 690 km $\times$ 885 km; Crisium, 590 km $\times$ 460 km; Humorum,

423 km × 460 km; Humboldtianum, 305 km × 410 km. Mare Crisium is longest in an East-West direction, though from Earth foreshortening makes it appear longest North-South. Of the irregular seas, the Oceanus Procellarum is much larger than the Mare Imbrium. On average the Oceanus Procellarum is only 1 km. lower than the mean sphere, whereas the deeper seas, such as Mare Crisium, are 4 km. lower. The Mare Orientale (discovered by Wilkins and Moore in 1938) is the only major sea which extends on to the Moon's far side, and is very badly placed for observation from Earth.

The largest crater is Bailly, diameter 295 km (depth 3·96 km). Next, insofar as the Moon's near side is concerned, is Clavius (232 km diamter, 4·91 km deep). Many of the large structures are much shallower – such as Ptolemæus (diameter 148 km, depth 1·2 km). Craters with central structures include Theophilus (101 km diameter, 4·4 km deep); Copernicus (97 km, 3·35 km.) and Tycho (84 km, 4·27 km). The great dark formation Plato has a depth of 2·4 km.

The deepest crater is thought to be Newton, with a floor 8·85 km below the crest of its wall; the wall rises 2·25 km above the outer surface.

The brightest crater on the Moon is Aristarchus, which often appears prominently when earthlit. **The darkest area** is the floor of Grimaldi.

The greatest ray-crater is Tycho, whose rays dominate the scene at or near full moon. Second to it is Copernicus, on the Oceanus Procellarum. Other important ray-centres include Kepler, Olbers, Anaxagoras, Thales and Proclus.

The most famous plateau on the Moon is Wargentin, which is lava-filled. All other known plateaux are much smaller and less regular.

The most famous rills on the Moon are those in the Mare Vaporum area (Hyginus, Ariadæus) and the Hadley Rill in the Apennine area, visited by the Apollo 15 astronauts; the Hadley Rill is 135 km long, 1 to 2 km wide, and 370 m deep. Other famous rills or rill-systems include those associated with Sirsalis, Bürg, Hesiodus, Triesnecker, Ramsden and Hippalus. Extending from Herodotus, near Aristarchus, is the imposing valley known as Schröter's Valley in honour of its discoverer, J. H. Schröter (in a way ·rather misleading, since the crater named after Schröter is in a completely different area).

Domes, up to 80 km in diameter, occur in various parts of the Moon – for instance near Arago in the Mare Tranquillitatis, Prinz in the

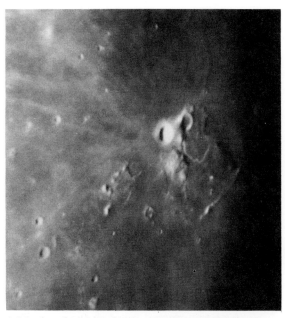

Aristarchus, the brightest crater on the Moon (Patrick Moore)

general area of Aristarchus, and on the floor of Capuanus. Many of them have summit pits; their slopes are gentle. Domes were first described in detail by the British selenographer R. Barker in 1932.

The first description of the 'grid system' (families of linear features, aligned in definite directions) was given by the American geologist J. E. Spurr in 1945; Spurr also invented the term. Most astronomers believe it to be of considerable significance in studies of the Moon's past history.

The origin of the lunar craters is still a matter for debate. There are various theories, ranging from the plausible to the eccentric! The main question is whether the craters were produced by internal action – that is to say, vulcanism – or by meteoritic impact. The impact theory was proposed by Gruithuisen in 1824; it was revived by G. K. Gilbert in 1892 and again by R. B. Baldwin in 1949. On the volcanic theory, it is generally assumed that the craters are of the caldera type. Probably both vulcanism and impact have played a part in the moulding of the lunar surface.

(Among the more erratic theories may be cited those of P. Fauth, who died in 1943 and supported the idea that the Moon is covered with ice; Weisberger, who in 1942 denied the existence of any lunar mountains or craters, and attributed the effects to storms and cyclones in a dense lunar atmosphere; and Ocampo, whose theory appeared in 1951. To Ocampo, the craters were the result of an atomic war between two races of Moon-men. The fact that some craters have central peaks while others do not shows, of course, that the warring factions used different kinds of bombs!)

The first reports of mascons, or **mass concentrations,** were made by P. Muller and W. L. Sjögren in 1968, from studies of the movements of the probe Orbiter 5. Exceptionally dense areas of material were found to lie below various regular features – Imbrium, Crisium, Smythii, Serenitatis, Humorum, Nectaris, Humboldtianum, Orientale, Æstuum and Grimaldi. On the other hand, the material below some of the other features, such as Copernicus, was of much lower density. (In 1968 B. Middlehurst and Patrick Moore had shown that TLP are often associated with features near the boundaries of the regular seas.)

The first attempted lunar probe was Able 1, launched by the Americans on 17 August 1958. It was not successful.

The first successful probe to the Moon was Luna 1 (2 January 1959). It by-passed the Moon, sending back valuable data.

The first probe to land on the Moon was Luna 2 (13 September 1959).

The first photographs of the Moon's far side were obtained from Luna 3, in October 1959.

The first good close-range pictures of the Moon were obtained from Ranger 7 (31 July 1964). The region, not far from Guericke, has been named the Mare Cognitum, or Known Sea, though the name has not come into general use.

The first successful soft landing on the Moon was made by Luna 9, launched on 31 January 1966.

The first lunar satellite was Luna 10, launched on 31 March 1966.

The first manned flight round the Moon was that of Apollo 8, in December 1968.

The first landing on the Moon was that of Apollo 11, on 20 July 1969. Early on 21 July Neil Armstrong stepped out on to the lunar surface, followed shortly afterwards by Edwin Aldrin.

The first unmanned probe to return samples from the Moon was Luna 16, launched on 12 September 1970.

The first 'crawler' on the Moon was Lunokhod 1. It was taken to the Moon by Luna 17, and landed there on 17 November 1970.

The first 'Moon Car' or LRV (Lunar Roving Vehicle) was taken to the Moon in Apollo 15 (30 July 1971). Astronauts Scott and Irwin used it to drive around in the Apennines area.

The first professional geologist to go to the Moon was Harrison ('Jack') Schmitt, in Apollo 17 (11 December 1972). It was Schmitt who found the famous 'orange soil' near the crater which was known unofficially as Shorty. The colour proved to be due to small glassy particles which were very ancient; the age is 3·7 to 3·8 æons. (One æon is taken to be a thousand million years.)

The last man on the Moon (so far!) was Eugene Cernan, in Apollo 17. When he re-entered the lunar module, following Dr Schmitt, the first stage in lunar manned exploration was over.

The Apollo results have changed many of our ideas about the Moon. There is a loose upper layer or **regolith,** from 1 to 20 metres deep (on average 4 to 5 metres in the maria, 10 in the highlands); this is a breccia containing many different ingredients. On the highlands, the fragments are chiefly plagioclase-rich rocks and broken crystals of plagioclase; on the maria, the rock fragments are basalt, with pyroxene, plagioclase etc. Below the regolith is a kilometre-thick layer of shattered bedrock. Under this is a layer of more solid rock going down to about 25 km. Beneath this is a layer of material 35 km thick, with properties similar to the anorthosites and the feldspar-rich gabbros of the highlands. The bottom of this layer is 60 km from the surface. At 60 km another kind of rock appears; there is a sharp boundary. These new rocks are pyroxene and olivine rich, and are dense; they go down to 150 km. Beneath this is the core region, which is metallic, and may be 1000 to 1500 km in diameter. A 1-tonne meteorite which hit the Moon on 17 July 1972 indicated, from the moonquake waves, that there is a zone 1000 to 1200 km deep inside the Moon where the rocks are hot enough to be molten. The crust on the far side of the Moon may be thicker than that on the Earth-turned side (75 to 100 km deep).

All the rocks brought home for analysis are igneous; the Apollo missions recovered 2196 samples, with a total weight of 381·69 kg, now divided into 35 600 samples. The youngest rocks seem to be about 3·1 æons old; the oldest, 4·7 æons – about the same as the age of the Moon itself. There are no sedimentary or metamorphic rocks. The main epoch of lava-flowing ended about 3 æons ago.

In the **lavas,** basalts are dominant. They contain more titanium than terrestrial lavas; over 10 per cent. in the Apollo 11 samples (3 to 5 per cent. in the Apollo 12 samples, as against 1 to 3 per cent. in terrestrial basalts). Small amounts of metallic iron were found. Generally, lunar rocks have one-tenth as much sodium and potassium as terrestrial rocks. A new mineral – an opaque oxide of iron, titanium and magnesium, not unlike ilmenite – has been named armacolite.

All the highland rocks are igneous, with twice as much calcium and aluminium as the Mare lavas, but less titanium, magnesium and iron. Plagioclase makes up at least half the highland rocks; there is little sodium or potassium, so that the volatiles must have been lost from the Moon at an early stage. The average age of highland rocks is from 4 to 4·2 æons. The so-called 'Genesis Rock' (Apollo 15) is anorthosite, with an age of 4 æons.

Breccias have been collected from every landing site on the Moon. These are complex rocks made up of shattered, crushed and sometimes melted pieces of rock. The fragments may be igneous rocks of various types, glasses, or bits of other breccias.

The Apollo 12 sample 12013 (collected by Conrad from the Oceanus Procellarum) is unique. It is about the size of a lemon, and contains 61 per cent of SiO_2, whereas the associated lavas have only 35 to 40 per cent of SiO_2. It also contains 40 times as much potassium, uranium and thorium, making it one of the most radioactive rocks found anywhere on the Moon. It is composed of a dark grey breccia, a light grey breccia and a vein of solidified lava.

(**Pyroxene,** a Ca-Mg-Fe silicate, is the most common mineral in lunar lavas, making up about half of most specimens; it forms yellowish-brown crystals a few centimetres in size. **Plagioclase** or feldspar, a Ca-Al silicate, forms elongated white crystals. **Anorthosite** contains over 90 per cent plagioclase; it may even be pure plagioclase. **Olivine,** a Mg-Fe silicate, is made up of pale green crystals a few millimetres in size, and is not uncommon. This is also true of **ilmenite,** an Fe-Ti oxide. It is notable that there are no hydrated materials on the Moon, so that evidently water has never existed there.)

Moonquakes have been recorded regularly; there are about 3000 per year, centred mainly from 600 to 800 km below the surface, though very shallow moonquakes also occur. The impacts from the crashed lunar modules of the Apollo missions showed that the outermost few kilometers of the Moon are made up of cracked and shattered rock, so that signals can echo back and forth for hours; it was even said that after an impact of a module the Moon 'rang like a bell'! Researches by J. Green, B. Middlehurst and Patrick Moore have confirmed that moonquakes are most common at the time of lunar perigee, when the Moon's crust is under maximum strain.

There is virtually no overall magnetic field, though it seems that one did exist in the remote past. However, there are areas of magnetized material, particularly in and near the far-side crater Van de Graaff.

All recording Apollo stations on the Moon were shut down on 30 September 1977, partly on the grounds of expense and partly because some of the equipment was showing signs of failure.

There is no Earth-type atmosphere, and the so-called lunar atmosphere is a collisionless gas in which hydrogen, helium, neon and argon have been identified; the density is negligible – about 2×10^5 molecules per cubic centimetre, which corresponds to what we normally call a very high laboratory vacuum. There is no trace of life, either past or present, and the quarantining of Apollo astronauts was dropped after the first three successful missions.

The origin of the Moon remains uncertain, but it is now thought that the Earth-Moon system should be regarded as a double planet rather than as a planet and a satellite (the Moon's orbit is always concave to the Sun.) The age of the Moon is about the same as that of the Earth. The Earthward-turned axis is 2·2 km longer than the rotational axis, so that there is a slight Earthward bulge. In 1968 the **north pole star** of the Moon was Omega Draconis; in 1977 it is 36 Draconis. The **south pole star** is Delta Doradûs.

No minor Earth satellites of natural origin seem to exist. Careful searches for them have been made, as in 1957 by Tombaugh (discoverer of Pluto), but without result. Clouds of loose material in the Moon's orbit, at the Lagrangian points, were reported by K. Kordylewski in 1961, but remain unconfirmed. It therefore seems safe to conclude that the Moon is the Earth's only attendant.

NAMED 'SEAS' ON THE NEAR SIDE OF THE MOON

Sinus Æstuum	The Bay of Heats
Mare Australe	The Southern Sea
Mare Crisium	The Sea of Crises
Palus Epidemiarum	The Marsh of Epidemics
Mare Fœcunditatis	The Sea of Fertility
Mare Frigoris	The Sea of Cold
Mare Humboldtianum	Humboldt's Sea
Mare Humorum	The Sea of Humours
Mare Imbrium	The Sea of Showers
Sinus Iridum	The Bay of Rainbows
Mare Marginis	The Marginal Sea
Sinus Medii	The Central Bay
Lacus Mortis	The Lake of Death
Palus Nebularum	The Marsh of Mists
Mare Nectaris	The Sea of Nectar
Mare Nubium	The Sea of Clouds
Mare Orientale	The Eastern Sea
Oceanus Procellarum	The Ocean of Storms
Palus Putredinis	The Marsh of Decay
Sinus Roris	The Bay of Dews
Mare Serenitatis	The Sea of Serenity
Mare Smythii	Smyth's Sea
Palus Somnii	The Marsh of Sleep
Lacus Somniorum	The Lake of the Dreamers
Mare Spumans	The Foaming Sea
Mare Tranquillitatis	The Sea of Tranquility
Mare Undarum	The Sea of Waves
Mare Vaporum	The Sea of Vapours

These 'seas' are not alike. Some, such as Mare Imbrium, are regular and largely mountain-bordered, whereas others, such as Mare Undarum, are discontinuous, and consist only of patches of dark mare material. Most of the seas are connected; for instance Palus Nebularum (a name omitted from some lists) is part of the Mare Imbrium, while Sinus Iridum is a bay leading out from the Mare Imbrium. Isolated seas are rare; the Mare Crisium is a good example. The Mare Orientale spreads on to the far side of the Moon. The name was given to it by H. P. Wilkins and the present writer, but owing to a policy decision by the International Astronomical Union, reversing lunar east and west, the *Eastern* Sea is now on the *western* limb of the Moon.

LUNAR CRATERS ON THE NEAR SIDE OF THE MOON

The following are named craters. The diameters are in many cases arbitrary, because so many of the craters are irregular in shape, but the values given here are of the right order. c.p. = central peak.

Name	Lat	Long	Diam, km	Notes
Abenezra	21 S	12 E	43	W. of Nectaris. Well-formed. Pair with Azophi.
Abulfeda	14 S	14 E	64	Abenezra area. Pair with Almanon.
Adams	32 S	69 E	64	E. of Vendelinus. Irregular walls.
Agatharchides	20 S	31 W	48	Humorum area. Remnant of c.p. Irregular walls.
Agrippa	4 N	11 E	48	Vaporum area. Regular; c.p. Pair with Godin.
Airy	18 S	6 E	35	Albategnius area: pair with Argelander. Irregular walls.

The surface of the Moon photographed
during the *Apollo 16* mission (Victor
Hasselblad AB)

Name	Lat	Long	Diam, km	Notes
Albategnius	12 S	4 E	129	Companion to Hipparchus. Terraced, irregular walls; c.p.
Alexander	40 N	14 E	105	N. end of Caucasus. Darkish floor. Low walls.
Alfraganus	6 S	19 E	19	N.W. of Theophilus. Very bright. Minor ray-centre.
Alhazen	18 N	70 E	32	Near border of Crisium.
Aliacensis	31 S	5 E	84	Walter area; pair with Werner. High walls. c.p.
Almanon	17 S	15 E	48	Pair with Abulfeda. Regular.
Alpetragius	16 S	4 W	43	Outside Alphonsus. High, terraced walls. Vast c.p. with summit pit.
Alphonsus	13 S	3 W	129	Ptolemæus group. Low c.p. Rills on floor.
Anaxagoras	75 N	10 W	52	N. polar area; distorts Goldschmidt. Major ray-centre. c.p.
Anaximander	66 N	48 W	87	Pythagoras area; pair with Carpenter, No c.p.
Anaximenes	75 N	45 W	72	Near Philolaus. Rather low walls.
Anděl	10 S	13 E	40	Highlands W. of Theophilus. Low, irregular walls.
Ångström	30 N	42 W	11	N. of Harbingers, in Imbrium.
Ansgarius	14 S	82 E	80	E. of Fœcunditatis. Distinct. Pair with La Peyrouse.
Apianus	27 S	8 E	63	Aliacensis area. High walls.
Apollonius	5 N	61 E	48	Uplands S. of Crisium. Well-formed.
Arago	6 N	21 E	29	On Tranquillitatis. Domes nearby.
Aratus	24 N	5 E	13	In Apennines. Bright, not regular.
Archimedes	30 N	4 W	75	On Imbrium. Very regular. No c.p.
Archytas	59 N	5 E	34	On Frigoris. Bright, distinct; c.p.
Argelander	17 S	6 E	42	Albategnius area; pair with Airy. c.p.
Ariadæus	5 N	17 E	15	Vaporum area. Associated with great rill.

Name	Lat	Long	Diam, km	Notes
Aristarchus	24 N	48 W	37	On Procellarum. Brightest crater on the Moon. Terraced walls; c.p.
Aristillus	34 N	1 E	56	Archimedes group. Fine c.p.
Aristoteles	50 N	18 E	97	Pair with Euduxus. High walls.
Arnold	67 N	38 E	81	N.W. of Democritus; N. of Frigoris. Low walls.
Arzachel	18 S	2 W	97	Ptolemæus group. High walls; c.p.
Asclepi	55 S	26 E	32	S. uplands, W. of Hommel. Distinct.
Atlas	47 N	44 E	89	N. of Somniorum; pair with Hercules. High walls; much interior detail.
Autolycus	31 N	1 E	36	Archimedes group. Regular, distinct.
Auwers	15 N	17 E	16	Foothills of Hæmus. Not very bright.
Auzout	10 N	64 E	31	Outside Crisium. Low c.p.
Azophi	22 S	13 E	43	W. of Nectaris; pair with Abenezra. Well-formed.
Babbage	58 N	52 W	145	Nr. Pythagoras. Irregular enclosure.
Baco	51 S	19 E	64	Licetus area. High walls; low c.p.
Baillaud	75 N	40 E	80	Uplands N.E. of Meton. Rather low walls.
Bailly	66 S	65 E	294	'Field of ruins'; southern uplands.
Baily	50 N	31 E	19	Frigoris area N. of Bürg.
Ball	36 S	8 W	40	On edge of Deslandres. High walls.
Barocius	45 S	17 E	80	Outside Maurolycus. High but broken walls.
Barrow	73 N	10 E	87	N. of W. C. Bond (N. polar uplands). Low, broken walls.
Bayer	51 S	35 W	52	Outside Schiller. High, terraced walls.
Beaumont	18 S	29 E	48	Bay on Nectaris.
Beer	27 N	9 W	11	On Imbrium. Pair with Feuillé.
Behaim	16 S	71 E	56	E. of Vendelinus. High walls, c. craters.
Bellot	12 S	48 E	19	Edge of Fœcunditatis. Bright floor.
Bernouilli	34 N	60 E	40	E. of Geminus. Fairly regular.
Berosus	33 N	70 E	61	E. of Cleomedes; pair with Hahn. Terraced walls.
Berzelius	37 N	51 E	24	Taurus area. Darkish floor; c.p.
Bessel	22 N	18 E	19	On Serenitatis. Associated with long ray.
Bettinus	63 S	45 W	66	Bailly area; one of a line. High walls; c.p.
Bianchini	49 N	34 W	40	In Jura Mts. Rather irregular; c.p.
Biela	55 S	52 E	74	Vlacq area. High walls; c.p.
Billy	14 S	50 W	42	S. edge Procellarum. Very dark floor. Pair with Hansteen.
Biot	23 S	51 E	16	In Fœcunditatis. Very bright.
Birmingham	64 N	10 W	106	N. of Frigoris. Low-walled; irregular.
Birt	22 S	9 W	18	On Nubium; W. of Straight Wall. Profile irregular. Rill to the W.
Blagg	1 N	2 E	5	In Sinus Medii. Quite distinct.
Blancanus	64 S	21 W	92	Near Clavius; pair with Scheiner. High walls.
Blanchinus	25 S	3 E	53	N. of Werner. Uneven walls; rough floor.
Bode	7 N	2 W	18	Outside Æstuum. Bright; minor ray-centre.
Boguslawsky	75 S	45 E	97	Southern uplands. High walls.
Bohnenberger	16 S	40 E	35	Edge of Nectaris. Low walls.
Bond, G. P.	32 N	36 E	23	E. of Posidonius. Fairly regular.
Bond, W. C.	64 N	3 E	160	N. of Frigoris. Old and broken.
Bonpland	8 S	17 W	58	Fra Mauro group (Nubium). Fairly regular.
Borda	25 S	47 E	42	W. of Petavius. Low walls.
Boscovich	10 N	11 E	43	On edge of Vaporum. Low walls; very dark floor.
Bouguer	52 N	36 W	24	In Jura uplands. Very distinct.
Boussingault	70 S	50 E	78	S. uplands. Made up of three rings.
Bouvard	37 S	87 W	129	Schickard area. Central ridge rising to a peak.
Brayley	21 N	37 W	16	On Procellarum. Low c.p.
Breislak	48 S	18 E	32	S.E. of Maurolycus. Fairly regular. No c.p.
Brenner	39 S	39 E	90	Broken. Adjoins Janssen (S. uplands).
Briggs	26 N	69 W	38	On Procellarum (Otto Struve area). Similar to Seleucus.
Brisbane	50 S	65 E	47	Australe area. Fairly regular.
Brown	46 S	18 W	40	N.E. of Longomontanus. Irregular.
Bruce	1 N	0	9	In Sinus Medii. Distinct.
Buch	39 S	18 E	48	Maurolycus area; adjoins Büsching. Regular.
Bullialdus	21 S	22 W	50	On Nubium. Massive, terraced walls; c.p.
Burckhardt	31 N	57 E	48	N. of Cleomedes. Member of complex group.
Bürg	45 N	28 E	48	N. of Mortis. Large c.p. with pit. Associated with major rill-system.
Burnham	14 S	7 E	29	Albategnius area. Low walls.
Büsching	38 S	20 E	58	Adjoins Buch; is less regular.
Byrgius	25 S	65 W	64	W. of Humorum, Byrgius A, on its E. crest, is a ray-centre.

Name	Lat	Long	Diam, km	Notes
Cabæus	85 S	20 W	140	S. polar uplands. Fairly high walls.
Calippus	39 N	11 E	31	N. end of Caucasus. Irregular.
Campanus	28 S	28 W	38	W. edge of Nubium; pair with Mercator. c.p.
Capella	8 S	36 E	48	Uplands N. of Nectaris. Large c.p. Cut by crater-valley.
Capuanus	34 S	26 W	56	Edge of Epidemiarum. Domes on floor.
Cardanus	13 N	73 W	52	On Procellarum; pair with Krafft. c.p.
Carlini	34 N	24 W	8	On Imbrium. Rather bright.
Carpenter	70 N	50 W	65	N. of Frigoris; adjoins Anaximander.
Carrington	44 N	62 E	25	Messala group.
Casatus	75 S	35 W	104	S. of Clavius; intrudes into Klaproth. High walls.
Cassini	40 N	5 E	58	Edge of Nebularum. Low walls. Contains deep crater, A.
Cassini, J. J.	68 N	16 W	50	Near Philolaus (N. uplands). Irregular ridge-bounded area.
Catharina	18 S	24 E	89	Theophilus group. Rough floor: no c.p.
Cauchy	10 N	39 E	13	On Tranquillitatis. Bright.
Cavalerius	5 N	67 W	64	In Hevel group. Central ridge.
Cavendish	25 S	54 W	52	W. of Humorum. Fairly high walls.
Cayley	4 N	15 E	13	Uplands W. of Tranquillitatis. Very bright.
Celsius	34 S	20 E	45	Rabbi Levi group. Elliptical.
Censorinus	0	32 E	5	Uplands S.E. of Tranquillitatis. Brilliant
Cepheus	41 N	46 E	48	Somniorum area; pair with Franklin.
Chacornac	30 N	32 E	48	Edge of Serenitatis, near Posidonius. Outline irregular.
Challis	78 N	9 E	56	N. polar area. Pair with Main.
Chevallier	45 N	52 E	47	Atlas area. Low walls.
Chladni	4 N	1 E	8	Sinus Medii area; abuts on Murchison. Bright.
Cichus	33 S	21 W	32	Just S. of Nubium. Well-formed.
Clairaut	48 S	14 E	64	Maurolycus area. Broken walls.
Clausius	37 S	44 W	24	Schickard area. Bright and distinct.
Clavius	56 S	14 W	232	S. uplands. Massive walls. Curved line of craters on floor. No c.p.
Cleomedes	27 N	55 E	126	Crisium area. Distorted by Tralles.
Cleostratus	60 N	74 W	70	Pythagoras area. Distinct.
Colombo	15 S	46 E	81	In Fœcunditatis. Irregular. Broken by large crater, A.
Condamine	53 N	28 W	48	Edge of Frigoris. Fairly regular.
Condorcet	12 N	70 E	72	Outside Crisium. Regular; no c.p.
Conon	22 N	2 E	21	In Apennine uplands. Fairly distinct.
Cook	18 S	49 E	42	Edge of Fœcunditatis. Darkish floor.
Copernicus	10 N	20 W	97	Great ray-crater. Massive walls. Central mountain group.
Crozier	14 S	51 E	24	Fœcunditatis area, S.E. of Colombo. c.p.
Crüger	17 S	67 W	48	S.W. of Procellarum. Very dark floor.
Curtius	77 S	5 E	81	Moretus area. Massive, terraced walls.
Cuvier	50 S	10 E	66	Licetus group. High walls; c.p.
Cyrillus	13 S	24 E	97	Theophilus group. Low c.p.
Cysatus	66 S	7 W	47	Moretus area. High walls.
Daguerre	12 S	34 E	32	In Nectaris. Very low walls.
Damoiseau	5 S	61 W	35	E. of Grimaldi. Very irregular.
Daniell	35 N	31 E	24	Adjoins Posidonius. No c.p.
Darney	15 S	24 W	17	On Nubium (Fra Mauro). Bright.
D'Arrest	2 N	15 E	30	W. of Sabine-Ritter pair. Low walls.
Darwin	20 S	69 W	130	Grimaldi area. Very irregular; low walls. Contains a large dome.
Da Vinci	9 N	45 E	30	Somnii area. Low walls.
Davy	12 S	8 W	32	E. edge of Nubium. Irregular walls.
Dawes	17 N	26 E	23	Between Serenitatis and Tranquillitatis. Distinct.
Debes	29 N	52 E	48	Outside Cleomedes. Fusion of 2 rings.
Dechen	46 N	67 W	12	On Procellarum (N.W. area). Not bright.
De Gasparis	26 S	50 W	48	W. of Humorum, S. of Mersenius. Fairly regular.
Delambre	2 S	18 E	52	Tranquillitatis area. High walls.
De la Rue	67 N	56 E	160	Adjoins Endymion to N.W. Very low, broken walls.
Delaunay	22 S	3 E	45	Albategnius area, near Faye. Irregular.
De l'Isle	30 N	35 W	22	On Procellarum; pair with Diophantus. c.p.
Delmotte	27 N	60 E	31	N. of Crisium. Not prominent.
Deluc	55 S	3 W	45	Clavius area. Moderate walls.
Dembowski	3 N	7 E	23	E. of Medii. Low walls.
Democritus	62 N	35 E	37	Highlands N. of Frigoris. Very deep.
Demonax	85 S	35 E	121	Boguslawsky area. Fairly regular.
De Morgan	3 N	15 E	8	Uplands W. of Tranquillitatis. Bright.
Descartes	12 S	16 E	48	N.E. of Abulfeda. Low, broken walls.

Crater Copernicus (Cdr. H. R. Hatfield)

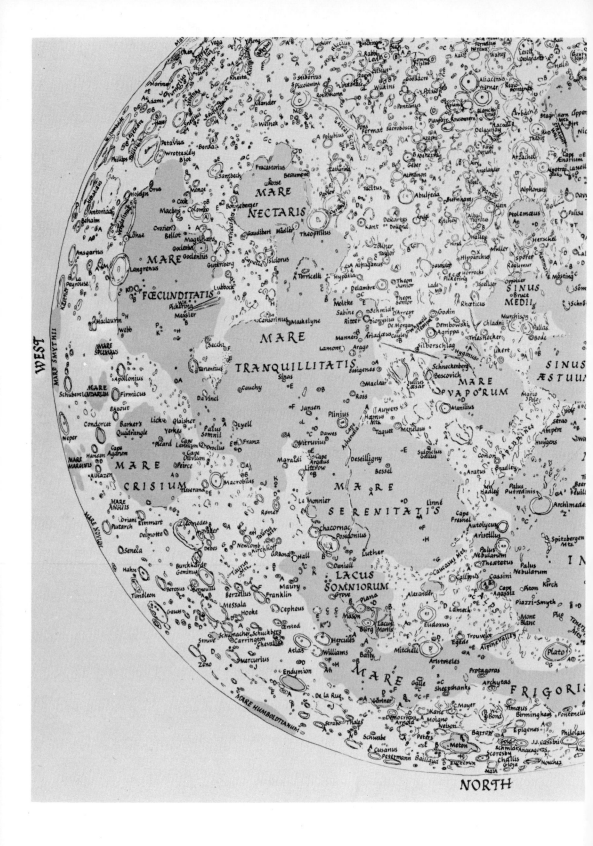

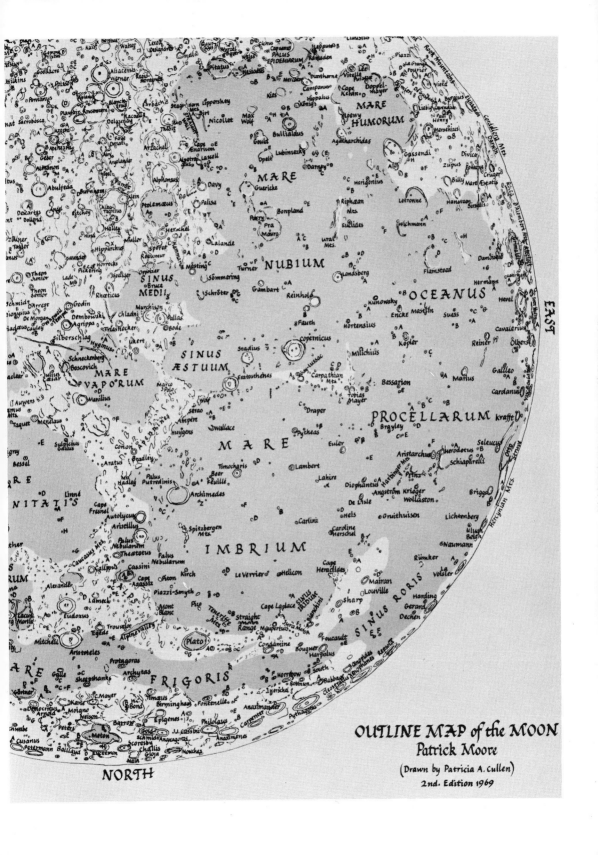

EAST

OUTLINE MAP of the MOON
Patrick Moore
(Drawn by Patricia A. Cullen)
2nd. Edition 1969

NORTH

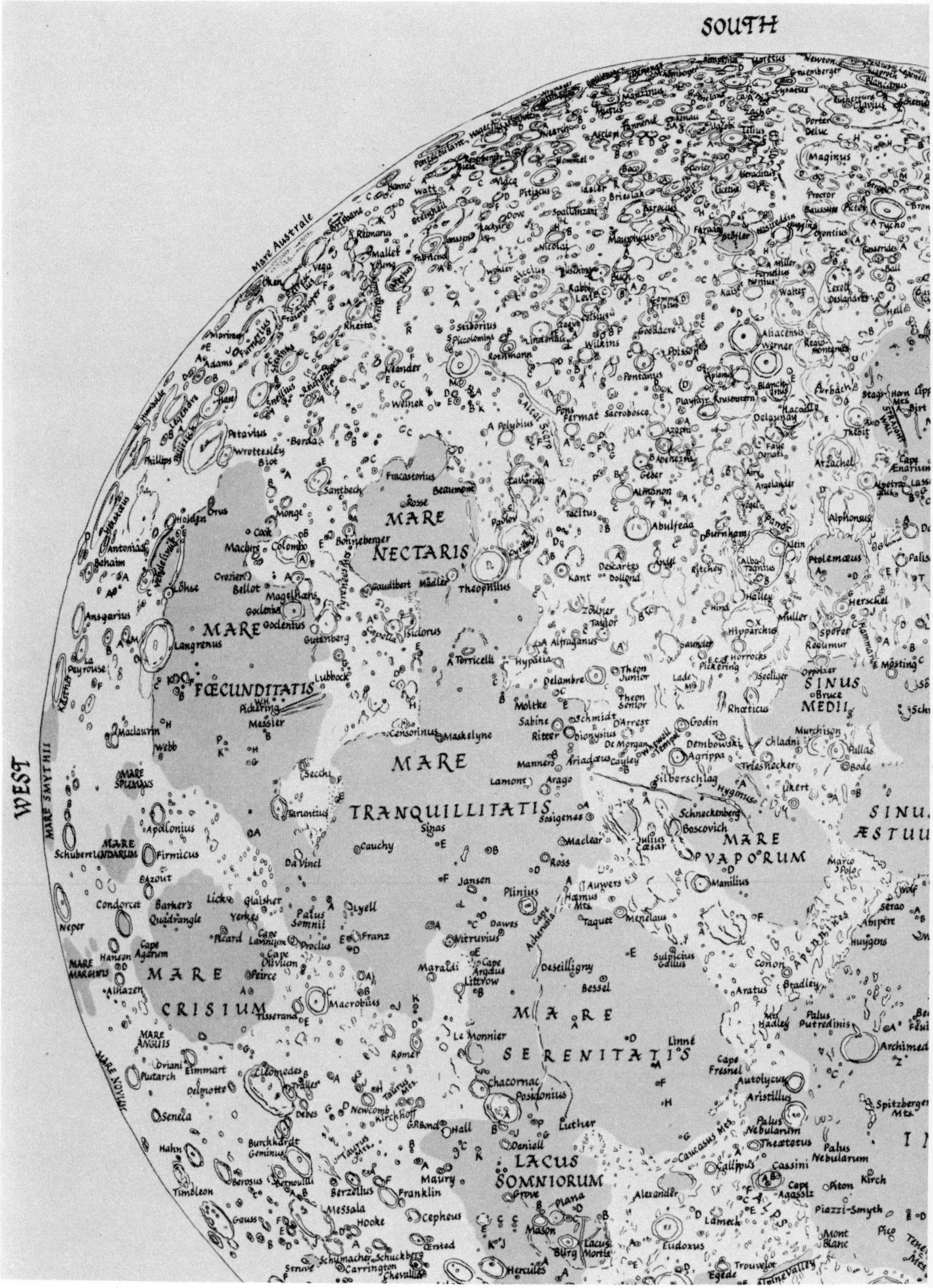

WEST

MARE AUSTRALE

MARE NECTARIS

MARE FOECUNDITATIS

MARE TRANQUILLITATIS

MARE VAPORUM

SINUS MEDII

SINUS AESTUUM

MARE CRISIUM

MARE SERENITATIS

LACUS SOMNIORUM

MARE SPUMANS

MARE UNDARUM

MARE SMYTHII

MARE MARGINIS

MARE ANGUIS

MARE NOVUM

SOUTH

EAST

MARE HUMORUM

MARE NUBIUM

OCEANUS PROCELLARUM

SINUS MEDII

SINUS AESTUUM

MARE VAPORUM

MARE IMBRIUM

SINUS IRIDUM

SINUS RORIS

PALUS EPIDEMIARUM

Palus Putredinis

Palus Nebularum

Maginus · Tycho · Clavius · Copernicus · Kepler · Archimedes · Aristarchus · Eratosthenes · Plato · Ptolemæus · Alphonsus · Arzachel · Hipparchus · Albategnius · Bullialdus · Gassendi · Letronne · Flamsteed · Grimaldi · Riccioli · Mersenius · Schickard · Wargentin · Longomontanus · Walter · Werner · Aliacensis · Abulfeda · Theophilus · Cyrillus · Catharina · Piccolomini · Fracastorius

Herschel · Lalande · Mösting · Sömmering · Gambart · Reinhold · Landsberg · Euclides · Herigonius · Encke · Kunowsky · Hortensius · Milichius · Tobias Mayer · Marius · Reiner · Cavalerius · Hevel · Olbers

Manilius · Menelaus · Julius Cæsar · Agrippa · Godin · Pallas · Bode · Murchison · Triesnecker · Hyginus · Ukert · Boscovich · Silberschlag

Conon · Aratus · Bradley · Hadley · Timocharis · Beer · Feuillée · Lambert · Euler · Pytheas · Lahire · Diophantus · Delisle · Carlini · Gruithuisen · Lichtenberg · Naumann · Rümker

Autolycus · Aristillus · Cassini · Theætetus · Kirch · Piazzi-Smyth · Le Verrier · Helicon · Mairan · Louville · Sharp · Mont Blanc · Tenerife Mts · Straight Range · Harding · Gerard · Oenopides

Caucasus Mts · Apennines · Carpathian Mts · Hercynian Mts · Harbinger Mts · Spitzbergen Mts

Cape Agassiz · Cape Fresnel · Cape Heraclides · Cape Laplace · Cape Kelvin · Cape Enarium

Copernicus · Stadius · Eratosthenes · Draper · Bessarion · Marco Polo · Wolf · Wallace · Huygens · Ampère · Conon

Ptolemæus · Alphonsus · Arzachel · Davy · Guericke · Bonpland · Parry · Fra Mauro · Gould · Opelt · Lubiniezky · Damoiseau · Byrgius · Fourier · Vieta · Doppelmayer · Vitello · Hippalus · König · Kies · Campanus · Mercator · Capuanus · Hesiodus · Pitatus · Cichus

Purbach · Regiomontanus · Thebit · Nicollet · Birt · Lippershey · Max Wolf

Ritchey · Albategnius · Halley · Hind · Müller · Spörer · Seeliger · Rhæticus · Réaumur · Oppolzer

De Morgan · Cayley · Whewell · D'Arrest · Dionysius · Dembowski · Arago · Manners · Sosigenes · Ross · Plinius

Schröter · Turner · Gay-Lussac · Kepler · Suess · Maestlin

Mairan · Sharp · Foucault · Bianchini · Maupertuis · Condamine · Fontenelle · Horrebow · Anaximenes · Philolaus

Aristarchus · Herodotus · Schiaparelli · Seleucus · Briggs · Wollaston · Angström · Krieger · Prinz

Theon Junior · Theon Senior · Taylor · Zöllner · Kant · Descartes · Andel · Abenezra · Geber · Almanon · Abulfeda · Burnham · Pons · Sacrobosco · Azophi · Apianus · Playfair · Krusenstern · Blanchinus · La Caille · Delaunay · Faye · Donati · Airy · Argelander · Parrot · Vogel

Newton · Moretus · Gruemberger · Cysatus · Curtius · Short · Bettinus · Kinau · Schmidt · Simpelius · Manzinus · Boguslawsky · Mutus · Hommel · Vlacq · Rosenberger · Nearch · Hagecius · Pitiscus · Baco · Barocius · Breislak · Maurolycus · Buch · Büsching · Riccius · Rabbi Levi · Zagut · Lindenau · Rothmann · Lockyer · Santbech · Piccolomini · Stiborius

Phocylides · Wargentin · Nasmyth · Inghirami · Baade · Bouvard · Drebbel · Lehmann · Piazzi · Lagrange · Fourier · Vieta · Henry · Rost · D'Alembert Mts · Leibnitz Mts · Doerfel Mts

Clausius · Elger · Hainzel · Epimenides · Mee · Hippalus · Agatharchides · Loewy · Puiseux · Liebig · Lacroix · Vitello · Cruger · Sirsalis · Rocca · Darwin

Maclear · Auwers · Sulpicius Gallus · Bessel · Linné · Palus Nebularum · Calippus · Alexander · Eudoxus · Lamèch · Trouvelot · Galle · Egede · C. Mayer · Sheepshanks · Bürg · Lacus Mortis

Name	Lat	Long	Diam, km	Notes
Deseilligny	21 N	21 E	9	On Serenitatis. Distinct.
Deslandres	32 S	61 W	186	Ruined enclosure W. of Walter.
De Vico	20 S	60 W	24	W. of Gassendi. Deep.
Dionysius	3 N	17 E	19	On edge of Tranquillitatis. Brilliant.
Diophantus	28 N	34 W	18	On Procellarum; pair with De l'Isle. c.p.
Dollond	10 S	14 E	10	W. of Theophilus. Borders a large "ghost".
Donati	21 S	5 E	35	S. of Albategnius; pair with Faye. Irregular. c.p.
Doppelmayer	28 S	41 W	68	Bay in Humorum. Remnant of c.p.
Dove	47 S	32 E	18	Janssen area. Low walls.
Draper	18 N	22 W	10	In Imbrium, S. of Pytheas. One of a pair.
Drebbel	41 S	49 W	29	Schickard area. Well-formed.
Drygalski	80 S	55 W	140	Adjoins Legentil (Bailly area).
Dubiago	4 N	70 E	48	Smythii area. Regular.
Dunthorne	30 S	32 W	18	Edge of Epidemiarum. Broad walls.
Egede	49 N	11 E	49	Near Alpine Valley. Low walls. Lozenge-shaped.
Eichstädt	23 S	80 W	52	Orientale area. Regular.
Eimmart	24 N	65 E	42	Near edge of Crisium. Regular.
Einstein	18 N	86 W	160	W. of Procellarum; Otto Struve area. Central crater.
Elger	35 S	30 W	24	Edge of Epidemiarum. Low-walled. Imperfect.
Encke	5 N	37 W	32	On Procellarum; Kepler area.
Endymion	55 N	55 E	117	Humboldtianum area. Darkish floor.
Epigenes	73 N	4 W	52	N. of Frigoris. Broad walls.
Epimenides	41 S	30 W	24	One of a pair E. of Hainzel.
Eratosthenes	15 N	11 W	61	End of Apennines. Very deep. c.p.
Euclides	7 S	29 W	12	Near Riphæan Mts. Lies on bright nimbus.
Eudoxus	44 N	16 E	64	S. of Frigoris. Pair with Aristoteles.
Euler	23 N	29 W	25	On Imbrium. Minor ray-centre.
Fabricius	43 S	42 E	89	Intrudes into Janssen. Rough floor. c.p.
Faraday	42 S	8 E	64	Intrudes into Stöfler. Irregular.
Fauth	6 N	20 W	13	On Procellarum, S. of Copernicus. Double crater (with A).
Faye	21 S	4 E	35	S. of Albategnius; pair with Donati. Irregular. c.p.
Fermat	23 S	20 E	40	Altai area. Distinct.
Fernelius	38 S	5 E	64	N. of Stöfler. Rather irregular.
Feuillé	27 N	10 W	13	On Imbrium. Twin with Beer
Firmicus	7 N	64 E	56	S. of Crisium. Dark floor. No c.p.
Flammarion	3 S	4 W	72	N.W. of Ptolemæus. Irregular enclosure.
Flamsteed	5 S	44 W	19	On Procellarum. Associated with 100-km 'ghost'.
Fontana	16 S	57 W	48	Between Billy and Crüger. c.p.
Fontenelle	63 N	19 W	45	N. edge of Frigoris. Deep, distinct.
Foucault	50 N	40 W	24	Jura Mts. area. Bright and deep.
Fourier	30 S	53 W	58	W. of Humorum. Terraced, with central crater.
Fracastorius	21 S	33 E	97	Bay at S. of Nectaris.
Fra Mauro	6 S	17 W	81	On Nubium; low walls. Trio with Bonpland and Parry.
Franklin	39 N	48 E	55	Regular. S.E. of Atlas.
Franz	16 N	40 E	34	Edge of Somnii. Low-walled.
Fraunhofer	39 S	59 E	48	S. of Furnerius. N.W. wall broken by craters.
Furnerius	36 S	60 E	129	In Petavius chain. Rather broken walls.
Galilaei	10 N	63 W	15	On Procellarum. Inconspicuous.
Galle	56 N	22 E	24	On Frigoris. Distinct.
Gambart	1 N	15 W	26	On Procellarum, 200 km S.S.E. of Copernicus. Regular. Low walls.
Gärtner	60 N	34 E	101	Bay on Frigoris; 'seaward' wall barely traceable.
Gassendi	18 S	40 W	89	Edge of Humorum; 'seaward' wall low. c.p. Many rills on floor.
Gaudibert	11 S	37 E	26	Edge of Nectaris. Low walls.
Gauricus	34 S	12 W	64	Pitatus group. Irregular outline.
Gauss	36 N	80 E	136	Along limb from Humboldtianum. High walls; c.p.
Gay-Lussac	14 N	21 W	24	N. of Copernicus. Irregular enclosure.
Geber	20 S	14 E	40	Between Almanon and Abenezra. Regular.
Geminus	35 N	57 E	90	Crisium area. Broad, terrraced walls; central hill.
Gemma Frisius	34 S	14 E	84	N. of Maurolycus. High but broken walls.
Gérard	44 N	75 W	87	W. of Roris. Fairly distinct.
Gioja	North polar		35	Fairly distinct and regular.
Glaisher	13 N	49 E	16	W. border of Crisium. Obscure.

Craters Aristoteles and Eudoxus (both lower left) with the Alps and Alpine Valley (Patrick Moore)

Name	Lat	Long	Diam, km	Notes
Goclenius	10 S	45 E	52	Edge of Fœcunditatis. Lava-flooded. Low c.p.
Godin	2 N	10 E	43	S. of Vaporum; pair with Agrippa. c.p.
Goldschmidt	75 N	0	109	E. of Anaxagoras (N. of Frigoris). Low walls; broken.
Goodacre	33 N	14 E	48	E. of Aliacensis. Low c.p.
Gould	19 S	17 W	42	'Ghost' in Nubium, E. of Bullialdus.
Grimaldi	6 S	68 W	193	W. of Procellarum. Low, irregular walls. No c.p. Darkest point on the Moon.
Grove	40 N	33 E	26	In Somniorum. Bright and deep.
Gruemberger	68 S	10 W	87	Moretus area. High walls; contains a very deep crater.
Gruithuisen	33 N	40 W	16	On Procellarum. Bright
Guericke	12 S	14 W	53	In Nubium (Fra Mauro area). Broken walls; irregular.
Gutenberg	8 S	41 E	72	Edge of Fœcunditatis, near Goclenius. Irregular.
Gyldén	5 S	1 E	48	N. of Ptolemæus; floor partly filled with lava. Crater-valley abuts on W.
Hagecius	60 S	46 E	81	Vlacq group. Wall broken by craters.
Hahn	31 N	74 E	56	E. of Cleomedes (Crisium area); pair with Berosus. Regular; c.p.
Haidinger	39 S	25 W	12	S. of Epidemiarum. Not conspicuous.
Hainzel	41 S	34 W	97	N. of Schiller. Compound; 2 coalesced rings.
Hall	34 N	37 E	25	Somniorum area, E. of Posidonius.
Halley	8 S	6 E	35	Hipparchus group; pair with Hind. Regular.
Hanno	57 S	73 E	64	Australe area. Darkish floor.
Hansen	44 S	83 E	36	Crisium area, near Agarum; similar to Alhazen. Regular.
Hansteen	12 S	52 W	52	S. edge of Procellarum; pair with Billy. Regular.
Harding	43 N	70 W	23	Roris area. Low walls.
Harpalus	53 N	43 W	52	Edge of Frigoris. Deep.
Hase	30 S	63 E	77	S. of Petavius. Rather irregular.
Heinsius	39 S	18 W	72	Tycho area. Irregular: 3 craters on its S. wall.
Heis	32 N	32 W	12	On Imbrium. Bright.
Hekatæus	23 S	84 E	180	S.E. of Vendelinus. Irregular walls; c.p.
Helicon	40 N	23 W	29	Iridum area. Pair with Le Verrier. Regular.
Hell	32 S	8 W	31	In W. of Deslandres. Low c.p.
Helmholtz	72 N	78 E	97	S. uplands; Boussingault area. Fairly regular.
Henry, Paul	24 S	57 W	40	W. of Humorum; distinct. Pair with Prosper Henry.
Henry, Prosper	24 S	59 W	45	Pair with Paul Henry.
Heraclitus	49 S	6 E	92	Cuvier-Licetus group, S. of Stöfler. Very irregular.
Hercules	46 N	39 E	72	Companion to Atlas. Bright walls; deep interior crater.
Herigonius	13 S	34 W	16	N.E. of Gassendi. Bright; c.p.
Hermann	1 S	57 W	16	On Procellarum, E. of Lohrmann. Bright.
Herodotus	23 N	50 W	37	Pair with Aristarchus; fairly regular. Associated with great valley.
Herschel	6 S	2 W	45	N. of Ptolemæus. Terraced walls; large c.p.
Herschel, Caroline	34 N	31 W	13	On Imbrium; group with Carlini and De l'Isle. Bright.
Herschel, John	62 N	41 W	145	N. of Frigoris. Ridge-bordered enclosure.
Hesiodus	29 S	16 W	45	Companion to Pitatus (S. of Nubium). A major rill runs S.W. from it.
Hevel	2 N	67 W	122	Grimaldi chain. Convex floor, containing rills. c.p.
Hind	8 S	7 E	26	Hipparchus group; pair with Halley. Regular.
Hippalus	25 S	30 W	61	Bay on Humorum; remnant of c.p. Associated with rills.
Hipparchus	6 S	5 E	145	E. of Ptolemæus; pair with Albategnius. Low, broken walls.
Holden	19 S	63 E	40	S. of Vendelinus. Deep.
Hommel	54 S	33 E	121	S. uplands. 2 large craters on floor.
Hooke	41 N	55 E	43	W. of Messala. Fairly regular.
Horrebow	59 N	41 W	32	Frigoris area, outside J. Herschel. Deep; pair with Robinson.
Horrocks	4 S	6 E	29	Within Hipparchus. Regular.
Hortensius	6 N	28 W	16	On Procellarum. Bright. Domes to N.
Huggins	41 S	2 W	48	Encroaches on Orontius. Irregular.
Humboldt, W.	27 S	81 E	193	E. of Petavius. Irregular. Rills on floor.
Hyginus	8 N	6 E	6	Depression on Vaporum. In great crater-rill.
Hypatia	4 S	23 E	48	S. of Tranquillitatis. Triangular; low walls.

Near side of the Moon (P. W. Foley)

Name	Lat	Long	Diam, km	Notes
Ideler	49 S	22 E	44	S.E. of Maurolycus. Distinct.
Inghirami	48 S	70 W	97	Schickard area. Regular; c.p.
Isidorus	8 S	33 E	48	Outside Nectaris; pair with Capella. Deep.
Jacobi	57 S	12 E	66	Heraclitus group. Fairly distinct.
Jansen	14 N	29 E	26	In Tranquillitatis. Low walls. Darkish floor.
Janssen	46 S	40 E	170	S. uplands. Great ruin, broken in N. by Fabricius.
Julius Cæsar	9 N	15 E	71	Vaporum area. Low, irregular walls. Very dark floor.
Kaiser	36 S	6 E	48	N. of Stöfler. Well-marked. No c.p.
Kane	63 N	26 E	45	N. of Frigoris. Fairly regular.
Kant	11 S	20 E	30	W. of Theophilus. Large c.p. with summit pit.
Kapteyn	13 S	70 E	35	E. of Langrenus. Not prominent.
Kästner	6 S	80 E	129	Smythii area. Distinct.
Kepler	8 N	38 W	35	On Procellarum. Major ray-centre.
Kies	26 S	23 W	42	On Nubium; Bullialdus area. Very low walls.
Kinau	60 S	15 E	42	Jacobi group. High walls; c.p.
Kirch	39 N	6 W	11	In Imbrium. Bright.
Kircher	67 S	45 W	74	Bettinus chain; Bailly area. Very high walls.
Kirchhoff	30 N	39 E	50	Cleomedes area; larger of 2 craters W. of Newcomb.
Klein	12 N	3 E	43	In Albategnius. Regular; c.p.
König	24 S	25 W	21	In Nubium; Bullialdus area. Deep.

Name	Lat	Long	Diam, km	Notes
Krieger	29 N	46 W	23	Aristarchus area. Distinct, but walls broken by craterlets.
Krusenstern	26 S	6 E	48	Werner area. Not prominent.
Kunowsky	3 N	32 W	21	On Procellarum, E. of Encke. Low central ridge.
Lacaille	24 S	1 E	53	Werner area. Irregular.
Lacroix	38 S	59 W	38	N. of Schickard. Fairly regular; c.p.
Lade	1 S	10 E	48	Highlands S. of Godin. Low walls.
Lagalla	45 S	23 W	63	Abuts on Wilhelm I. Low-walled, irregular.
Lagrange	33 S	72 W	165	W. of Humorum. Low walls; rough floor.
Lalande	4 S	8 W	24	Ptolemæus area. Low c.p.
Lambert	26 N	21 W	29	In Imbrium. Bright. Central crater.
Lamé	15 S	65 E	81	On N.E. wall of Vendelinus. Well-formed.
Lamèch	43 N	13 E	95	Eudoxus area. V. low, irregular walls.
Lamont	5 N	23 E	62	On Tranquillitatis; Arago area. Low walls.
Landsberg	0	26 W	42	On Nubium. Massive walls; c.p.
Langrenus	9 S	61 E	137	E. edge of Fœcunditatis. Massive walls; c.p. Petavius chain.
La Peyrouse	10 S	78 E	72	E. of Fœcunditatis; pair with Ansgarius. Distinct.
Lassell	16 S	8 W	23	In Nubium; Alphonsus area. Low walls.
La Voisier	36 N	70 W	71	W. of Procellarum. Well-formed.
Lee	31 S	41 W	42	S. edge of Humorum; damaged by lava.
Legendre	29 S	70 E	74	W. Humboldt area. Central ridge.
Legentil	73 S	80 E	140	S. of Bailly. Distinct.
Lehmann	40 S	56 W	45	Schickard area. Irregular.
Le Monnier	26 N	31 E	55	Bay on Serenitatis; smooth floor.
Lepaute	33 S	34 W	14	Edge of Epidemiarum. Distinct.
Letronne	10 S	43 W	113	Bay at S. edge of Procellarum; N. wall destroyed. Low c.p.
Le Verrier	40 N	20 W	25	Imbrium area; pair with Helicon. Distinct.
Lexell	36 S	4 W	63	Edge of Deslandres; N. wall reduced. Remnant of c.p.
Licetus	47 S	6 E	74	Cuvier-Heraclitus group (Stöfler area). Fairly regular; c.p.
Lichtenberg	32 N	68 W	19	Edge of Procellarum. Minor ray-centre.
Lick	12 N	53 E	35	Edge of Crisium. Incomplete.
Liebig	24 S	48 W	45	W. of Humorum. Moderate walls.
Lilius	54 S	6 E	52	Jacobi group. High walls; c.p.
Lindenau	32 S	25 E	56	Rabbi Levi group. Terraced walls.
Linné	28 N	12 E	11	On Serenitatis. Craterlet surrounded by nimbus.
Lippershey	26 S	10 W	6	On Nubium; Pitatus area. Distinct.
Littrow	22 N	31 E	35	Edge of Serenitatis. Irregular.
Lockyer	46 S	37 E	48	Intrudes into Janssen. Bright walls.
Loewy	23 S	33 W	23	E. edge of Humorum. Distinct.
Lohrmann	1 S	67 W	45	Between Grimaldi and Hevel. c.p.
Lohse	14 S	60 E	32	On N. wall of Vendelinus. Deep; c.p.
Longomontanus	50 S	21 W	145	Clavius area. Complex walls; much floor detail.
Louville	44 N	47 W	35	In Jura Mts. Low walls; darkish floor.
Lubbock	4 S	42 E	13	W. edge of Fœcunditatis. Fairly bright.
Lubiniezky	18 S	24 W	45	In Nubium, Bullialdus area. Low walls.
Luther	33 N	24 E	8	In Serenitatis (N. part). Distinct.
Lyell	14 N	41 E	43	W. edge of Sumnii. Darkish floor.
Maclaurin	2 S	68 E	45	W. of Smythii. Concave floor; uneven walls.
Maclear	11 N	20 E	18	On Tranquillitatis. Darkish floor.
McClure	15 S	50 E	23	In Fœcunditatis (Goclenius area). Bright
Macrobius	21 N	46 E	68	Crisium area: high walls, compound c.p.
Mädler	11 S	30 E	32	On Nectaris. Irregular.
Maestlin	5 N	41 W	50	On Procellarum, near Encke. Obscure.
Magelhæns	12 S	44 E	40	Edge of Fœcunditatis; pair with A. Darkish floor.
Maginus	50 S	6 W	177	Clavius area; irregular walls. Obscure near Full Moon.
Main	81 N	9 E	48	N. polar area; twin with Challis.
Mairan	42 N	43 W	40	Jura Mts. area. Bright; not perfectly regular.
Mallet	45 S	53 E	50	Rheita Valley area; not conspicuous.
Manilius	15 N	9 E	36	On edge of Vaporum. c.p.; brilliant walls.
Manners	5 N	20 E	16	On Tranquillitatis (Arago area). Bright.
Manzinus	68 S	25 E	90	Boguslawsky area. High, terraced walls.
Maraldi	19 N	35 E	37	In N. of Tranquillitatis. Distinct.
Marco Polo	15 N	2 W	19	Apennines area. Irregular; darkish floor.
Marinus	50 S	75 E	48	Australe area. Distinct, with c.p.
Marius	12 N	51 W	42	In Procellarum; pair with Reiner. Low c.p.
Marth	31 S	29 W	13	In Epidemiarum. Concentric crater.
Maskelyne	2 N	30 E	24	In Tranquillitatis. Low c.p.

Name	Lat	Long	Diam, km	Notes
Mason	43 N	30 E	31	Bürg area; companion to Plana.
Maupertuis	50 N	27 W	43	In Juras. Irregular mountain enclosure.
Maurolycus	42 S	14 E	109	E. of Stöfler. Rough floor; central peak group.
Maury	37 N	40 E	18	Atlas area. Bright and deep.
Mayer, C.	63 N	17 E	39	N. of Frigoris. Rhomboidal.
Mayer, T.	16 N	29 W	35	In Carpathian Mts. c.p.
Mee	44 S	35 W	114	Abuts on Hainzel. Low, broken walls.
Menelaus	16 N	16 E	32	In Hæmus Mts. Brilliant. Interior peak.
Mercator	29 S	26 W	38	Pair with Campanus (Humorum area). Noticeably dark floor.
Mercurius	56 N	65 E	53	Humboldtianum area. Low c.p.
Mersenius	21 S	49 W	72	W. of Humorum. Convex floor. Rills nearby.
Messala	39 N	60 E	128	Humboldtianum area. Oblong; low, broken walls.
Messier	2 S	48 E	13	On Fœcunditatis. Twin with Messier A (= W. H. Pickering). 'Comet' ray to W.
Metius	40 S	44 E	81	Janssen group. Pair with Fabricius. Distinct.
Meton	74 N	25 E	170	N. polar area, near Scoresby. Compound formation; smooth floor.
Milichius	10 N	30 W	13	On Procellarum. Bright. Dome to the W.
Miller	39 S	1 E	48	Orontius group. Fairly distinct.
Mitchell	50 N	20 E	19	Abuts on Aristoteles. Distinct.
Möltke	1 S	24 E	7	N. edge of Tranquillitatis. Distinct.
Monge	19 S	48 E	32	Edge of Fœcunditatis. Rather irregular.
Montanari	46 S	20 W	87	Longomontanus area. Distorted.
Moretus	70 S	8 W	105	Southern uplands. Massive walls; very high c.p.
Mösting	1 S	6 W	26	Medii area. Mösting A, to the S.S.E., is a bright craterlet used as a reference point on older maps.
Müller	7 S	3 E	24	Ptolemæus area. Fairly regular.
Murchison	5 N	0	56	Edge of Medii. Low-walled; irregular.
Mutus	63 S	30 E	81	S. uplands. 2 large craters on floor.
Nasireddin	41 S	0	48	Orontius group. Fairly distinct.
Nasmyth	52 S	53 W	74	Phocylides group. Fairly regular; no c.p.
Naumann	35 N	62 W	10	In N. of Procellarum. Bright walls.
Neander	31 S	40 E	48	Rheita Valley area. Well-formed.
Nearch	58 S	39 E	61	Vlacq area. Craterlets on floor.
Neison	68 N	28 E	45	Meton area. Regular; no c.p.
Neper	7 N	83 E	113	Marginis/Smythii area. Deep.
Neumayer	62 S	75 E	80	Boussingault area. Distinct.
Newcomb	30 N	44 E	52	Cleomedes area. S. wall broken by crater.
Newton	78 S	20 W	113	Moretus area. Unusually deep; irregular.
Nicolai	42 S	26 E	43	Janssen area. Regular.
Nicollet	22 S	12 W	16	In Nubium, W. of Birt. Distinct.
Nöggerath	49 S	45 W	35	Schiller area. Low walls.
Nonius	35 S	4 E	32	Stöfler area. Fairly regular.
Œnopides	57 N	65 W	68	Limb area, near Roris. High walls.
Œrsted	43 N	47 E	40	Edge of Somniorum. Rather irregular.
Oken	44 S	78 E	80	Australe area. Darkish floor.
Olbers	7 N	78 W	64	Grimaldi area. Major ray-centre.
Opelt	16 S	18 W	43	'Ghost' in Nubium, E. of Bullialdus.
Oppolzer	2 S	1 W	48	Edge of Medii. Low walls
Orontius	40 S	4 W	84	N.E. of Tycho; one of a group. Irregular.
Palisa	9 S	7 W	32	Alphonsus area. On edge of larger ring.
Palitzsch	28 S	64 E	97 × 32	Outside Petavius to the E. Really a crater-chain.
Pallas	5 N	2 W	47	Medii area; adjoins Murchison. c.p.
Palmieri	29 S	48 W	48	Humorum area. Floor darkish.
Parrot	15 S	3 E	64	Albategnius area. Very irregular, compound structure.
Parry	8 S	16 W	42	On Nubium; Fra Mauro group. Fairly regular.
Peirce	18 N	53 E	19	In Crisium. Fairly regular.
Peirescius	46 S	71 E	64	Australe area. Rather irregular.
Pentland	64 S	12 E	72	S. uplands, near Curtius. High walls; c.p.
Petavius	25 S	61 E	170	One of a great chain. High walls; c.p.; interior rill.
Phillips	26 S	78 E	120	W. of W. Humboldt. Central ridge.
Philolaus	75 N	33 W	74	Frigoris area; pair with Anaximenes. Deep regular.
Phocylides	54 S	58 W	97	Schickard group. Much floor-detail.
Piazzi	36 S	68 W	90	Schickard area. Very broken walls.
Piazzi Smyth	42 N	3 W	10	On Imbrium. Bright.

Mare Serenitatis (Cdr H. R. Hatfield)

Name	Lat	Long	Diam, km	Notes
Picard	15 N	55 E	34	Largest crater on Crisium. C. hill.
Piccolomini	30 S	32 E	80	End of Altai Scarp. High walls; c.p.
Pickering	3 S	7 E	16	Hipparchus area. Distinct.
Pictet	43 S	7 W	48	Closely E. of Tycho. Fairly regular.
Pitatus ·	30 S	14 W	80	S. edge of Nubium. Passes connect it with Hesiodus.
Pitiscus	51 S	31 E	80	S. uplands. Walls narrow in places.
Plana	42 N	28 E	39	Bürg area; pair with Mason. Darkish floor.
Plato	51 N	9 W	97	Edge of Imbrium; Alps area. Regular; smooth, very dark floor.
Playfair	23 S	9 E	43	Abenezra area. Fairly regular.
Plinius	15 N	24 E	48	Between Serenitatis and Tranquillitatis. C. craters.
Plutarch	25 N	75 E	64	N.E. of Crisium. c.p.
Poisson	30 S	11 E	72	Aliacensis area. Very irregular; compound.
Polybius	22 S	26 E	32	Theophilus/Catharina area. Distinct.
Pons	25 S	22 E	32	Altai area. Very thick walls.
Pontanus	28 S	15 E	45	Altai area. Regular; no c.p.
Pontécoulant	69 S	65 E	97	S. uplands. High walls.
Posidonius	32 N	30 E	96	Edge of Serenitatis. Narrow walls. Much floor detail.
Porter	56 S	10 W	40	On wall of Clavius. Very distinct. c.p.
Prinz	26 N	44 W	50	In Procellarum, N.E. of Aristarchus. Incomplete. Domes nearby.
Proclus	16 N	47 E	29	W. of Crisium. Bright; low c.p. Ray-crater.
Proctor	46 S	5 W	43	Maginus area. Fairly regular.
Protagoras	56 N	7 E	19	In Frigoris. Bright, regular.
Ptolemæus	14 S	3 W	148	Trio with Alphonsus and Arzachel. Floor smooth, but with details including a large crater, A.
Puiseux	28 S	39 W	23	On Humorum, near Doppelmayer. Very low walls.
Purbach	25 S	2 W	120	Walter group. Rather irregular.
Pythagoras	65 N	65 W	113	N.W. of tridum. High, massive walls; c.p.
Pytheas	21 N	20 W	19	On Imbrium. Bright; c.p. Minor ray-centre.
Rabbi Levi	35 S	24 E	80	One of a group of 6, S.W. of Piccolomini.
Ramsden	33 S	32 W	24	Edge of Epidemiarum. Rills nearby.
Réaumur	2 S	1 E	45	Medii area. Low walls.
Regiomontanus	28 S	0	129 × 105	Prominent, though distorted between Walter and Purbach. c.p.
Reichenbach	30 S	48 E	48	Rheita area; irregular. Crater-chain to S.E. ("Reichenbach Valley").
Reimarus	46 S	55 E	45	Rheita Valley area. Irregular.
Reiner	7 N	55 W	32	On Procellarum; pair with Marius. Low c.p.
Reinhold	3 N	23 W	48	On Procellarum, S.W. of Copernicus. Pair with Reinhold B, to the N.E.
Repsold	50 N	70 W	140	W. of Roris. One of a group.
Rhæticus	0	5 E	45	Medii area. Low walls.
Rheita	37 S	47 E	68	W. of Furnerius; sharp crests. Associated with great crater-chain ("Rheita Valley").
Riccioli	3 S	75 W	160	Companion to Grimaldi. Broken walls; very dark patches on floor. No c.p.
Riccius	37 S	26 E	80	Rabbi Levi group. Broken walls, rough floor.
Ritchey	11 S	9 E	26	E. of Albategnius. Broken walls,
Ritter	2 N	19 E	31	On Tranquillitatis; pair with Sabine. c.p.
Robinson	59 N	46 W	27	Frigoris area; similar to Horrebow. Distinct.
Rocca	15 S	72 W	97	S. of Grimaldi. Irregular walls.
Rømer	25 N	37 E	37	Taurus area. Very massive c.p. with summit pit.
Rosenberger	55 S	43 E	80	Vlacq group. Darkish floor; c.p.
Ross	12 N	22 E	29	On Tranquillitatis. c.p.
Rosse	18 S	35 E	16	On Nectaris. Bright.
Rost	56 S	34 W	55	Schiller area; pair with Weigel. Fairly regular.
Rothmann	31 S	28 E	42	Altai area. Fairly deep and regular.
Rümker	41 N	58 W	49	On Roris. Semi-ruined plateau.
Rutherfurd	61 S	12 W	40	On wall of Clavius. Distinct; c.p.
Sabine	2 N	20 E	31	On Tranquillitatis; pair with Ritter. c.p.
Sacrobosco	24 S	17 E	84	Altai area. Irregular.
Santbech	21 S	44 E	70	E. of Fracastorius. Darkish floor.
Sasserides	39 S	9 W	97	N. of Tycho. Irregular enclosure.
Saunder	4 S	9 E	43	E. of Hipparchus. Low walls.
Saussure	43 S	4 W	50	N. of Maginus. Interrupts larger ring.
Scheiner	60 S	28 W	113	Clavius area; pair with Blancanus. High walls; interior craterlet.
Schiaparelli	23 N	59 W	29	On Procellarum. Distinct.

The Straight Wall, showing the immense crater-chain of Walter, Regiomontanus and Purbach (Cdr. H. R. Hatfield)

Name	Lat	Long	Diam, km	Notes
Schickard	44 S	54 W	202	Rather low walls; great walled plain.
Schiller	52 S	39 W	180 × 97	Schickard area. Fusion of 2 rings.
Schmidt	1 N	19 E	13	In Tranquillitatis; Sabine/Ritter area. Bright.
Schömberger	76 N	30 E	65	Boguslawsky area. Regular.
Scoresby	77 N	15 E	58	Polar uplands. Deep; prominent. c.p.
Schröter	3 N	7 W	32	Medii area. Low walls.
Schubert	3 N	70 E	74	In Smythii area. Distinct.
Schumacher	42 N	55 E	40	Messala area. Fairly distinct.
Secchi	2 N	43 E	23	In Fœcunditatis. Bright walls; c.p.
Seeliger	2 S	1 E	45	N. of Hipparchus. Irregular.
Segner	59 S	48 W	74	Schiller area; pair with Zucchius. Distinct.
Seleucus	21 N	66 W	45	On Procellarum. Terraced walls; c.p.
Sharp	46 N	40 W	35	In Jura Mts. Deep; small c.p.
Sheepshanks	59 N	17 E	19	N. of Frigoris. Fairly regular.
Short	76 S	.5 W	70	Moretus group. Deep; high walls.
Shuckburgh	43 N	53 E	47	Cepheus group. Fairly regular.
Silberschlag	6 N	13 E	13	Near Ariadæus. Bright.
Simpelius	75 S	15 E	80	Moretus area. Deep.
Sinas	9 N	32 E	8	On Tranquillitatis. Not prominent.
Sirsalis	13 S	60 W	32	Just outside Procellarum; twin with Sirsalis A. Associated with great rill.
Snellius	29 S	56 E	80	Furnerius area; pair with Stevinus. High walls; c.p.
Sömmering	0	7 W	27	Medii area. Low walls.
Sosigenes	9 N	18 E	23	Edge of Tranquillitatis. Bright; low c.p.
South	57 N	50 W	98	Frigoris area. Ridge-bounded enclosure.
Spallanzani	46 S	25 E	25	W. of Janssen. Low walls.
Spörer	4 S	2 W	35	N. of Ptolemæus and Herschel. Partly lava-filled.
Stadius	11 N	14 W	70	Ghost ring E. of Copernicus. Pitted.
Steinheil	50 S	48 E	70	S.W. of Janssen. Twin with Watt.
Stevinus	33 S	54 E	70	Furnerius area; pair with Snellius. High walls; c.p.
Stiborius	34 S	32 E	44	S. of Piccolomini (Altai area). c.p.
Stöfler	41 S	6 E	145	S.E. of Walter; broken by Faraday. Smooth darkish floor.
Strabo	62 N	55 E	47	Near De la Rue. Minor ray-centre.
Street	46 S	10 W	50	S. of Tycho. Fairly regular; no c.p.
Struve	43 N	65 E	18	Adjoins Messala; lies on dark patch.
Struve, Otto	25 N	75 W	160	—each of 2 ancient rings; W. edge of Procellarum.
Suess	4 N	48 W	15	On Procellarum, W. of Encke. Obscure.
Sulpicius Gallus	20 N	12 E	13	On Serenitatis. Very bright.
Sven Hedin	5 N	75 W	98	W. of Hevel. Broken walls; irregular.
Tacitus	16 S	19 E	40	Catharina area. Polygonal; 2 floor-craters.
Tannerus	56 S	22 E	24	S. uplands, near Mutus.
Taquet	17 N	19 E	10	Just on Serenitatis, near Menelaus. Bright.
Taruntius	6 N	48 E	60	On Fœcunditatis. Concentric crater, with c.p. Rather low walls.
Taylor	5 S	17 E	40	Delambre area. Rather elliptical.
Tempel	4 N	12 E	8	Uplands W. of Tranquillitatis. Bright.
Thales	59 N	41 E	39	Near Strabo. Major ray-centre.
Theætetus	37 N	6 E	26	On Nebularum. Low c.p.
Thebit	22 S	4 W	60	Edge of Nubium, near Arzachel. W. wall broken by A, which is itself broken by F— a characteristic trio.
Theon Junior	2 S	16 E	16	Near Delambre: pair with Theon Senior. Bright.
Theon Senior	1 S	15 E	17	Pair with Theon Junior. Bright.
Theophilus	12 S	26 E	101	Edge of Nectaris. Very deep; massive walls; c.p. Chain with Cyrillus and Catharina.
Timæus	63 N	1 W	34	Edge of Frigoris. Bright.
Timocharis	17 N	13 W	35	On Imbrium. Central crater. Minor ray-centre.
Timoleon	35 N	75 E	130	Adjoins Gauss. Fairly distinct.
Tisserand	21 N	48 E	33	Crisium area. Regular; no c.p.
Torricelli	5 S	29 E	19	On Nectaris. Irregular, compound structure.
Tralles	28 N	53 E	48	On wall of Cleomedes. Very deep.
Triesnecker	4 N	4 E	23	Vaporum area. Associated with major rill-system.
Trouvelot	49 N	6 E	10	Alpine Valley area. Rather bright.
Turner	2 S	13 W	13	On Nubium, near Gambart. Deep.
Tycho	43 S	11 W	84	Brilliant southern ray-centre—greatest on the Moon. Terraced walls; c.p.
Ukert	8 N	1 E	23	Edge of Vaporum. Rills nearby.
Ulugh Beigh	29 N	85 W	70	W. of Procellarum. High walls; c.p.

Mare Humboldtianum—photographed by *Mariner 10* on its way to Venus (NASA)

Name	Lat	Long	Diam, km	Notes
Vasco da Gama	15 N	85 W	80	W. of Procellarum. Central ridge.
Vega	45 N	63 E	80	Australe area. Deep.
Vendelinus	16 S	62 E	165	Petavius chain. Irregular and broken.
Vieta	29 S	57 W	52	W. of Humorum. Low c.p.
Vitello	30 S	38 W	38	S. edge of Humorum. Concentric crater.
Vitruvius	18 N	31 E	31	Between Serenitatis and Tranquillitatis. Low but rather bright walls.
Vlacq	53 S	39 E	90	S.W. of Janssen; one of a group of six. Deep, with c.p.
Vogel	15 S	6 E	60 × 25	S.E. of Albategnius. Chain of 4 craters.
Wallace	10 N	9 W	29	In Imbrium; Apennine area. Imperfect ring; very low walls
Walter	33 S	1 E	129	Outside Nubium; massive walls, interior peak and craters. Trio with Regiomontanus and Purbach.
Wargentin	50 S	60 W	89	Schickard group. The famous plateau.
Watt	50 S	51 E	72	S.W. of Janssen. Twin with Steinheil.
Webb	1 S	60 E	26	Edge of Fœcunditatis. Darkish floor; c.p. Minor ray-centre.
Weigel	58 S	39 W	55	Schiller area; pair with Rost. Fairly regular.
Weinek	28 S	37 E	30	N.E. of Piccolomini (Altai area). Darkish floor.
Weiss	32 S	20 W	60	Pitatus group. Very irregular.
Werner	28 S	3 E	66	Pair with Aliacensis. Very regular; high walls, c.p.
Whewell	4 N	14 E	8	Uplands W. of Tranquillitatis. Bright.
Wichmann	8 S	38 W	13	On Procellarum. Associated with large ghost-crater.
Wilhelm I	43 S	20 W	97	Longomontanus area. Uneven walls.
Wilkins	30 S	20 E	64	Rabbi Levi group. Irregular.
Williams	42 N	37 E	45	Somniorum area. Not prominent.
Wilson	69 S	33 W	74	Bettinus chain, near Bailly. Deep and regular. No c.p.
Wöhler	38 S	31 E	45	Near Rabbi Levi group. Fairly regular.
Wolf	23 S	17 W	32	In S. of Nubium. Irregular; low walls.
Wollaston	31 N	47 W	11	N. of Harbinger Mts. Bright.
Wrottesley	24 S	57 E	55	Petavius group. Twin-peaked c. mountain.
Wurzelbauer	34 S	16 W	80	Pitatus group. Irregular walls; much floor-detail.
Xenophanes	57 N	77 W	108	Limb near Roris. High walls; c.p.
Yerkes	15 N	52 E	25	On W. edge of Crisium. Irregular. Low walls.
Young	42 S	51 E	45	Rheita Valley area. Irregular.
Zach	61 S	5 E	52	E. of Clavius. Fairly regular and deep.
Zagut	32 S	22 E	80	Rabbi Levi group. Irregular.
Zöllner	8 S	19 E	40	N.W. of Theophilus. Elliptical.
Zucchius	61 S	50 W	63	Schiller area; pair with Segner. Distinct.
Zupus	17 S	52 W	26	S. of Billy. Low walls; irregular. Very dark floor.

Mare Nectaris (Patrick Moore)

Recently-named Lunar Craters

Most of the craters in this list are relatively small. The names have however been approved by the International Astronomical Union.

Name	Lat	Long	Former designation
Abbot	6 N	55 W	Apollonius K
Banting	27 N	16 E	Linné E
Bowen	18 N	9 E	Manilius A
Brackett	18 N	24 E	
Cajal	13 N	31 E	Jansen F
Cameron	6 N	46 E	Taruntius C
Carmichael	20 N	40 E	Macrobius A
Clerke	22 N	29 E	Littrow B
Curtis	16 N	57 E	
Daly	5 N	57 E	Apollonius A
Daubrée	16 N	15 E	Menelaus S
Eckert	18 N	58 E	
Franck	23 N	36 E	Rømer K

Name	Lat	Long	Former designation
Freud	26 N	52 W	
Galen	22 N	5 E	Aratus A
Hadley	25 N	3 E	Hadley C
Haldane	1 S	84 E	
Hill	21 N	41 E	Macrobius B
Hornsby	24 N	12 E	Aratus CB
Houtermans	9 S	87 E	
Humason	31 N	57 W	Lichtenberg G
Huxley	20 N	5 W	Wallace B
Joy	25 N	7 E	Hadley A
Kiess	6 S	84 E	
Knox-Shaw	5 N	80 E	Banachiewicz F
Kreiken	9 S	85 E	
Lawrence	8 N	43 E	Taruntius M
Lucian	15 N	37 E	Maraldi B
Nielsen	32 N	52 W	Wollaston C
Peek	3 N	87 E	
Runge	2 S	87 E	
Sarabhai	25 N	21 E	Bessel A
Shapley	10 N	57 W	Picard H
Spurr	26 N	3 W	Archimedes M
Tacchini	5 N	86 E	Neper K
Tebbutt	10 N	54 E	Picard G
Theophrastus	18 N	39 E	Maraldi M
Väisälä	26 N	48 W	Aristarchus A
Very	26 N	25 E	Le Monnier B
Watts	9 N	46 E	Taruntius D
Widmanstätten	6 S	86 E	
Yangel	17 N	5 E	Manilius F
Zinner	27 N	59 W	Schiaparelli B

Mare Nubium (Cdr H. R. Hatfield)

Named Capes and promontories

Acherusia	17 N	22 E	Serenitatis: E. end of Hæmus Mountains.
Ænarium	19 S	7 W	Nubium: Straight Wall area.
Agarum	14 N	66 E	Crisium, near Condorcet.
Agassiz	42 N	2 E	Imbrium, near Cassini.
Argæus	19 N	29 E	Serenitatis; near Littrow.
Banat	17 N	26 W	Carpathians, near T. Mayer.
Deville	46 N	0	Imbrium: Alpine area, near Agassiz.
Fresnel	29 N	5 E	Serenitatis; end of Apennines.
Heraclides	40 N	34 W	Sinus Iridum.
Kelvin	27 S	33 W	Humorum, near Hippalus.
Laplace	46 N	26 W	Sinus Iridum.

Mountain Ranges

Alps	N. border of Imbrium.
Altai Scarp	SW of Nectaris, from Piccolomini.
Apennines	Bordering Imbrium.
Carpathians	Bordering Imbroium, to the S.
Caucasus	Separating Serenitatis and Nebularum.
Cordillera	Limb range: Grimaldi/Darwin area.
Hæmus	S. border of Serenitatis.
Harbinger	Clumps of hills in Imbrium (Aristarchus area).
Jura	Bordering Iridum.
Percy	NW. border of Humorum. Not a major range.

Pyrenees	Clumps of hills bordering Nectaris, to the E.
Riphæans	In Nubium. Short range.
Rook	Limb range, associated with Orientale.
Spitzbergen	In Imbrium, N of Archimedes. Mountain clump.
Stag's-Horn	At S. end of Straight Wall (Nubium, near Thebit).
Straight Range	In Imbrium: Plato area. Very regular.
Taurus	Mountain clumps E. of Serenitatis.
Teneriffe	In Imbrium, S. of Plato. Mountain clumps.
Ural	Extension of the Riphæans.

Named Peaks

Ampère	20 N	4 W	In Apennines.
Blanc	45 N	1 E	In Alps.
Bradley	22 N	1 E	In Apennines.
Hadley	27 N	5 E	In Apennines.
Huygens	20 N	3 W	In Apennines.
La Hire	28 N	25 W	In Imbrium (Lambert area).
Pico	46 N	9 W	In Imbrium, S of Plato (edge of old ring).
Piton	41 N	1 W	In Imbrium, near Agassiz.
Schneckenberg	9 N	9 E	'Spiral mountain', near Hyginus Rill.
Serao	17 N	6 W	In Apennines.
Wolff	17 N	7 W	In Apennines.

NOMENCLATURE OF THE FAR SIDE OF THE MOON

Names allotted to features on the Moon's far side are listed here. Older maps show, on the Moon's near side, the Leibnitz Mountains and the D'Alembert Mountains; these have proved to be not true ranges, and the names have been deleted so that they could be re-allotted. The name of Porter, in honour of the American astronomer Russell W. Porter, was suggested for a far-side crater at latitude 56°S, longitude 110°W; but following a suggestion by the present writer, it was officially transferred to a crater inside Clavius, on the near side of the Moon. Early maps of the far side showed a mountain range, recorded by the Russian probe Luna 3 in 1959, and named the Soviet Mountains.

The name has been deleted, for the excellent reason that the mountains do not exist. (The feature recorded was nothing more than a bright ray.)

The following craters are actually on the near side of the Moon, but have only been named recently, and are extremely difficult to study from Earth because they are so foreshortened:

Name	Lat		Long		Name	Lat		Long	
Aston	32	N	88	W	Hubble	22	N	87	E
Baade	47	S	83	W	Jansky	8	N	87	E
Balboa	20	N	84	W	Krasnov	31	S	89	W
Barnard	32	N	86	E	Liapunov	27	N	88	E
Boole	65	N	89	W	Lyot	48	S	88	W
Bunsen	41	N	85	W	Nicholson	27	S	85	W
Catalan	46	S	87	W	Petrov	61	S	88	E
Dalton	18	N	86	W	Pettit	28	S	86	W
Gibbs	19	S	85	W	Rynin	37	N	86	E
Graff	43	S	88	W	Schlüter	7	S	84	W
Gum	40	S	89	E	Shaler	33	S	88	W
Hamilton	44	S	83	E	Voskresensky	28	N	88	W
Hartwig	7	S	82	W	Wright	31	S	88	W
Hayn	63	N	85	E					

FEATURES ON THE FAR SIDE OF THE MOON

Features on the Moon's far side are named according to recommendations from the International Astronomical Union.

Name	Lat		Long		Name	Lat		Long	
Abbe	58	S	175	E	Belopolsky	18	S	128	W
Abul Wafa	2	N	117	E	Belyayev	23	N	143	E
Aitken	17	S	173	E	Bergstrand	19	S	176	E
Al-Biruni	18	N	93	E	Berkner	25	N	105	W
Alden	24	S	111	E	Berlage	64	S	164	W
Alekhin	68	S	130	W	Bhabha	56	S	165	W
Alter	19	N	108	W	Birkeland	30	S	174	E
Amici	10	S	172	W	Birkhoff	59	N	148	W
Amundsen	83	S	103	W	Bjerknes	38	S	113	E
Anders	42	S	144	W	Blazhko	31	N	148	W
Anderson	16	N	171	E	Bobone	29	N	131	W
Antoniadi	69	S	173	W	Boltzmann	55	S	115	W
Apollo	37	S	153	W	Bolyai	34	S	125	E
Appleton	37	N	158	E	Borman	37	S	142	W
Arrhenius	55	S	91	W	Bose	54	S	170	W
Artamonov	26	N	104	E	Boss	46	N	90	E
Artemev	10	N	145	W	Boyle	54	S	178	E
Avicenna	40	N	97	W	Bragg	42	N	103	W
Avogadro	64	N	165	E	Brashear	74	S	172	W
					Bredikhin	17	N	158	W
					Brianchon	77	N	90	W
Babcock	4	N	94	E	Bridgman	44	N	137	E
Backlund	16	S	103	E	Brouwer	36	S	125	W
Baldet	54	S	151	W	Brunner	10	S	91	E
Banachiewicz	51	N	135	W	Buffon	41	S	134	W
Barbier	24	S	158	E	Buisson	1	S	113	E
Barringer	29	S	151	W	Butlerov	9	N	109	W
Bartels	24	N	90	W	Buys-Ballot	21	N	175	E
Becquerel	41	N	129	E					
Běcvár	2	S	125	E					
Beijerinck	13	S	152	E	Cabannes	61	S	171	W
Belkovich	60	S	92	E	Cajori	48	S	168	E
Bell	22	N	96	W	Campbell	45	N	152	E
Bellingshausen	61	S	164	W	Cannizzaro	55	N	100	W

Name	Lat		Long	
Cantor	38	N	118	E
Carnot	52	N	144	W
Carver	43	S	127	E
Cassegrain	52	S	113	E
Ceraski	49	S	141	E
Chaffee	39	S	155	W
Chamberlin	59	S	96	E
Champollion	37	N	175	E
Chandler	44	N	171	E
Chang Heng	19	N	112	E
Chant	41	S	110	W
Chaplygin	6	S	150	E
Chapman	50	N	101	W
Chappell	62	N	150	W
Charlier	36	N	132	W
Chaucer	3	N	140	W
Chauvenet	11	S	137	E
Chebyshev	34	S	133	W
Chrétien	33	S	113	E
Clark	38	S	119	E
Coblentz	38	S	126	E
Cockcroft	30	N	164	W
Compton	55	N	104	E
Comrie	23	N	113	W
Comstock	21	N	122	W
Congreve	0		168	W
Cooper	53	N	176	E
Coriolis	0		172	E
Coulomb	54	N	115	W
Cremona	68	N	93	E
Crocco	47	S	150	E
Crommelin	68	S	147	W
Crookes	11	S	165	W
Curie	23	S	92	E
Cyrano	20	S	157	E
Dædalus	6	S	180	
D'Alembert	52	N	164	E
Danjon	11	S	123	E
Dante	25	N	180	
Das	14	S	152	W
Davisson	38	S	175	W
Dawson	67	S	134	W
Debye	50	N	177	W

Name	Lat	Long	Name	Lat	Long	Name	Lat	Long
De Forest	76 S	162 W	Guthnick	48 S	94 W			
Dellinger	7 S	140 E	Guyot	11 N	117 E			
Delporte	16 S	121 E						
Denning	16 S	143 E				Name	Lat	Long
De Roy	55 S	99 W	Hagen	56 S	135 E			
Deutsch	24 N	111 E	Hansky	10 S	97 E	Lacchini	41 N	107 W
De Vries	20 S	177 W	Harriott	33 N	114 E	Lamarck	57 S	158 E
Dewar	3 S	166 E	Hartmann	3 N	135 E	Lamb	43 S	101 E
Dirichlet	10 N	151 W	Harvey	19 N	147 W	Lampland	31 S	131 E
Donner	31 S	98 E	Hatanaka	29 N	122 W	Landau	42 N	119 W
Doppler	13 S	160 W	Hayford	13 N	176 W	Lane	9 S	132 E
Douglass	35 N	122 W	Healy	32 N	111 W	Langemak	10 S	119 E
Dreyer	10 N	97 E	Heaviside	10 S	167 E	Langevin	44 N	162 E
Drude	39 S	91 W	Hedin	68 N	123 E	Langmuir	36 S	129 W
Dryden	33 S	157 W	Helberg	22 N	102 W	Larmor	32 N	180
Dufay	5 N	170 E	Henderson	5 N	152 E	Lauritsen	27 S	96 E
Dugan	65 N	103 E	Hendrix	48 S	161 W	Leavitt	46 S	140 W
Dunér	45 N	179 E	Henyey	13 N	152 W	Lebedev	48 S	108 E
Dyson	61 N	121 W	Hertz	13 N	104 E	Lebedinsky	8 N	165 W
Dziewulski	21 N	99 E	Hertzsprung	0	130 W	Leeuwenhoek	30 S	179 W
			Hess	54 S	174 E	Leibnitz	38 S	178 E
			Heymans	75 N	145 W	Lemaître	62 S	150 W
Edison	25 N	99 E	Hilbert	18 S	108 E	Lenz	3 N	102 W
Ehrlich	41 N	172 W	Hippocrates	71 N	146 W	Leonov	19 N	148 E
Eijkman	62 S	141 W	Hirayama	6 S	93 E	Leucippus	29 N	116 W
Einthoven	5 S	110 E	Hoffmeister	15 N	137 E	Levi-Civita	24 S	143 E
Ellerman	26 S	121 W	Hogg	34 N	122 E	Lewis	19 S	114 W
Ellison	55 N	108 W	Hohmann	18 S	94 W	Ley	43 N	154 E
Elvey	9 N	101 W	Holetschek	28 S	151 E	Lindblad	70 N	99 W
Emden	63 N	176 W	Houzeau	18 S	124 W	Lobachevsky	10 N	113 E
Engelhardt	5 N	159 W	Hutton	37 N	169 E	Lodygin	18 S	147 W
Eötvös	36 S	134 E				Lomonosov	27 N	98 E
Erro	6 N	98 E	Ibn Yunis	14 N	91 E	Lorentz	34 N	100 W
Esnault-Pelterie	47 N	142 W	Icarus	6 S	173 W	Love	6 S	129 E
Espin	28 N	109 E	Idelson	81 S	114 E	Lovelace	82 N	107 W
Evans	10 S	134 W	Ingalls	4 S	153 W	Lovell	39 S	149 W
Evdokimov	35 N	153 W	Ingenii, Mare	34 S	163 E	Lowell	13 S	103 W
Evershed	36 N	160 W	Innes	28 N	119 E	Lucretius	9 S	121 W
			Izsak	23 S	117 E	Lundmark	39 S	152 E
Fabry	43 N	100 E				Lütke	17 S	123 E
Fechner	58 S	125 E	Jackson	22 N	163 W	Lyman	65 S	162 E
Fenyi	45 S	105 W	Jeans	53 S	91 W			
Feoktistov	31 N	140 E	Jenner	42 S	96 E	Mach	18 N	149 W
Fermi	20 S	122 E	Joffe	15 S	129 W	Maksutov	41 S	169 W
Fersman	18 N	126 W	Joliot	26 N	94 E	Malyi	22 N	105 E
Firsov	4 N	112 E	Joule	27 N	144 W	Mandelstam	4 N	156 E
Fitzgerald	27 N	172 W	Jules Verne	36 S	146 E	Marci	22 N	169 W
Fizeau	58 S	133 W				Marconi	9 S	145 E
Fleming	15 N	109 E	Kamerlingh Onnes	15 N	116 W	Mariotte	29 S	140 W
Focas	34 S	94 W	Karpinsky	73 N	166 E	Maunder	14 S	94 W
Foster	23 N	142 W	Kearons	12 S	113 W	Maxwell	30 N	98 E
Fowler	43 N	145 W	Keeler	10 S	162 E	McKeller	16 S	171 W
Freundlich	25 N	171 E	Kékulé	16 N	138 W	McLaughlin	47 N	93 W
Fridman	13 S	127 W	Khwolson	14 S	112 E	McMath	15 N	167 W
Froelich	80 N	110 W	Kibaltchitch	2 N	147 W	McNally	22 N	127 W
Frost	37 N	119 W	Kidinnu	36 N	123 E	Mees	14 N	96 W
			Kimura	57 S	118 E	Meggers	24 N	123 E
Gadomski	36 N	147 W	King	5 N	120 E	Meitner	11 S	113 E
Gagarin	20 S	150 E	Kirkwood	69 N	157 W	Mendel	49 S	110 W
Galois	16 S	153 W	Kleimenov	33 S	141 E	Mendeleev	6 N	141 E
Gamow	65 N	143 E	Klute	37 N	142 W	Merrill	75 N	116 W
Ganswindt	79 S	110 E	Koch	43 N	150 E	Mesentsev	72 N	129 W
Garavito	17 N	131 E	Kohlschütter	15 N	154 E	Meshcerski	12 N	125 E
Gavrilov	17 N	131 E	Kolhörster	10 N	115 W	Metchnikiff	11 S	149 W
Geiger	14 S	158 E	Komarov	25 N	153 E	Michelson	6 N	121 W
Gerasmović	23 S	124 W	Kondratyuk	15 S	115 E	Milanković	77 N	170 E
Gernsback	36 S	99 E	Konstantinov	20 N	159 E	Millikan	47 N	121 E
Ginzel	14 N	97 E	Kopff	17 S	90 W	Mills	9 N	156 E
Giordano Bruno	36 N	103 E	Korolev	5 S	157 W	Milne	31 S	113 E
Glasenapp	2 S	138 E	Kostinsky	14 N	118 E	Mineur	25 N	162 W
Golitzyn	25 S	105 W	Kovalevskaya	31 N	129 W	Minkowski	56 S	145 W
Golovin	40 N	161 E	Kovalsky	22 S	101 E	Mitra	18 N	155 W
Grachev	3 S	108 W	Kramers	53 N	128 W	Möbius	16 N	101 E
Green	4 N	133 E	Krasovsky	4 N	176 W	Mohorovičić	19 S	165 W
Gregory	2 N	127 E	Krylov	9 S	157 W	Moiseev	9 N	103 E
Grigg	13 N	130 W	Kugler	53 S	104 E	Montgolfier	47 N	160 W
Grissom	45 S	160 W	Kulik	42 N	155 W	Moore	37 N	178 W
Grotrian	66 S	128 E	Kuo Shou Ching	8 N	134 W	Morozov	5 N	127 E
Gullstrand	45 N	130 W	Kurchatov	38 N	142 E	Morse	22 N	175 W
						Moseley	23 N	95 W

Name	Lat	Long
Moscoviense, Mare	27 N	147 E
Moulton	61 S	97 E
Nagaoka	20 N	154 E
Nassau	25 S	177 E
Nernst	36 N	95 W
Neujmin	27 S	125 E
Niépce	72 N	120 W
Nijland	33 N	134 E
Nikolayev	35 N	151 E
Nishina	45 S	171 W
Nobel	15 N	101 W
Nöther	66 N	114 W
Numerov	71 S	161 W
Nušl	32 N	167 E
Obruchev	39 S	162 E
O'Day	31 S	157 E
Ohm	18 N	114 W
Olcott	20 N	117 E
Omar Khayyám	58 N	102 W
Oppenheimer	35 S	166 W
Oresme	43 S	169 E
Orlov	26 S	175 W
Östwald	11 N	122 E
Paneth	63 N	95 W
Pannekoek	4 S	140 E
Papaleski	10 N	164 E
Paracelsus	23 S	163 E
Paraskevopoulos	50 N	150 E
Parenago	26 N	109 W
Parkhurst	34 S	103 E
Parsons	37 N	171 W
Paschen	14 S	141 W
Pasteur	12 S	105 E
Pauli	45 S	137 E
Pavlov	29 S	142 E
Pawsey	44 N	145 E
Pease	13 N	106 W
Perelman	24 S	106 E
Perepelkin	10 S	128 E
Perkin	47 N	176 W
Perrine	42 N	129 W
Petrie	45 N	108 E
Petropavlovsky	37 N	115 W
Petzval	63 S	113 W
Pirquet	20 S	140 E
Pizzetti	35 S	119 E
Planck	57 S	135 E
Plaskett	82 N	175 E
Plummer	25 S	155 W
Pogson	42 S	111 E
Poincaré	57 S	161 E
Poinsot	79 N	147 W
Polzunov	26 N	115 E
Popov	17 N	99 E
Poynting	17 N	133 W
Prager	4 S	131 E
Prandtl	60 S	141 E
Priestly	57 S	108 E
Purkyne	1 S	95 E
Quételet	43 N	135 W
Racah	14 S	180
Raimond	14 N	159 W
Ramsay	40 S	145 E
Rasumov	39 N	114 W
Rayet	45 N	114 E
Rayleigh	67 S	179 E
Riccó	75 N	177 E

Name	Lat	Long
Riedel	49 S	140 W
Riemann	40 N	96 E
Rittenhouse	74 S	107 E
Ritz	15 S	92 E
Roberts	71 N	175 W
Robertson	22 N	105 W
Roche	42 S	135 E
Rowland	57 N	163 W
Rozhdestvensky	86 N	155 W
Rumford	29 S	170 W
Rydberg	47 S	96 W
Safarik	10 N	177 E
Saha	2 S	103 E
Sänger	4 N	102 E
St. John	10 N	150 E
Sanford	32 N	139 W
Sarton	49 N	121 W
Scaliger	27 S	109 E
Schaeberle	26 S	117 E
Schjellerup	69 N	157 E
Schlesinger	47 N	138 W
Schliemann	2 S	155 E
Schneller	42 N	164 W
Schönfeld	45 N	98 W
Schorr	19 S	90 E
Schrödinger	75 S	133 E
Schuster	4 N	147 E
Schwarzschild	71 N	120 E
Seares	74 N	145 E
Sechenov	7 S	143 W
Segers	47 N	128 E
Seidel	33 S	152 E
Seyfert	29 N	114 E
Shajn	33 N	172 E
Sharonov	13 N	173 E
Shatalov	24 N	140 E
Shi Shen	76 N	105 E
Siedentopf	22 N	135 E
Sierpinski	27 S	155 E
Sisakian	41 N	109 E
Sklodowska	18 S	159 E
Slipher	50 N	160 E
Smoluchowski	60 N	96 W
Sniadecki	22 S	169 W
Sommerfeld	65 N	161 W
Spencer Jones	13 N	166 E
Spiru Haret	59 S	176 W
Stark	25 S	134 E
Stebbins	65 N	143 W
Stefan	46 N	109 W
Stein	7 N	179 E
Steklov	37 S	105 W
Steno	33 N	162 E
Sternberg	19 N	117 W
Stetson	40 S	119 W
Stoletov	45 N	155 W
Stoney	56 S	156 W
Størmer	57 N	145 E
Stratton	6 S	165 E
Strömgren	22 S	133 W
Subbotin	29 S	135 E
Sumner	37 N	109 E
Sundman	11 N	91 W
Swann	52 N	112 E
Szilard	34 N	106 E
Teisserenc de Bort	32 N	137 W
ten Bruggencate	9 S	134 E
Terashkova	28 N	147 E
Tesla	38 N	125 E
Thiel	40 N	134 W
Thiessen	75 N	169 W
Thomson	32 S	166 E
Tikhomirov	25 N	162 E
Tikhov	62 N	172 E
Tiling	52 S	132 W
Timiryazev	5 S	147 W
Titius	27 S	101 E

Name	Lat	Long
Titov	28 N	150 E
Trümpler	28 N	168 E
Tsander	5 N	149 W
Tsiolkovskii	21 S	129 E
Tsu Chung-chi	17 N	144 E
Tyndall	35 S	117 E
Valier	7 N	174 E
Van de Graaff	27 S	172 E
Van den Bergh	31 N	159 W
Van der Waals	44 S	119 E
Van Gent	16 N	160 E
Van Gu	11 S	139 W
Van Maanen	36 N	127 E
Van Rhijn	52 N	145 E
Van't Hoff	62 N	133 W
Van Wijk	63 S	119 E
Vashakidze	44 N	93 E
Vavilov	1 S	139 W
Vening Meinesz	0	163 E
Ventris	5 S	158 E
Vernadsky	23 N	130 E
Vesalius	3 S	115 E
Vestine	34 N	94 E
Vetchinkin	10 N	131 E
Vilev	6 S	144 E
Volterra	57 N	131 E
Von der Pahlen	25 S	133 W
Von Kármán	45 S	176 E
Von Neumann	40 N	153 E
Von Zeipel	42 N	142 W
Walker	26 S	162 W
Waterman	26 S	128 F
Watson	63 S	124 W
Weber	50 N	124 W
Wegener	45 N	113 W
H. G. Wells	41 N	122 E
Wexler	69 S	90 E
Weyl	16 N	120 W
White	48 S	149 W
Wiechart	84 S	165 E
Wiener	41 N	146 E
Wilsing	22 S	155 W
Winkler	42 N	179 W
Winlock	35 N	160 W
Woltjer	45 N	160 W
Wood	44 N	121 W
Wyld	1 S	98 E
Yablochkov	61 N	127 E
Yamamoto	59 N	161 E
Zeeman	75 S	135 W
Zelinsky	29 S	167 E
Zernike	18 N	168 E
Zhiritsky	25 S	120 E
Zhukovsky	7 N	167 W
Zinger	57 N	176 E
Zsigmondy	59 N	105 W
Planck Rima:	65 S	129 E
	to 54 S	125 E
Schrödinger Rima:	62 S	99 E
	to 71 S	114 E

(The name 'Hedin' commemorates the Swedish scientist Sven Hedin. The name of Sven Hedin, used for a large enclosure near Grimaldi which is visible from Earth, has been deleted from some maps, but is retained here because lunar observers continue to use it.)

Far side of the Moon (NASA)

ECLIPSES OF THE MOON

Eclipses of the Moon are caused by the Moon's entry into the cone of shadow cast by the Earth. At the mean distance of the Moon, the diameter of the shadow cone is approximately 9170 km; the shadow is 1 367 650 km long on average. Totality may last for up to 1h 44m.

Lunar eclipses may be either total or partial. If the Moon misses the main cone, and merely enters the zone of 'partial shadow' or penumbra to either side, there is a slight dimming; but a penumbral eclipse is difficult to detect with the naked eye. Of course, the Moon must pass through the penumbra before entering the main cone or umbra.

During an eclipse the Moon becomes dim, often coppery. The colour during eclipse depends upon conditions in the Earth's atmosphere; thus the eclipse of 19 March 1848 was so 'bright' that lay observers refused to believe that an eclipse was happening at all. On the other hand, it is reliably reported that during the eclipses of 18 May 1761 and 10 June 1816 the Moon became completely invisible to the naked eye. The French astronomer A. Danjon has given an 'eclipse scale' from 0 (dark) to 4 (bright) and has attempted to correlate this with solar activity, though the evidence is still far from conclusive.

Ancient eclipse records are naturally rather uncertain. It has been claimed that an eclipse seen from the Middle East can be dated to 3450 BC; the eclipse of 1361 BC is more definite. Ptolemy gives the dates of observed eclipses as 721 BC and 720 BC. In *The Clouds*, the Greek playwright Aristophanes alludes to an eclipse seen from Athens on 9 October, 425 BC. There was certainly a lunar eclipse in August 413 BC which had unfortunate reults for Athens, since it caused Nicias, the commander of the Athenian expedition to Sicily, to delay the evacuation of his army; the astrologers advised him to stay where he was 'for thrice nine days'. When he eventually tried to embark his forces, he found that he had been blockaded by the Spartans. His fleet was destroyed, and the expedition annihilated – a reverse which led directly to the final defeat of Athens in the Peloponnesian War. According to Polybius, an eclipse in September 218 BC so alarmed the Gaulish mercenaries in the service of Attalus I of Pergamos that they refused to continue a military advance. On the other hand, Christopher Columbus was able to turn the eclipse of AD 1504 to his advantage. He was anchored off Jamaica, and the local inhabitants refused to supply his men with food; Columbus threatened to extinguish the Moon – and when the eclipse duly took place, the natives were so alarmed that there was no further trouble.

Obviously, a lunar eclipse can happen only at full moon: a solar eclipse, at new moon. In the original edition of the famous novel *King Solomons Mines*, H. Rider Haggard described a full moon, a solar eclipse and another full moon on successive days. When the mistake was pointed out, he altered the second edition, turning the solar eclipse into a lunar one!

LUNAR ECLIPSES, 1923-1977

*=total. For partial eclipses, the percentage maximum phase is given.

1923	March 2	38	1950	April 2	*
1923	August 25	17	1950	September 26	*
1924	February 20	*	1952	February 10/11	9
1924	August 14	*	1952	August 5	54
1925	February 8/9	74	1953	January 29/30	*
1925	August 4	75	1953	July 26	*
1927	June 15	*	1954	January 18/19	*
1927	December 8	*	1954	July 15/16	41
1928	June 3	*	1955	November 29	13
1928	November 27	*	1956	May 24	97
1930	April 13	11	1956	November 18	*
1930	October 7	3	1957	May 13/4	*
1931	April 2	*	1957	November 7	*
1931	September 26	*	1958	May 3	2
1932	March 22	97	1959	March 24	27
1932	September 14	98	1960	March 13	*
1934	January 30	12	1960	September 5	*
1934	July 26	67	1961	March 2	81
1935	January 19	*	1961	August 26	99
1935	July 16	*	1963	July 6/7	71
1936	January 8	*	1963	December 30	*
1936	July 4	27	1964	January 24/5	*
1937	November 18	15	1964	December 19	*
1938	May 14	*	1965	June 13/14	18
1938	November 7/8	*	1967	April 24	*
1939	May 3	*	1967	October 18	*
1939	October 28	99	1968	April 13	*
1941	March 13	33	1968	October 6	*
1941	September 5	· 6	1970	February 21	5
1942	March 2/3	*	1970	August 17	41
1942	August 26	*	1971	February 10	*
1943	February 20	77	1971	August 6	*
1943	August 15	88	1972	January 30	*
1945	June 25	60	1972	July 26	55
1945	December 18/19	*	1973	December 10	11
1946	June 14	*	1974	June 4	83
1946	December 8	*	1974	November 29	*
1947	June 3	2	1975	May 25	*
1948	April 23	3	1975	November 18/19	*
1949	April 13	*	1976	May 13	13
1949	October 6/7	*			

LUNAR ECLIPSES
1977-2000

Date (Mid-eclipse)	Magni- tude, per cent. (* = total)	Length of Totality h	Length of Totality m	Date (Mid-eclipse)	Magni- tude, per cent. (* = total)	Length of Totality h	Length of Totality m
4 April 1977	21	—		17 August 1989	*	1	38
24 March 1978	*	1	30	9 February 1990	*	0	46
16 September 1978	*	1	22	6 August 1990	68	—	
13 March 1979	89	—		21 December 1991	9	—	
6 September 1979	*	0	52	15 June 1992	69	—	
17 July 1981	58	—		10 December 1992	*	1	14
9 January 1982	*	1	24	4 June 1993	*	1	38
6 July 1982	*	1	42	29 November 1993	*	0	50
30 December 1982	*	1	6	25 May 1994	28	—	
25 June 1983	·34	—		15 April 1995	12	—	
4 May 1985	*	1	10	4 April 1996	*	1	24
28 October 1985	*	0	42	27 September 1996	*	1	12
24 April 1986	*	1	8	24 March 1997	93	—	
17 October 1986	*	1	14	16 September 1997	*	1	6
7 October 1987	1	—		28 July 1999	42	—	
27 August 1988	30	—		21 January 2000	*	1	24
20 February 1989	*	1	16	16 July 2000	*	1	42

The tracks of men on the Moon—
a photograph of the *Apollo 12* mission
(NASA)

LUNAR PROBES, 1958–1977

American, pre-Ranger

Name	Launch date	Results
Able 1	17 August 1958	Failed after 77 s, at 20 km (explosion of lower stage of launcher).
Pioneer 1	11 October 1958	Reached 113 000 km.
Pioneer 2	9 November 1958	Failed when third stage failed to ignite.
Pioneer 3	6 December 1958	Reached 106 000 km.
Pioneer 4	3 March 1959	Passed within 60 000 km of the Moon on 5 March
Able 4	26 November 1959	Failure soon after lift-off.
Able 5A	25 October 1960	Total failure.
Able 5B	15 December 1960	Exploded 70 s after lift-off.

American Rangers

Ranger 1	23 August 1961	Failed to go anywhere near the Moon.
Ranger 2	18 November 1961	Total failure.
Ranger 3	26 January 1962	Missed the Moon by 37 000 km (28 January).
Ranger 4	23 April 1962	Instruments and guidance failure. Probably landed on the night side of the Moon on 26 April.
Ranger 5	18 October 1962	Missed the Moon by over 630 km.
Ranger 6	30 January 1954	Hit the Moon (2 February), but no pictures received.
Ranger 7	28 July 1964	Hit the Moon on 31 July, in the Mare Nubium. Sent back 4308 photographs before impact.
Ranger 8	17 February 1965	Hit the Moon on 20 February, in the Mare Tranquillitatis. Sent back 7137 photographs before impact.
Ranger 9	21 March 1965	Hit the Moon on 24 March; interior of Alphonsus. Sent back 5814 photographs before impact.

American Surveyors

Surveyor 1	30 May 1966	Landed N. of Flamsteed; returned 11 150 photographs.
Surveyor 2	20 September 1966	Guidance failure. Crash-landed S.E. of Copernicus.
Surveyor 3	17 April 1967	Landed in Oceanus Procellarum, 612 km E. of Surveyor 1, and close to the site of the later Apollo 12 landing. 6315 photographs returned; soil physics studies carried out.
Surveyor 4	14 July 1967	Failure. Crash-landed in Sinus Medii.
Surveyor 5	8 September 1967	Landed in Mare Tranquillitatis, 25 km from the later Apollo 11 site. 18 000 photographs returned; soil physics studied.
Surveyor 6	7 November 1967	Landed in Sinus Medii. 30 000 photographs returned; soil physics, etc.
Surveyor 7	17 January 1968	Landed on N. rim of Tycho. 21 000 photographs returned, plus a great deal of miscellaneous data.

American Orbiters

Orbiter 1	10 August 1966	Successful photographic probe.
Orbiter 2	7 November 1966	Successful photographic probe.
Orbiter 3	25 February 1967	Successful photographic probe.
Orbiter 4	4 May 1967	Successful photographic probe.
Orbiter 5	1 August 1967	Successful photographic probe. Final controlled impact, 31 January 1968.

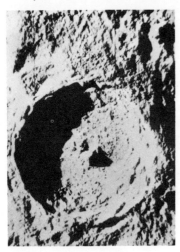

Crater Tycho, with its central peak casting a prominent shadow, photographed from *Surveyor* 7 (NASA)

American Apollo Probes

Number	Landing date	Crew	Site	EVA duration, hours	Distance covered, km
Apollo 11	20 July 1969	Armstrong Aldrin Collins	Mare Tranquillitatis Lat 00°67′N Long 23°49′E	2·2	—
Apollo 12	19 November 1969	Conrad Bean Gordon	Oceanus Procellarum Lat 03°12′S Long 23°23′W	7·6	1·4
Apollo 13	11 April 1970 (launch)	Lovell Haise Swigert	No landing; failure on outward trip. Splashdown 17 April	—	—
Apollo 14	31 January 1971	Shepard Mitchell Roosa	Fra Mauro Lat 03°40′S Long 17°28′E	9·2	3·4
Apollo 15	30 July 1971	Scott Irwin Worden	Hadley-Apennines, Lat 26°06′N Long 03°39′E	18·3	28
Apollo 16	21 April 1972	Young Duke Mattingly	Descartes Lat 08°60′S Long 15°31′E	20·1	26
Apollo 17	11 December 1972	Cernan Schmitt Evans	Taurus-Littrow Lat 20°10′N Long 30°46′E	22	29

Russian Luna and Zond Probes

Name	Launch date	Results
Luna 1	2 January 1959	Passed within 5955 km of the Moon on 4 January.
Luna 2	12 September 1959	Landed on the Moon, 13 September.
Luna 3	4 October 1959	Went round the Moon and photographed the far side. Approached the Moon to 6200 km.
Luna 4	2 April 1963	Missed the Moon by 8529 km; contact lost.
Luna 5	9 May 1965	Crash-landed on the Mare Nubium on 12 May. Unsuccessful soft-lander.
Luna 6	8 June 1965	Missed the Moon by 161 000 km (11 June).
Zond 3	18 July 1965	Photographed far side of the Moon; 25 pictures taken. Minimum distance from Moon, 9219 km. Pictures sent back from 2 200 000 km on 27 July.

Name	Launch date	Results
Luna 7	4 October 1965	Crashed in Oceanus Procellarum. Unsuccessful soft-lander.
Luna 8	3 December 1965	Crashed in Oceanus Procellarum. Unsuccessful soft-lander.
Luna 9	31 January 1966	Successful soft-lander; 100 kg capsule landed on the Moon (Oceanus Procellarum). Photographs obtained.
Luna 10	31 March 1966	First lunar satellite; contact maintained for 2 months (460 orbits). Minimum distance from the Moon, 350 km. Valuable data obtained.
Luna 11	24 August 1966	Lunar satellite. Contact maintained until 1 October. Minimum distance from the Moon, 159 km.
Luna 12	22 October 1966	Lunar satellite. Pictures showed craters down to 15 m. Contact lost on 19 January 1967.
Luna 13	21 December 1966	Soft landing in Oceanus Procellarum. Contact maintained until 27 December. Soil density, etc. studied.
Luna 14	7 April 1968	Lunar satellite. Minimum distance from Moon, 160 km. Valuable data obtained.
Zond 5	14 September 1968	Went round the Moon (minimum distance 1950 km) and returned to Earth on 21 September. Plants, seeds, insects and tortoises carried.
Zond 6	10 November 1968	Went round the Moon (minimum distance 2420 km) and filmed the far side; returned to Earth on 17 November.
Luna 15	13 July 1969	Unsuccessful attempt to return lunar samples. Crash-landed in Mare Crisium, 21 July.
Zond 7	7 August 1969	Circum-lunar flight. Lunar far side photographed from 2000 km; colour pictures of Earth and Moon secured. Returned to Earth.
Luna 16	12 September 1970	Landed in Mare Fœcunditatis (lat 0°41′S, long 56°18′E). Secured 100 g of material, and after 26½ hours lifted off. Returned to Earth; capsule recovered in Kazakhstan on 24 September.
Zond 8	20 October 1970	Circumlunar flight; colour pictures of Moon and Earth. Returned to Earth, 27 October, in Indian Ocean.
Luna 17	10 November 1970	Carried Lunokhod 1 to the Moon; landed 17 November in Mare Imbrium.
Luna 18	2 September 1971	Unsuccessful soft-lander. Contact lost during descent manœuvre to Mare Fœcunditatis.
Luna 19	28 September 1971	Lunar satellite. Contact maintained for over a year and over 4000 lunar orbits; studies of mascons, lunar gravitational field, etc., as well as solar flares.
Luna 20	14 February 1972	Successful sample-recovery probe. Landed 120 km N. of Luna 16's impact point, south of the Mare Crisium. Returned on 25 February.
Luna 21	8 January 1973	Carried Lunokhod 2 to the Moon; landed near Le Monnier, 180 km from Apollo 17's site, on 16 January.
Luna 22	29 May 1974	Successful lunar orbiter.
Luna 23	28 October 1974	Landed in the southern part of Mare Crisium, 6 November. Unsuccessful.
Luna 24	9 August 1976	Landed on 18 August in Mare Crisium, latitude 12°45′N, long 62°12′E. Drilled down to 2 metres, and obtained samples. Lifted off on 19 August, and returned to Earth on 22 August.

Russian Lunokhods

Name	Carrier	Weight, kg	Site	Notes
Lunokhod 1	Luna 17	756	Mare Imbrium	Operated for 11 months after its arrival on 17 November 1970. Area photographed exceeded 80 000 m² Over 200 panoramic pictures and 20 000 photographs returned. Distance travelled, 10·5 km.
Lunokhod 2	Luna 21	850	Le Monnier	Operated until mid-May 1973; landed 16 January 1973. 86 panoramic pictures and 80 000 television pictures obtained. Distance travelled, 37 km. On 3 June the Soviet authorities announced that the programme had ended—perhaps prematurely.

MERCURY

Mosaic of photographs of Mercury
taken by *Mariner 10* (NASA)

MERCURY DATA

Mean distance from the Sun:
  57·9 million km = 0·387 a.u.
Maximum distance from the Sun:
  69·7 million km = 0·467 a.u.
Minimum distance from the Sun:
  45·9 million km = 0·306 a.u.
Sidereal period: 87·969 days
Rotation period: 58·6461 days
Mean orbital velocity: 47·87 km/s
Axial inclination: Negligible
Orbital inclination: 7° 00′ 15″·5
Orbital eccentricity: 0·206
Diameter: 4880 km
Apparent diameter from Earth:
  max. 12″·9, min 4″·5
Reciprocal mass, Sun = 1 : 6 000 000
Density, water = 1 : 5·5
Mass, Earth = 1 : 0·055
Volume, Earth = 1 : 0·056
Escape velocity: 4·3 km/s
Surface gravity, Earth = 1 : 0·38
Mean surface temperature: +350°C
  (day), −170°C (night)
Oblateness: Negligible
Albedo: 0·06
Maximum magnitude: −1·9
**Mean diameter of Sun, seen from
  Mercury:** 1°22′40″

The innermost planet is Mercury; it is also the **smallest** of the principal planets, with the possible exception of Pluto. It was once thought that a still closer-in planet must exist, and it was even named (Vulcan), but it is now known that no planet of appreciable size moves within the orbit of Mercury.

Elongations of Mercury, 1977–2000

Western	*Eastern*
1977 Jan. 28, May 27, Sept. 21.	1977 Apr. 10, Aug. 8, Dec. 3.
1978 Jan. 11, May 9, Sept. 4, Dec. 24.	1978 Mar. 24, July 22, Nov. 16.
1979 Apr. 21, Aug. 19, Dec. 7.	1979 Mar. 8, July 3, Oct. 29.
1980 Apr. 2, Aug. 1, Nov. 19.	1980 Feb. 19, June 14, Oct. 11.
1981 Mar. 16, July 14, Nov. 3.	1981 Feb. 2, May 27, Sept. 23.
1982 Feb. 26, June 26, Oct. 17.	1982 Jan. 16, May 8, Sept. 6, Dec. 30.
1983 Feb. 8, June 8, Oct. 1.	1983 Apr. 21, Aug. 19, Dec. 13.
1984 Jan. 22, May 19, Sept. 14.	1984 Apr. 3, July 31, Nov. 25.
1985 Jan. 3, May 1, Aug. 28.	1985 Mar. 17, July 14, Nov. 8.
1986 Apr. 13, Aug. 11, Nov. 30.	1986 Feb. 28, June 25, Oct. 21.
1987 Mar. 26, July 25, Nov. 13.	1987 Feb. 12, June 7, Oct. 4.
1988 Mar. 8, July 6, Oct. 26.	1988 Jan. 26, May 19, Sept. 15.
1989 Feb. 18, June 18, Oct. 10.	1989 Jan. 9, May 1, Aug. 29, Dec. 23.
1990 Feb. 1, May 31, Sept. 24.	1990 Apr. 13, Aug. 11, Dec. 6.
1991 Jan. 14, May 12, Sept. 7, Dec. 27.	1991 Mar. 27, July 25, Nov. 19.
	1992 Mar. 9, July 6, Oct. 31.
1992 Apr. 23, Aug. 21, Dec. 9.	1993 Feb. 21, June 17, Oct. 14.
1993 Apr. 5, Aug. 4, Nov. 22.	1994 Feb. 4, May 30, Sept. 26.
1994 Mar. 19, July 17, Nov. 6.	1995 Jan. 19, May 12, Sept. 9.
1995 Mar. 1, June 29, Oct. 20.	1996 Jan. 2, Apr. 23, Aug. 21, Dec 15.
1996 Feb. 11, June 10, Oct. 3.	1997 Apr. 6, Aug. 4, Nov. 28.
1997 Jan 24., May 22, Sept. 16.	1998 Mar. 20, July 17, Nov. 11.
1998 Jan. 6, May 4, Aug. 31, Dec. 20.	1999 Mar. 3, June 28, Oct. 24.
1999 Apr. 16, Aug. 14, Dec. 2.	2000 Feb. 15, June 9, Oct. 6.
2000 Mar. 28, July 27, Nov. 15.	

The discovery of Mercury must have been prehistoric. The oldest observation of the planet which has come down to us is dated 15 November 265 BC, when the planet was one lunar diameter away from a line joining the stars Delta and Beta Scorpii. This information has been given by the last great astronomer of Classical times, Ptolemy (*c.* AD 120–180). Plato (*Republic*, X, 14) commented upon the yellowish colour of Mercury, though most naked-eye observers will describe it as being white. Mercury can actually become brighter than any star, but can never be seen against a really dark sky.

The phases of Mercury were first detected by Hevelius, in the first half of the 17th century.

The first prediction of a transit of Mercury across the face of the Sun was made by Kepler, for the transit of 7 November 1631. His prediction enabled Gassendi to observe the transit.

The following transits occurred between 1631 and 1978:

1631 Nov. 7	1756 Nov. 7	1878 May 8
1644 Nov. 8	1769 Nov. 9	1881 Nov. 8
1651 Nov. 2	1776 Nov. 2	1891 May 8
1661 May 3	1782 Nov. 12	1894 Nov. 10
1664 Nov. 4	1786 May 3	1907 Nov. 12
1677 Nov. 7	1789 Nov. 5	1914 Nov. 6
1690 Nov. 10	1799 May 7	1924 May 8
1697 Nov. 3	1802 Nov. 9	1927 Nov. 10
1707 May 6	1815 Nov. 12	1937 May 11
1710 Nov. 5	1822 Nov. 5	1940 Nov. 11
1723 Nov. 9	1832 May 5	1953 Nov. 14
1736 Nov. 11	1835 Nov. 7	1957 May ,6
1740 May 2	1845 May 8	1960 Nov. 7
1743 Nov. 5	1848 Nov. 9	1970 May 9
1753 May 6	1861 Nov. 12	1973 Nov. 10
	1868 Nov. 5	

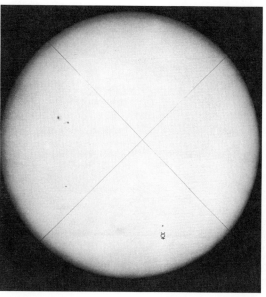

Transit of Mercury across the disk of the Sun on 7 November 1914 (RAS)

The following transits will occur between 1978 and 2000:

1986 Nov. 13, Mid-transit 4h 08m UT.
1993 Nov. 6, Mid-transit 3h 58m UT.
1999 Nov. 15, Mid-transit 21h 42m UT.

Transits can occur only in May and November. May transits occur with Mercury near aphelion; at November transits Mercury is near perihelion, and November transits are the more frequent in the approximate ratio of 7 to 3. The longest transits (those of May) may last for almost 9 hours.

The first serious telescopic observations of Mercury were made in the late 18th century by Sir William Herschel, who, however, could make out no surface detail. At about the same time Mercury was studied by J. H. Schröter, who recorded some surface patches, and who believed that he had detected high mountains. It seems, however, that these results were illusory.

The first attempted map of Mercury was compiled by G. V. Schiaparelli, from Milan, using 0m.218 and 0m.49 refractors between 1881 and 1889. He recorded various dark markings, and believed the rotation period to be synchronous – that is to say, equal to Mercury's revolution

period. This would mean that part of the planet would be in permanent daylight and another part in permanent night, with an intervening 'twilight zone' over which the Sun would rise and set. Schiaparelli was also the first observer to study Mercury in daylight, when both it and the Sun were high above the horizon.

The best pre-Mariner map was compiled by E. M. Antoniadi, using the 0·38m Meudon refractor. The map was published in 1934, and various features were named. Like Schiaparelli, Antoniadi believed the rotation period to be synchronous, and he also believed in local obscurations, due to material suspended in a thin Mercurian atmosphere. These conclusions are now known to be wrong.

The first (and probably only!) observer to draw 'canal-like' features on Mercury was P. Lowell, at Flagstaff (Arizona) in 1896. These features are completely illusory.

The first disproof of the synchronous rotation theory was obtained in 1962 by W. E. Howard and his colleagues at Michigan, who measured the long-wavelength radiations from Mercury and found that the dark side was much warmer than it would be if it never received any

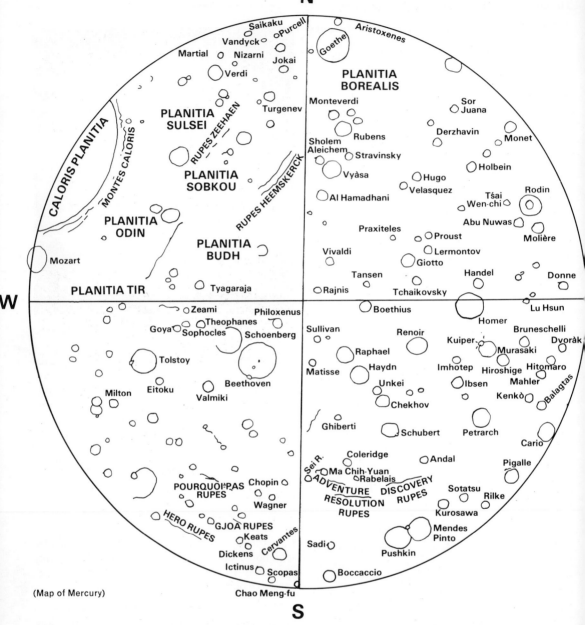

(Map of Mercury)

(Map of Mercury)

direct sunlight. In 1965 the shorter rotation period was confirmed by radar methods (R. Dyce and G. Pettengill, at Arecibo in Puerto Rico). The true period is 2/3 of the revolution period, and when Mercury is best placed for observation from Earth the same face is always presented to us.

The first (and so far, the only) Mercury Probe, Mariner 10, made three active passes of the planet: in March and September 1974 and March 1975. It was launched on 3 November 1973, and obtained some photographs of the Moon before by-passing Venus (5 February 1974) and making its first rendezvous with Mercury. **The first crater to be identified on Mercury** during the initial encounter was the bright ray-crater now named Kuiper. Closest approach took place on 29

March, and altogether 647 pictures were obtained. The second closest approch was that of 21 September 1974, and the third occurred on 16 March 1975, by which time the equipment was deteriorating. Contact was finally lost on 24 March 1975, though no doubt the probe is still in solar orbit and continues to make periodical approaches to Mercury.

The atmosphere of Mercury was first detected from Mariner 10. The ground pressure is about 10^{-9} millibars, and the main constituent is helium, suggesting that the weak magnetic field traps helium nuclei from the solar wind. Earlier spectroscopic reports of an atmosphere (initially by H. Vogel in 1871) are now known to have been erroneous.

The magnetic field of Mercury was also detected by Mariner 10. The first encounter yielded a value of from 350 to 700 gammas at the surface, or about 1 per cent of the Earth's magnetic field, but at the third and closest encounter the estimated strength of the magnetic field was increased to 1/30 that of the Earth. This means that the magnetic field of Mercury is stronger than that of either Venus, Mars or the Moon. The field is dipolar, with two equal magnetic poles of opposite polarity aligned with the rotational axis of the planet.

The densest planets are Mercury and the Earth, each with a specific gravity of about 5·5. Mercury has apparently a dense iron core, about 3600 km in diameter (somewhat larger than the whole of the Moon) and containing 80 per cent of the planet's mass.

The Mercurian craters were first detected from Mariner 10, as were other features: plains (**planitia**), mountains (**montes**), valleys (**valles**), scarps (**dorsa**) and ridges (**rupes**). The south pole of Mercury lies inside the crater Chao Meng Fu. It has been agreed that the 20th-degree meridian of Mercury passes through the centre of the 1·5 km crater Hun Kal, 0°·58 south of the Mercurian equator (the name stands for 20 in the language of the Maya, who used a base-20 number system).

The most imposing information on Mercury is the Caloris Basin (Caloris Planitia), so called because it lies near a 'hot pole', the highest-temperature region of Mercury (there are two 'hot poles'). The Caloris Basin is 1300 km in diameter, and is bounded by a ring of smooth mountain blocks rising 1 to 2 km above the surrounding surface. Unfortunately only half of the Caloris Basin was recorded from Mariner 10; at each

Surface of Mercury near the south pole photographed by *Mariner 10* at a range of 54 600 km

encounter the same regions were on view, so that our knowledge of the surface is still very incomplete.

There is an obvious similarity between the surface features of Mercury and those of the Moon, though Mercury lacks the broad lunar-type maria. The distribution of the craters is of the same general type – when a crater breaks into another, it is generally the smaller formation which intrudes into the larger; there are pairs, groups and lines of craters, and some of the formations have central peaks while others do not. Ray-craters also exist. There are, naturally, definite differences, but it seems likely that the lunar and the Mercurian craters were formed by the same process. Arguments as to vulcanism versus impact still continue!

Mercury is certainly a hostile world in every way. With its excessively thin atmosphere and its extreme temperatures, it is unsuited to any form of life, and there seems little chance of its being visited in the foreseeable future, though no doubt many unmanned probes – both orbiters and landers – will be sent there during the coming decades.

NAMED FORMATIONS ON MERCURY

The naming of Mercurian features is not yet complete, and, of course, less than half the total surface was covered from Mariner 10 – our sole source of information at the time of writing (1978). In general craters have been named after persons who have made great contributions to human culture; plains (**planitia**) from the names of Mercury in various languages (for instance, the Suisei Planitia is the Japanese form); valleys (**valles**) after radar installations; ridges (**rupes**) after famous ships of exploration and discovery, and scarps (**dorsa**) after astronomers who have been particularly associated with the observation of Mercury. There are a few exceptions to these general rules. Caloris Basin is 'the hot basin' because it lies near one of the two hot poles; the late Gerard P. Kuiper is honoured by having the first-identified crater on Mercury named after him; and as we have seen, Hun Kal is the Mayan word for 'twenty', since, by definition, the 20th meridian on Mercury passes through its centre.

Apart from the Caloris Basin, the largest circular structure is Beethoven (625 km in diameter). There are various ray-craters; those named are Copley, Kuiper, Mena, Tansen, and Snorri.

No doubt the nomenclature of Mercury will be extended in the near future. Meantime, here is the 1977 list, as approved by the International Astronomical Union. The positions and crater diameters are approximate only.

CRATERS

Name	Latitude	Longitude	Diameter, km
Abu Nuwas	+18	021	115
Africanus Horton	−51	042	120
Al Hamadhani	+39	090	170
Al Jāhiz	+02	022	95
Amru al-Qays	+13	176	50
Andal	−47	036	90
Aristoxenes	+82	011	65
Asvogasha	+11	021	80
Bach	−69	103	225
Balagtas	−22	014	100
Beethoven	−20	124	625
Bello	−19	121	150
Bernini	−80	136	145
Boccaccio	−81	030	135
Boethius	−01	074	130
Bramante	−46	062	130
Brunelleschi	−09	023	140
Byron	−08	033	100
Callicrates	−65	032	65
Camões	−71	070	70
Carducci	−36	090	75
Cario	−28	010	160
Cervantes	−75	122	200
Chao Meng-Fu	−88	132	150

Name	Latitude	Longitude	Diameter, km
Chekhov	−36	062	180
Chiang K'ui	+15	103	105
Chopin	−65	124	100
Chu Ta	+03	106	100
Coleridge	−55	067	110
Copley	−39	087	30
Deprez	+81	092	40
Derzhavin	+45	037	145
Dickens	−73	157	75
Donne	+03	014	90
Dürer	+22	120	190
Dvořák	−10	013	80
Eitoku	−22	158	105
Equiano	−39	031	80
Futabatei	−17	084	55
Gauguin	+67	097	75
Ghiberti	−48	080	100
Giotto	+13	056	150
Gluck	+38	019	85
Goethe	+80	044	340
Goya	−07	153	135
Guido d'Arezzo	−38	019	50
Handel	+04	034	150
Harunobu	+16	141	100
Haydn	−28	070	230
Hesiod	−58	036	90
Hiroshige	−13	027	140
Hitomaro	−16	016	105
Holbein	+36	029	85
Holberg	−67	060	66
Homer	−01	037	320
Horace	−69	050	48
Hugo	+39	048	190
Hun Kal	−0·6	020	1·5
Ibsen	−24	036	160
Ictinus	−79	175	110
Imhotep	−18	39	160
Jokai	+73	136	85
Judah Ha-Levi	+12	108	85
Kālidāsā	−18	178	110
Keats	−71	156	110
Kenkō	−21	017	90
Khansa	−59	052	100
Kuan Han-ching	+29	053	155
Kuiper	−11	032	40
Kurosawa	−52	023	180
Leopardi	−73	185	69
Lermontov	+16	049	160
Li Ching Chao	−77	075	60
Li Po	+18	035	120
Lu Hsun	+01	024	95
Lysippus	+02	133	150
Ma Chih-Yuan	−61	077	170
Machaut	−02	083	105
Mahler	−19	019	100
Mansart	+74	120	75
Mark Twain	−11	139	140
Marti	−76	169	63
Martial	+69	178	45
Matisse	−24	090	210
Melville	+22	010	135
Mena	+0·2	125	20
Mendes Pinto	−61	019	170
Mickiewicz	+24	103	115
Milton	−26	175	175
Mistral	+05	054	100
Mofolo	−38	29	90
Molière	+16	018	140
Monet	+44	010	250
Monteverdi	+64	077	130
Mozart	+08	191	225
Murasaki	−12	031	125
Myton	+71	080	30
Nampeyo	−40	051	40
Neumann	−38	034	100
Nizami	+72	165	70
Ovid	−70	022	40
Petrarch	−30	027	160
Phidias	+09	150	155
Philoxenus	−08	112	95
Pigalle	−38	012	130

Name	Latitude	Longitude	Diameter, km
Po Chu-I	−07	166	60
Polygnotus	00	069	130
Po-ya	−46	019	90
Praxiteles	+27	060	175
Proust	+20	047	140
Puccini	−65	046	110
Purcell	+61	148	80
Pushkin	−66	023	200
Rabelais	−54	063	130
Rajnis	+05	097	85
Rameau	−54	038	50
Raphael	−21	077	350
Renoir	−18	052	220
Repin	−19	063	95
Rilke	−46	013	70
Rodin	+22	019	270
Rubens	+60	074	180
Rublev	−15	158	125
Rudaki	−04	052	120
Sadi	−79	055	60
Saikaku	+73	177	80
Schönberg	−16	136	30
Schubert	−47	055	160
Scopas	−81	185	95
Sei	−64	090	130
Shevchenko	−53	047	130
Sholem Aleichem	+51	077	190
Sinan	+16	030	140
Snorri	−09	084	20
Sophocles	−07	146	145
Sor Juana	+49	024	80
Sōtatsu	−48	019	130
Spitteler	−68	059	65
Stravinsky	+51	073	170
Sullivan	−17	087	135
Tansen	+05	072	25
Tchaikovsky	+08	051	160
Thākur	−04	064	115
Theophanes	−04	143	50
Tintoretto	−48	023	60
Titian	−03	043	115
Tolstoy	−15	165	400
Ts'ai Wen-chi	+24	023	120
Ts'ao Chan	−13	142	110
Tsurayuki	−62	023	80
Tung Yuan	+74	055	60
Turgenev	+66	135	110
Tyagaraja	+04	149	100
Unkei	−31	063	110
Vālmiki	−24	142	220
Van Gogh	−76	139	95
Vandyck	+77	165	100
Velasquez	+37	54	120
Verdi	+65	169	150
Vivaldi	+15	086	210
Vyāsa	+49	080	275

Name	Latitude	Longitude	Diameter, km
Wagner	−68	114	135
Wang Meng	+10	104	170
Wergeland	−37	057	35
Wren	+25	036	215
Yeats	+10	035	90
Yun Sun-do	−73	110	60
Zeami	−03	148	125

PLANITIA

	Latitude	Longitude	
Borealis	+75	085	
Budh	+18	150	
Caloris	+30	199	
Odin	+24	171	
Sobkou	+39	128	
Suisei	+59	157	
Tir	+03	177	

VALLES

	Latitude	Longitude	
Arecibo	−27	029	
Goldstone	−15	032	
Haystack	+04	046	
Simeiz	−13	066	

RUPES

	Latitude	Longitude	
Adventure	−64	063	
Astrolabe	−42	071	
Discovery	−54	038	
Endeavour	+38	031	
Fram	−57	094	
Gjoa	−65	163	
Heemskerck	+27	125	
Hero	−57	173	
Mirni	−37	040	
Pourquoi Pas	−58	156	
Resolution	−62	052	
Santa Maria	+06	020	
Victoria	+50	032	
Vostok	−38	019	
Zarya	−42	022	
Zeehaen	+50	158	

DORSA

	Latitude	Longitude	
Antoniadi	+28	030	
Schiaparelli	+24	164	

MONTES

	Latitude	Longitude	
Caloris	{ +22	180	
	+40	180	

VENUS

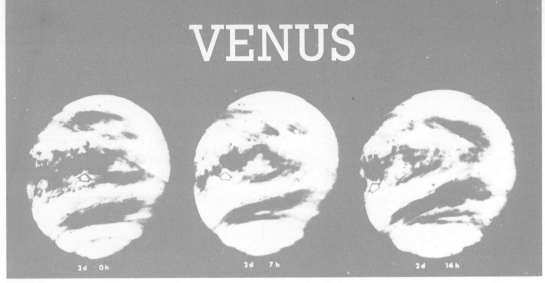

Venus photographed by *Mariner 10*. Rotation is shown by the arrowed feature (NASA)

The brightest planet is Venus, second in order of distance from the Sun.

Phenomena of Venus, 1977–2000

E. Elongation	Inferior Conjunction	W. Elongation	Superior Conjunction
1977 Jan. 24	1977 Apr. 6	1977 June 15	1978 Jan. 22
1978 Aug. 29	1978 Nov. 7	1979 Jan. 18	1979 Aug. 25
1980 Apr. 5	1980 June 15	1980 Aug. 24	1981 Apr. 7
1981 Nov. 11	1982 Jan. 21	1982 Apr. 1	1982 Nov. 4
1983 June 16	1983 Aug. 25	1983 Nov. 4	1984 June 15
1985 Jan. 22	1985 Apr. 3	1985 June 13	1986 Jan. 19
1986 Aug. 27	1986 Nov. 5	1987 Jan. 15	1987 Aug. 23
1988 Apr. 3	1988 June 13	1988 Aug. 22	1989 Apr. 5
1989 Nov. 8	1990 Jan. 10	1990 Mar. 30	1990 Nov. 1
1991 June 13	1991 Aug. 22	1991 Nov. 2	1992 June 13
1993 Jan. 19	1993 Apr. 1	1993 June 10	1994 Jan. 17
1994 Aug. 25	1994 Nov. 2	1995 Jan. 13	1995 Aug. 20
1996 Apr. 1	1996 June 10	1996 Aug. 19	1997 Apr. 2
1997 Nov. 6	1998 Jan. 16	1998 Mar. 27	1998 Oct. 30
1999 June 11	1999 Aug. 20	1999 Oct. 30	2000 June 11

DATA

Mean distance from the Sun:
108·2 million km = 0·723 a.u.
Maximum distance from the Sun:
109 million km = 0·728 a.u.
Minimum distance from the Sun:
107·4 million km = 0·718 a.u.
Sidereal period: 224·701 days
Rotation period: 243 days
Mean orbital velocity: 35·02 km/s
Axial inclination: 178°
Orbital inclination: 3° 23′ 39″·8
Orbital eccentricity: 0·007
Diameter: 12 104 km
Apparent diameter from Earth:
max 65″·2, min 9″·5, mean 37″·3
Reciprocal mass, Sun = 1: 408 520
Density, water = 1: 5·25
Mass, Earth = 1: 0·815
Volume, Earth = 1: 0·86
Escape velocity: 10·36 km/s
Surface gravity, Earth = 1: 0·903
Mean surface temperature: cloud-tops
−33°C, surface +480°C
Oblateness: 0
Albedo: 0·76
Maximum magnitude: −4·4
Mean diameter of Sun, seen from Venus: 44′15″

The first man to find that Venus is at its brightest when in the crescent stage was Edmond Halley, in 1721. This is because when more of the illuminated hemisphere faces the Earth, Venus is further away from us.

The discovery of Venus must, of course, have been prehistoric. The most ancient observations known to us are Babylonian, and are recorded on the Venus Tablet found by Sir Henry Layard at Konyunjik, now to be seen in the British Museum. Homer (*Iliad*, XXII, 318) refers to Venus as 'the most beautiful star set in the sky'.

Venus is **the only planet referred to by Napoleon Bonaparte**! According to the French astronomer F. Arago, Napoleon was visiting Luxembourg when he saw that the crowd was paying more attention to the sky than to him; it was noon, but Venus was visible, and Napoleon himself saw it. Not surprisingly, his followers referred to it as being the star of 'the Conqueror of Italy'.

The first men to refer to shadows cast by Venus were the Greek astronomer Simplicius, in his *Commentary on the Heavens of Aristotle*, and the Roman writer Pliny around AD 60.

The first man to record the phases of Venus telescopically was Galileo, in 1610. This was an important observation, since according to the old Ptolemaic theory, with the Earth in the centre of the planetary system, Venus could never show a

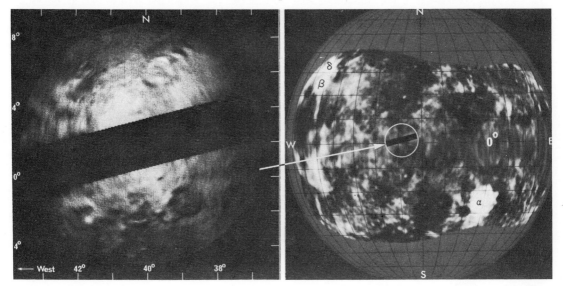

Radar map of part of the surface of Venus—it was not possible to achieve a clear view of the central area (NASA)

full cycle of phases from new to full. The observation therefore strengthened Galileo's faith in the 'Copernican' or Sun-centred system. (The phases had not previously been mentioned specifically, though exceptionally keen-sighted people can see the crescent form with the naked eye.)

The first prediction of a transit of Venus across the disk of the Sun was Johannes Kepler, who in 1627 predicted a transit for 6 December 1631. This transit was not actually observed.

The first observation of a transit of Venus was made by two English amateurs, J. Horrocks and W. Crabtree, on 24 November 1639 (O.S.).

The first suggestion of using a transit of Venus to measure the length of the astronomical unit was Edmond Halley. The method was used during the transits of 1761, 1769, 1874 and 1882, but proved to be inaccurate, and is now obsolete. It depended upon timing the exact moment of the beginning of the transit, but when Venus passes on to the Sun it seems to draw a strip of blackness after it, and when this strip disappears the transit is already in progress. The effect, termed the 'Black Drop.' is caused by Venus' atmosphere.

The last transit of Venus was that of 1882. The next will be in 2004.

The only occultation of Mars by Venus to be observed was that of 3 October 1590, seen by M. Möstlin at Heidelberg.

The only occultation of Mercury by Venus to be observed was that of 17 May 1737, seen by J. Bevis at Greenwich.

The last occultation of a first-magnitude star by Venus was that of 7 July 1959, when Venus occulted Regulus. The phenomenon was widely observed, and led to a determination of the height of Venus' atmosphere. (The writer of this book observed the occultation from Selsey in Sussex, using a 12 in reflecting telescope.)

The first observation of the Ashen Light was made on 9 January 1643 by G. Riccioli. The Ashen Light is the faint visibility of the night hemisphere of Venus. It may be due to electrical phenomena in the upper atmosphere of Venus, though some astronomers dismiss it as being nothing more than a contrast effect.

The first markings on Venus were reported in 1645 by F. Fontana. However, Fontana was using a small-aperture, long-focus refractor, and there is no doubt that his 'markings on Venus' were illusory.

The first map of Venus was drawn by F. Bianchini in 1727. However, the markings seen by Bianchini, like those reported by Fontana, were certainly illusory. Bianchini believed that he had charted seas and continents on the surface of Venus!

The first observer to report an atmosphere around Venus was the Russian astronomer M. V. Lomonosov, during the transit of 1761.

The first really serious observer of Venus telescopically was J. H. Schröter. His series of observations began in 1779, but it was not until 28 February 1788 that he observed markings, which he correctly interpreted as being atmospheric. On 11 December 1798 he believed that he had identified a mountain 43 km high, but this was obviously erroneous. In 1793 he was the first to find that the observed and theoretical phases of Venus do not agree exactly; when Venus is waning, dichotomy (half-phase) is earlier than predicted by a few days, while when Venus is waxing dichotomy is late. The phenomenon is now generally called Schröter's Effect (a term due to the writer of this book!). It is due to effects of Venus' atmosphere.

The first report of a satellite of Venus was made by G. D. Cassini on 18 August 1686. Other reports followed, the last being that of Montbaron, at Auxerre, on 29 March 1764. It is now certain that no satellite exists, and that the observers were deceived by 'telescopic ghosts'.

The first man to estimate the rotation period of Venus was G. D. Cassini in 1666–7; he gave a period of 23h 21m. Many other estimates were made in later years. but there was no general agreement.

The first man to suggest a synchronous rotation period was G. V. Schiaparelli, in 1890. This would mean that the rotation period and the sidereal period of Venus were identical – 224·7 Earth-days – in which case Venus would keep the same hemisphere turned towards the Sun all the time. However, this did not seem to fit the facts, and in 1954 G. P. Kuiper proposed a period of 'a few weeks'.

The first estimate of the rotation period by using spectroscopic methods (the Doppler shift) was made by R. S. Richardson in 1956. Richardson concluded that the rotation was very slow, and retrograde – that is to say, opposite in sense to that of the Earth. This has proved to be correct. The true rotation period is 243 days, so that Venus is the only planet with a rotation period longer than its sidereal period. (The 'solar day' on Venus is equal to 118 Earth-days.) Venus is also the only planet to have a genuinely retrograde rotation, though it is true that the axial tilt of Uranus is slightly more than a right angle (98 degrees). However, during the 1960s, French observers

established that the upper clouds have a rotation period of only 4 days; this was confirmed by photographs obtained from the Mariner 10 probe in 1974.

The first substance to be detected spectroscopically in the atmosphere of Venus was carbon dioxide, identified by W. S. Adams and T. Dunham at Mount Wilson in 1932.

The first reliable temperature measurements of the upper clouds of Venus were made in 1923–8 by E. Pettit and S. B. Nicholson, using a thermocouple attached to the 100 in Hooker reflector at Mount Wilson. They gave a value of —38°C for the day side and —33°C for the dark side, which is in good agreement with modern values.

The first good infra-red and ultra-violet photographs of Venus were taken by F. E. Ross in 1923. The infra-red pictures showed no detail, but vague features were shown in ultra-violet, indicating high-altitude cloud phenomena.

The first suggestion of formaldehyde clouds on Venus was made by R. Wildt in 1937. This is no longer accepted, and it seems more likely that the clouds contain quantities of sulphuric acid. Hydrogen chloride has also been detected; it was found, along with hydrogen fluoride, by W. S. Benedict of the University of Maryland.

The first suggestion that Venus might be entirely water-covered was made by F. L. Whipple and D. H. Menzel in 1954; they believed that the clouds were made up of H_2O. The 'marine theory' was disproved by the Mariner 2 results of 1962. Venus is too hot for liquid water to exist on its surface.

The first radio measurements of Venus at centimetre wavelengths were made by Mayer and his colleagues in the United States in 1958. They indicated a very high temperature for the surface of Venus, and this has been fully confirmed.

The first reliable radar contact with Venus was made in 1961 by the team at the Lincoln Laboratory, United States (the success first claimed in 1958 proved to be premature). Several other groups made radar contact with the planet at about the same time. The measurements led to a better determination of the length of the astronomical unit, and in 1964 the International Astronomical Union officially adopted a revised value of 149 600 000 km.

The first Venus probe was Venera 1 (USSR), launched on 12 February 1961. It was not successful, as contact with it was lost at a distance of 7 500 000 km from Earth.

The first successful Venus probe was Mariner 2 (USA), launched on 26 August 1962, which bypassed Venus at 35 000 km. on 14 December 1962, confirming the high surface temperature and the virtual absence of any magnetic field.

The first probe to land on Venus was Venera 3 (USSR), on 1 March 1966. It failed to return any data, because it was crushed by the intense pressure as it parachuted down through the atmosphere of Venus.

The first probe to transmit information during the descent through Venus' atmosphere was Venera 4 (USSR), on 18 October 1967.

The first probe to transmit information from Venus' surface was Venera 7 (USSR), on 15 December 1970. It survived for 23 minutes before being put out of action by the intensely hostile conditions.

The first close-range photographs of the clouds of Venus were obtained from Mariner 10 (USA) on 5 February 1974, as the probe bypassed Venus en route for Mercury.

The first picture to be sent back from the surface of Venus was transmitted by Venera 9 (USSR) on 21 October 1975. Venera 10 landed on 25 October 1975, and also sent back one picture.

VENUS PROBES, 1961-1978
The Mariners were American; the other probes, Russian

Name	Launch	Arrival	Closest approach, km	Results
Venera 1	12 February 1961	19 May 1961	100,000	Contact lost at 7 500 000 km from Earth
Mariner 1	22 July 1962	—	—	Total failure
Mariner 2	26 August 1962	14 December 1962	35,000	Fly-by. Data transmitted
Zond 1	2 April 1964	?	?	Contact lost within a few weeks
Venera 2	12 November 1965	27 February 1966	24,000	In solar orbit. No Venus data received
Venera 3	16 November 1965	1 March 1966	Landed	Crushed by Venus atmosphere. No data received
Venera 4	12 June 1967	18 October 1967	Landed	Data transmitted during 94-minute descent
Mariner 5	14 June 1967	19 October 1967	4,000	Fly-by. Data transmitted
Venera 5	5 January 1969	16 May 1969	Landed	Data transmitted during descent
Venera 6	10 January 1969	17 May 1969	Landed	Data transmitted during descent
Venera 7	17 August 1970	15 December 1970	Landed	Lander transmitted for 23 minutes after arrival
Venera 8	26 March 1972	22 July 1972	Landed	Lander transmitted for 50 minutes after arrival
Mariner 10	3 November 1973	5 February 1974	5800	Pictures of upper clouds, plus other data. Went on to Mercury
Venera 9	8 June 1975	21 October 1975	Landed	Transmitted for 53 minutes after arrival. One picture received
Pioneer Venus 1	20 May 1978	4 December 1978	145	In orbit. Sending back information
Pioneer Venus 2	8 August 1978	9 December 1978	Landed	Multi-probe—five probes landed. Data transmitted back
Venera 11	9 September 1978	21 December 1978	Landed	Transmitted confirmatory observations about atmospheric and surface for about 60 minutes
Venera 12	14 September 1978	25 December 1978	Landed	Transmitted confirmatory observations about atmospheric and surface for about 60 minutes

Venus is an intensely hostile planet. The atmospheric pressure at the surface is thought to be about 90 000 millibars, which is some 90 times as great as the pressure of the Earth's atmosphere at sea-level. Our knowledge of the surface structure depends upon the Venera 9 and 10 pictures. The Venera 9 landscape was described as 'a heap of stones', several dozen centimetres in diameter, with rather sharp edges; the Venera 10 panorama was smoother, and gave the impression of being an older plateau. Wind velocities were low, and the light level was described as being much the same as that in Moscow at noon on a cloudy winter day.

It has been suggested that Venus once had a thinner atmosphere and lower temperature than it has now, but solar heat trapped by the atmosphere raised the temperature; any water was evaporated, and eventually the surface temperature became so high that carbon dioxide was driven out of carbonates in the rocks, producing the conditions which we find today. In view of the intense heat and also the high pressure of the carbon-dioxide atmosphere, plus the sulphuric acid in the clouds, we may safely discount the idea of any Earth-type life on Venus.

THE EARTH

DATA

Mean distance from the Sun:
149·5979 million km (1 a.u.)
Minimum distance from the Sun:
147 million km (0·9833 a.u.)
Maximum distance from the Sun:
152 million km (1·0167 a.u.)
Perihelion (1977): 3 January.
Aphelion: 5 July
Equinoxes (1977): 20 March, 17h 43m;
23 September, 03h 30m
Solstices (1977): 21 June, 12h 14m;
21 December, 23h 24m
Obliquity of the ecliptic, 1977:
23°·44227 (23° 27′)
Sidereal period: 365·256 days
Rotation period: 23h 56m 04s
Mean orbital velocity: 29·79 km/s
Orbital inclination: 0 (by definition)
Orbital eccentricity: 0·016719
Equatorial diameter: 12 756 km
Reciprocal mass, Sun = 1: 328 900
Density, water = 1: 5·517
Escape velocity: 11·18 km/s
Oblateness: 0·003
Mean surface temperature: 22°C
Albedo: 0·36
Mass: 5·976 × 10^{24} kg

The Earth is the largest and most massive of the inner group of planets.

The Earth and the Moon photographed by *Voyager 1* (NASA)

MARS

Mean distance from the Sun:
227·94 million km (1·524 a.u.)
Maximum distance from the Sun:
249·1 million km (1·666 a.u.)
Minimum distance from the Sun:
206·7 million km (1·381 a.u.)
Sidereal period: 686·980 days (= 668·60 sols)
Synodic period: 779·9 days
Rotation period: 24h 37m 22·6s (= 1 sol)
Mean orbital velocity: 24·1 km/s
Axial inclination: 23° 59′
Orbital inclination: 1° 50′ 59″·4
Orbital eccentricity: 0·093
Diameter: 6787 km
Apparent diameter from Earth:
max. 25″·7, min 3″·5
Reciprocal mass, Sun = 1: 3 098 700
Density, water = 1: 3·94
Mass, Earth = 1: 0·107
Volume, Earth = 1: 0·150
Escape velocity: 5·03 km/s
Surface gravity, Earth = 1: 0·380
Mean surface temperature: −23°C
Oblateness: 0·009
Albedo: 0·16
Maximum magnitude: −2·8
Mean diameter of Sun, seen from Mars: 21′

The only red planet is Mars, which comes fourth in order of distance from the Sun, and is the first planet beyond the orbit of the Earth.

Opposition Dates, 1977–2000

Date of Opposition	Closest Approach to Earth	Apparent diameter, ″	Magnitude	Constellation
1978 Jan. 22	1978 Jan. 19	14·3	−1·1	Gemini/Cancer
1980 Feb. 25	1980 Feb. 26	13·8	−1·0	Leo
1982 Mar. 31	1982 Apr. 5	14·7	−1·2	Virgo
1984 May 11	1984 May 19	17·5	−1·8	Libra
1986 July 10	1986 July 16	23·1	−2·4	Sagittarius
1988 Sept. 28	1988 Sept. 22	23·7	−2·6	Pisces
1990 Nov. 27	1990 Nov. 20	17·9	−1·7	Taurus
1993 Jan. 7	1993 Jan. 3	14·9	−1·2	Gemini
1995 Feb. 12	1995 Feb. 11	13·8	−1·0	Leo
1997 Mar. 17	1997 Mar. 20	14·2	−1·1	Virgo
1999 Apr. 24	1999 May 1	16·2	−1·5	Virgo

The closest oppositions occur with Mars near its perihelion. During this period the closest approach is that of 1988, when Mars will approach the Earth to 36 300 000 miles or 58 400 000 km. The closest approach during the past half-century has been that of August 1971 (minimum distance 56 200 000 km).

The greatest distance between Mars and the Earth, with Mars at superior conjunction, may amount to 400 000 000 km.

The longest interval between successive oppositions is 810 days; the **shortest interval** is 764 days. The mean synodic period (see Data Table) is 779·9 days.

The least favourable oppositions occur with Mars at aphelion, as in 1980 (least distance from Earth, 101 320 000 km).

The minimum phase of Mars as seen from Earth is 85 per cent. (At opposition, the phase is, of course, virtually 100 per cent.)

Across Argyre—a mosaic of photographs taken by *Mariner 9* at a range of 18 000 km (NASA)

Mars photographed by *Mariner 7* at a range of 471 750 km on 8 April 1969 (NASA)

Mars' north polar cap—a mosaic of
photographs taken by *Viking Orbiter 2*
at a range of 4025 km (NASA)

Martian Seasons: The seasons are of the same general type as those of the Earth, since the axial tilt is very similar and the Martian day (sol) is not a great deal longer (1 sol = 1·029 days). The lengths of the seasons are as follows:

	Days	Sols
S. spring (N. autumn)	146	142
S. summer (N. winter)	160	156
S. autumn (N. spring)	199	194
S. winter (N. summer)	182	177
	687	669

Southern summer occurs near perihelion. Therefore, climates in the southern hemisphere of Mars show a wider range of temperature than for the north. The effect is much greater than in the case of the Earth, partly because there is no sea on Mars and partly because of the greater eccentricity of the Martian orbit. At perihelion, Mars receives 44 per cent more solar radiation than at aphelion.

The discovery of Mars was unquestionably prehistoric. The planet was recorded in Egypt, China and Assyria. Later, its redness led to its being named in honour of the War-God, Ares (Mars); the study of the Martian surface is still officially known as 'areography'.

The first observation of an occultation of Mars by the Moon was made by Aristotle (384–322 BC). The exact date of the phenomenon is not known.

The first precise observation of the position of Mars dates, according to Ptolemy, back to 17 January 272 BC, when Mars was observed to be close to the star Beta Scorpii.

The best pre-telescopic observations of the movements of Mars were made by the Danish astronomer Tycho Brahe, from his island observatory at Hven between 1576 and 1596. It was these observations which enabled Kepler, in 1609, to publish his first Laws of Planetary Motion, showing that the planets move round the Sun rather than the Earth.

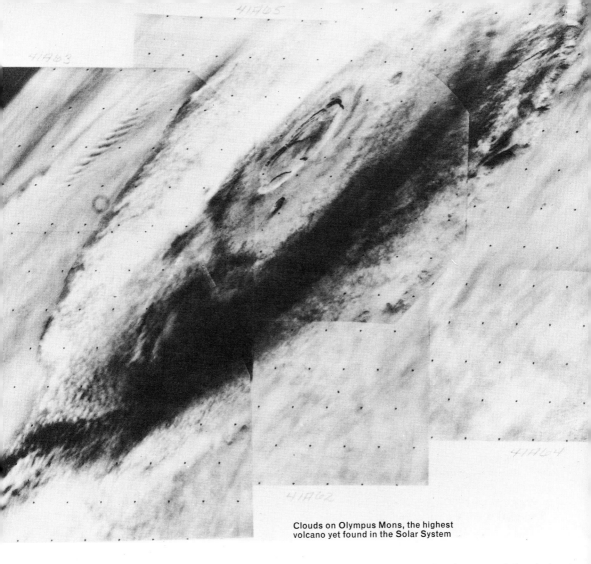

Clouds on Olympus Mons, the highest volcano yet found in the Solar System

The first telescopic observations of Mars were made by Galileo in 1610. No surface details were seen.

The first detection of the phase of Mars was also due to Galileo in 1610, and recorded by him in a letter written to Father Castelli on 30 December of that year.

The first telescopic drawing of Mars was made by F. Fontana, in Naples, in 1636. Mars was shown as spherical, and 'in its centre was a dark cone in the form of a pill'. This feature was, of course, an optical effect. Fontana's second drawing (24 August 1638) was similar.

The first marking to be recorded on Mars was the Syrtis Major, by C. Huygens on 28 November 1659, at 7 p.m. The Syrtis Major is easily recognizable (though exaggerated in size). The sketch has been very useful in confirming the constancy of Mars' rotation period.

The first reasonably accurate estimate of the rotation period of Mars was also due to Huygens. On 1 December 1659 he recorded that the period was about 24 hours.

The first really accurate measurement of the rotation period was made by G. D. Cassini in 1666; his value was 24h 40m.

The first record of the polar caps was also due to Cassini in 1666. It has been suggested that a cap was seen by Huygens in 1656, but his surviving drawing is very inconclusive. (However, Huygens undoubtedly saw the south polar cap in 1672.)

The discovery that the polar caps do not coincide with the geographical (or areographical!) **poles** was made by G. Maraldi, in 1719. In 1704 Maraldi had studied the caps, and had also given a rotation period of 24h 39m, which was very near the truth.

The first known panic due to Mars was that of 1719. Mars was at perihelic opposition, and was so bright that people mistook it for a red comet which might well collide with the Earth.

The first identification of the polar caps with ice and snow was suggested by W. Herschel, from his observations made between 1777 and 1783. Herschel also measured the rotation period, and his observations were later re-worked by W. Beer and J. H. Mädler, yielding a period of 24h 37m 23s·7, which is only one second in error.

The first comment upon the Martian atmosphere was made by Herschel in 1783. From observations of the close approach of a star to Mars, he rightly concluded that the atmosphere could not be very extensive.

The first good determination of the axial inclination of Mars was also due to Herschel at the same period. His value was 28 degrees, which is only 4 degrees too great. The Martian north polar star is at present Deneb (Alpha Cygni), but the inclination varies between 35 degrees and 14 degrees in a period of about 50 000 years, so that precessional effects are important in any long-term consideration of the changing Martian climate.

The first suggestion of signalling to the inhabitants of Mars was made by the great German mathematician K. F. Gauss, about 1802; the plan was to draw vast geometrical patterns in the Siberian tundra. In 1819 J. von Littrow, of Vienna, proposed to use signal fires lit in the Sahara. Later (1874) Charles Cros, in France, put forward a scheme to focus the Sun's heat on to the Martian deserts by means of a huge burning-glass; the glass could be swung around to write messages in the deserts!

The first suggestion of clouds in the atmosphere of Mars was made by the French astronomer H. Flaugergues in 1811.

The first reasonably good charts of Mars were made by W. Beer and J. H. Mädler in 1830–32, from Berlin. They were also the first to report a dark band round the periphery of a shrinking polar cap. This band was later seen by many observers, and in the late 19th century P. Lowell

attributed it, wrongly, to moistening of the ground owing to the melting polar ice.

The first report that the southern polar cap has a greater range of size than the northern was made by the French astronomer F. Arago in 1853. (Flaugergues, in 1811, had suggested that this should be so, because the temperature range is more extreme in the south.)

The first observations of 'white' clouds on Mars were made by A. Secchi in 1858.

The first suggestion that the dark areas are vegetation tracts rather than seas was made in 1860 by E. Liais, a French astronomer who spent much of his life in Brazil. This was also the view of G. V. Schiaparelli, who in 1863 pointed out that the dark areas did not show the Sun's reflection as they would be expected to do if they were sheets of water. The vegetation theory was regarded as probable up to the space-probe era.

The first reported detection of water vapour in the Martian atmosphere was by J. Janssen in 1867. Taking his equipment to the top of Mount Etna, to get above the densest and wettest part of the Earth's atmosphere, Janssen compared the spectrum of Mars with that of the airless Moon, and believed that he could see differences. It is now known that these results were spurious.

The first system of Martian nomenclature was that of R. A. Proctor in 1867. The features were named in honour of past and contemporary observers of the planet. The system was superseded in 1877 by that of Schiaparelli.

The first detection of the alleged canal network was due to G. V. Schiaparelli in 1877, when he recorded 40 canals. Streaks had been recorded earlier by various observers, including Beer and Madler (1830–1840) and the Rev W. R. Dawes (1864), but Schiaparelli's work marked the beginning of the 'canal controversy'. He also produced a new map, and as the zero for Martian longitudes he adopted the feature which had also been selected by Beer and Mädler; it is now called Sinus Meridiani, or the Meridian Bay. Following the space-probe results from 1965, the zero was fixed as the centre of a small crater in the Sinus Meridiani; this has been named Airy-0, in honour of the 19th-century Astronomer Royal, Sir George Airy (the zero for terrestrial longitudes passes through the Airy transit circle at the old Royal Observatory in Greenwich Park.) On this map Schiaparelli introduced a new system of nomenclature, which has only recently been modified in view of the Mariner and Viking results.

The surface of Mars photographed by *Viking Lander 2*. The rock in the foreground is about 250 mm across (NASA)

The first reports of the twinning or 'gemination' of some canals were made by Schiaparelli in 1879. The first observers to support Schiaparelli's canal network were Perrotin and Thollon, at the Nice Observatory in France, in 1886. Subsequently canals became fashionable, and P. Lowell, who built the observatory which bears his name at Flagstaff and observed Mars between 1896 and 1916, believed the network to be artificial. Only since the close-range photography of Mars by space-probes has it been shown that the canal system is completely illusory.

The first astronomer to use the term 'deserts' for the bright regions of Mars was W. H. Pickering, in 1886. In 1892 Pickering observed features in the dark regions as well as the 'deserts', and this more or less killed the theory that the dark areas were seas.

The first observations of craters on Mars seem to have been made by E. E. Barnard in 1892, using the Lick Observatory 36 in refractor. However, Barnard did not publish his results, for fear of ridicule. Craters were also seen by J. Mellish in 1917, with the 40-in Yerkes refractor, but these observations also remained unpublished. The first prediction of craters on Mars was made by D. L. Cyr in 1944, but it was not until the flight of Mariner 4, in 1965, that the existence of craters was proved.

The first suggestion that the polar caps were made of solid carbon dioxide was due to A. C. Ranyard and Johnstone Stoney in 1898. The theory was more or less disregarded until the early space-probe results, beginning with Mariner 4 in 1965, but was then favoured until the missions of Vikings 1 and 2 in 1976. It is now known that the main caps are of ice, many metres thick, with a thin surface layer of solid carbon dioxide which disappears during Martian summer. However, the caps may not be identical, and the south cap may contain carbon dioxide ice as well as water ice.

The first prize offered for communicating with extraterrestrial beings was the Guzman Prize, announced in Paris on 17 December 1900. The sum involved was 100 000 francs – but Mars was excluded, because it was felt that contact with the Martians would be too easy!

The first good measurements of the surface temperature of Mars were made in 1909, by Nicholson and Pettit at Mount Wilson and by Coblentz and Lampland at Flagstaff. They found that Mars has a mean surface temperature of − 28°C, as against +15°C for the Earth.

The first report of the 'Violet Layer' was made by Lampland in 1909. This was supposed to block out the short-wavelength radiations, and prevent them from reaching the Martian surface, except on occasions when the layer cleared away. It could not be seen visually. Following the space-

A panorama over dunes photographed by *Viking Lander 1* (NASA)

probe results, its existence is now regarded as doubtful.

The first analysis of the Martian atmosphere by using the Doppler method was made by W. S. Adams and T. Dunham in 1933. When Mars is approaching the Earth, the spectral lines should be shifted toward the short-wave end of the band; when Mars is receding, the shift should be to the red. It was hoped that in this way any lines due to gases in the Martian atmosphere could be disentangled from the lines produced by the same gases in the Earth's atmosphere. From their results, Adams and Dunham concluded that the amount of oxygen over Mars was less than 1/10 of 1 per cent of the amount existing in the atmosphere of the Earth.

The first reasonably good determination of the atmospheric pressure on Mars was made by the Russian astronomer N. Barabaschev in 1934. He gave a value of 50 millibars. Later estimates increased this to between 80 and 90

millibars. However, the Mariner and Viking results have shown that the pressure is nowhere as high as 10 millibars.

The first reported spectroscopic detection of carbon dioxide in the Martian atmosphere was due to G. P. Kuiper in 1947. However, it was then thought that the atmosphere must consist chiefly of nitrogen. The Mariner and Viking results show that this is wrong. The atmosphere of Mars is made up of 95 per cent carbon dioxide (CO_2), 2·7 per cent nitrogen (N_2), 0·15 per cent oxygen (O_2), 1·6 per cent argon, and smaller amounts of gases such as krypton and xenon.

The first reported detection of organic matter in the spectra of the Martian dark areas was announced by W. M. Sinton in 1959. The results were later found to be spurious.

The first attempt to send a probe to Mars was made by the Russians in 1962 (Mars-1). The first successful probe was the US Mariner 4, in 1965.

MARS PROBES, 1962-1977

Name	Date of launch	Date of encounter	Nearest approach (km)	Remarks
Mars 1 (USSR)	1 November 1962	?	190,000?	Contact lost at 106 000 000 km from Earth
Mariner 3 (USA)	5 November 1964	—	—	In solar orbit. Contact lost soon after launch
Mariner 4 (USA)	28 November 1964	14 July 1965	10,000	Successful. Returned 21 pictures, plus miscellaneous data
Zond 2 (USSR)	30 November 1964	August 1965?	?	Contact lost on 2 May 1965
Mariner 6 (USA)	24 February 1969	31 July 1969	3390	Flew over equator of Mars. Returned 76 pictures
Mariner 7 (USA)	27 March 1969	4 August 1969	3500	Flew over southern hemisphere of Mars. Returned 126 pictures
Mariner 8 (USA)	8 May 1971	—	—	Total failure; crashed in the Atlantic immediately after launch
Mars 2 (USSR)	19 May 1971	27 November 1971	Landed	Orbiter achieved Mars orbit; lander crashed on the surface
Mars 3 (USSR)	28 May 1971	2 December 1971	Landed	Orbiter returned data. Lander lost contact 20 seconds after arrival; no useful data received
Mariner 9 (USA)	30 May 1971	13 November 1971	1395	Orbiter. Returned 7329 pictures. Contact finally lost on 27 October 1972
Mars 4 (USSR)	21 July 1973	10 February 1974	?	Failed to orbit; missed Mars
Mars 5 (USSR)	25 July 1973	12 February 1974	?	In Mars orbit
Mars 6 (USSR)	5 August 1973	12 March 1974	Landed	Contact lost at touchdown on Mars
Mars 7 (USSR)	9 August 1973	9 March 1974	?	Failed to orbit; missed Mars
Viking 1 (USA)	20 August 1975	19 June 1976 (in orbit)	Landed	Landed 20 July 1976, lat. 22°·4N, long. 47°·5W (in Chryse)
Viking 2 (USA)	9 September 1975	7 August 1976 (in orbit)	Landed	Landed 3 September 1976 lat. 48°N, long. 226°W (in Utopia). Viking 2 lander is 7420 km north-east of Viking 1 lander

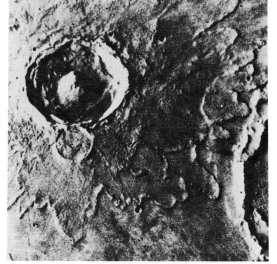

Crater Yuty photographed by *Viking Orbiter 1*

The latest nomenclature for Martian features involves naming the craters in honour of men and women who have been associated with the study of Mars. Other classes of features are as follows:

Catena: crater-chain; a line or chain of craters (e.g. Tithonius Catena).

Chasma: canyon; elongated, steep-sided depression (e.g. Coprates Chasma).

Dorsum (ridge); irregular, elongated elevation (e.g. Argyre Dorsum).

Fossa (ditch); long, narrow, shallow depressions (e.g. Claritas Fossæ).

Labyrinthus (valley complex). The only example is Noctis Labyrinthus.

Mensa (mesa). Flat-topped prominence with steep edges (e.g. Nilosyrtis Mensæ).

Mons (mountain). Large elevation (e.g. Ascræus Mons).

Patera Irregular crater or a complex one with scalloped edges (e.g. Alba Patera).

Planitia (plain). Smooth, low area. often a basin (e.g. Hellas Planitia).

Planum (plateau). Smooth, high area (e.g. Solis Planum).

Tholus (hill). Isolated peak or hill (e.g. Uranius Tholus).

Vallis (valley). Sinuous channel, often with tributaries (e.g. Valles Marineris).

Vastitas; extensive plain (e.g. the north-polar circumpolar plain, Vastitas Borealis).

The greatest Martian dust-storm of the present century may have been that of late 1971, when Mariner 9 reached the vicinity of Mars, but the storm of 1909 was of the same magnitude, and there have been others in 1911, 1924, 1956 and 1973. It seems that major storms are most often seen when Mars is near perihelic opposition – that is to say, at its closest both to the Sun and to the Earth.

The most conspicuous dark area on Mars is the Syrtis Major, formerly known as the Hourglass Sea or Kaiser Sea (or even the Atlantic Canal, though there is no connection with Lowell's canal network). The Syrtis Major is V-shaped, and was drawn by Huygens more than 300 years ago. It was formerly thought to be a vegetation-filled sea-bed, but is now known to be a plateau. It lies in the equatorial region of Mars; the most prominent dark marking further north is the Acidalia Planitia, formerly known as Mare Acidalium.

The brightest area on Mars, excluding the polar caps, may often be Hellas, once thought to be a snow-covered plateau and now known to be a basin. However, Hellas is very variable in brightness.

The highest volcano on Mars is Olympus Mons, formerly known as Nix Olympica (it is visible from Earth as a bright point). It is 25 km high, and has a base measuring 600 km, so that it is the largest known volcano in the Solar System; by contrast, the Hawaiian volcanoes on Earth, which also are shield volcanoes, have a height of only 9 km at most above their bases on the ocean floor. (Heights on Mars are bound to be somewhat arbitrary, since there is no sea-level to act as a standard.)

The largest caldera on a Martian volcano is that of Arsia Mons; the caldera is 140 km in diameter. Other large calderas crown the volcanoes Olympus Mons (65 km), Ascræus Mons (50 km) and Pavonis Mons (45 km); all these volcanoes lie in the Tharsis area. Arsia Mons has a height of 14 km.

The deepest basin on Mars is Hellas, which measures 2200 km by 1800 km. and is about 3 km deep (reckoned from the level on Mars where the atmospheric pressure is 6·1 millibars).

The longest valley system on Mars is that of the Valles Marineris, in the Coprates-Tithonius region and not very far from the giant volcanoes of the Tharsis ridge. The system can be traced for a length of 4000 km, with a maximum width of 200 km, and a greatest depth of about 6 km.

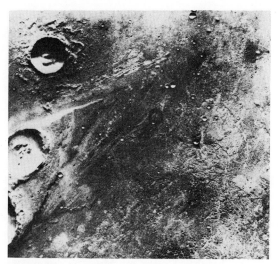

Chryse region photographed by *Viking Orbiter 1* showing an 'island' in a channel complex which could be evidence of past water flows (NASA)

above the rim. There are various other sinuous channels over 1000 km long, giving every impression of having been cut by running water.

The greatest canyon system on Mars is Noctis Labyrinthus (formerly known as Noctis Lacus). The canyons are from 10 to 20 km wide, with slopes of from 10 to 15 degrees. The whole system covers 120 000 km^2.

The highest atmospheric pressure so far measured on Mars is 8·9 millibars, on the floor of Hellas. The pressure at the top of Olympus Mons is below 3 millibars. When Viking 1 landed in Chryse, the pressure was approximately 7 millibars. A decrease of 0·012 millibar per sol was subsequently measured, due probably to carbon dioxide condensing out of the atmosphere to be deposited on the south polar cap, but this is·a seasonal phenomenon, and ceased while Vikings 1 and 2 were still operating.

The maximum temperature of the aeroshell of Viking 1 during descent to Mars was 1500°C; that of Viking 2 was similar.

The first analysis of the surface material was made from the lander of Viking 1: 16 per cent iron, 15 per cent to 30 per cent silicon, 3 per cent to 8 per cent calcium, 2 per cent to 7 per cent aluminium, 0·25 per cent to 1·5 per cent titanium. Results from Viking 2 were similar. The general composition seems to be not very different from that of the lunar maria.

The first temperature measurements from the lander of Viking 1 ranged between —86°C after dawn to a maximum of —31°C near noon. The results from Viking 2 were similar.

The minimum summer temperature of the north polar cap was found to be —73°C; at this temperature carbon dioxide would evaporate. At this time the temperature of the southern cap was found to be —137°C.

The first attempts to detect Martian life from the surface were made by Viking 1. There were three main experiments (pyrolitic release, labelled release, and gas exchange) but all were inconclusive; no certain evidence of Martian life has been found. The experiments were repeated from Viking 2, with equally inconclusive results.

The maximum windspeeds on Mars may, it is estimated, reach 400 km/h, but winds in that thin atmosphere will have little force. Windspeeds measured from the Vikings were light, little exceeding 20 km/h.

All our ideas about Mars have been revolutionised by the Mariner and Viking results. The dark areas are no more than albedo features, and are not depressed seabeds; neither are they vegetation-covered. The crust of Mars may be about 200 km deep; below this comes the mantle, and below this again is the core. No magnetic field has been detected. Much of the water on Mars is locked up in the permanent polar caps. If all the water in the atmosphere were condensed, it would cover the Martian surface with a layer only 1/100 mm deep, but the release of all the water in the ice-caps would produce a layer 10 m deep.

The atmospheric pressure is at present too low for liquid water to exist on the surface, though the presence of apparently water-cut channels indicates that the pressure was much greater some tens of thousands of years ago; evidently there are marked climatic changes, due possibly to the fact that when the axial inclination is at a suitable value the volatiles in the caps may be evaporated, leaving no permanent residual ice-caps such as those which exist today. Traces of carbon monoxide, ozone and hydrogen have been detected in the atmosphere; the height of the Martian tropopause is perhaps 120 km, with carbon dioxide or H_2O wave clouds rising to 24 km, water ice clouds to a maximum of 8 km, and dust-storms up to over 40 km (though even during major storms, the tops of the Tharsis volcanoes often protrude above the dusty level). No doubt future probes to Mars will produce further unexpected results!

	Long., deg.	Lat., deg.
Lunæ	70 to 60	+05 to +20
Ophir	61 to 55	−09 to −12
Solis	98 to 88	−20 to −30
Syria	105 to 100	−10 to −18
Sinai	90 to 70	−10 to −20

Patera

	Long., deg.	Lat., deg.
Alba	110	+40
Amphitrites	299	−59
Apollinaris	186	−08
Biblis	124	+02
Hadriaca	267	−31
Orcus	181	+14
Pavonis	121	+03
Tyrrhena	253	−22
Uranius	93	+26

Tholus

	Long., deg.	Lat., deg.
Albor	210	+19
Australis	323	−57
Ceraunius	97	+24
Hecates	210	+32
Hippalus	89	+76
Iaxartes	15	+72
Jovis	117	+18
Kison	358	+73
Ortygia	8	+70
Tharsis	91	+14
Uranius	98	+26

Vallis

	Long., deg.	Lat., deg.
Al Qahira	202 to 194	−23 to −15
Ares	23 to 14	+02 to +10
Auqakuh	298	+28
Huo Hsing	295 to 292	+32 to +28
Maadim	183	−27 to −20
Mangala	151	−10 to −4
Marineris	96 to 45	−05 to −15
Nirgal	44 to 36	−32 to −27
Kaseli	70 to 56	+21
Shalbatana	45	+01 to +15
Simud	40 to 37	00 to +14
Tiu	32	+10 to +18

Vastitas

	Long., deg.	Lat., deg.
Borealis	continuous	+55 to +67

Craters

	Long., deg.	Lat., deg.
Adams	197	+31
Agassiz	83	−70
Airy	0	−0·5
Antoniadi	299	+22
Arago	330	+10
Arrhenius	237	−40
Bakhuysen	344	−23
Baldet	295	+23
Barabashov	69	+47
Barnard	298	−61
Becquerel	8	+22
Beer	8	−15
Bianchini	97	−64
Bjerknes	189	−43
Boeddicker	197	−15
Bond	36	−33
Bouguer	333	−19
Brashear	120	−54
Briault	270	−10
Burroughs	243	−72
Burton	156	−14
Campbell	195	−54
Cassini	328	+24
Cerulli	338	+32
Chamberlain	124	−66
Charlier	169	−69
Clark	134	−56
Coblentz	91	−55
Columbus	166	−29
Comas Solá	158	−20

	Long., deg.	Lat., deg.
Catena		
Coprates	66 to 56	−15
Ganges	71 to 67	−02 to −03
Tithonia	98 to 80	−06

Chasma

	Long., deg.	Lat., deg.
Australis	270	−80 to −88
Borealis	65 to 30	+85
Candor	78 to 73	−04 to −06
Capri	52 to 32	−14 to −03
Coprates	68 to 54	−11 to −14
Eos	51 to 32	−16 to −17
Ganges	52 to 48	−08
Hebes	81 to 73	+01 to −01
Ius	98 to 80	−07
Juventæ	61	−04
Melas	78 to 70	−08 to −12
Ophir	77 to 64	−03 to −09
Tithonia	90 to 80	−04

Dorsum

	Long., deg.	Lat., deg.
Argyre	70	−61 to −65

Fossa

	Long., deg.	Lat., deg.
Alba	117 to 109	+38 to +49
Cerauniæ	107	+25
Claritas	108 to 105	−19 to −32
Elysium	225 to 219	+28 to +26
Hephæstus	240 to 233	+22 to +18
Mareotis	85 to 69	+41 to +48
Medusæ	162	−08
Memnonia	158 to 140	−22 to −15
Nili	284 to 279	+20 to +26
Sirenum	163 to 138	−36 to −27
Tantalus	105 to 99	+34 to +47
Tempe	80 to 62	+35 to +46
Thaumasia	100 to 80	−36 to −40

Labyrinthus

	Long., deg.	Lat., deg.
Noctis	110 to 92	−05 to −08

Mensa

	Long., deg.	Lat., deg.
Deuteronilus	346 to 340	+42 to +45
Nilosyrtis	290	+32
Protonilus	315	+38

Mons

	Long., deg.	Lat., deg.
Arsia	121	−09
Ascræus	104	+12
Elysium	213	+25
Olympus	133	+18
Pavonis	113	+01

Montes

	Long., deg.	Lat., deg.
Charitum	50 to 32	−57
Hellesponti	315	−45 to −48
Nereidum	57 to 43	−48 to −38
Phlegra	195	+31 to +46
Tharsis	125 to 101	−12 to +16

Planitia

	Long., deg.	Lat., deg.
Acidalia	30	+48
Arcadia	155	+48
Amazonis	160	+13
Argyre	43	−49
Chryse	45	+17
Elysium	210	+15
Hellas	290	−45
Isidis	270	+15
Syrtis	290	+15
Utopia	235	+35

Planum

	Long., deg.	Lat., deg.
Auroræ	52 to 48	−10 to −11
Hesperia	258 to 242	−10 to −35

	Long., deg.	Lat., deg.
Copernicus	169	−50
Crommelin	10	+5
Cruls	197	−43
Curie	5	+29
Daly	22	−66
Dana	32	−73
Darwin	20	−57
Dawes	322	−9
Denning	326	−18
Douglass	70	−52
Du Martheray	266	−6
Du Toit	46	−72
Eddie	218	+12
Eiriksson	174	−19
Escalante	245	0
Eudoxus	147	−44
Fesenkov	87	+22
Flammarion	312	+26
Flaugergues	341	−17
Focas	347	+34
Fontana	73	−64
Fournier	287	−4
Gale	222	−6
Galilei	27	+6
Galle	31	−51
Gilbert	274	−68
Gill	354	+16
Gledhill	273	−53
Graff	206	−21
Green	8	−52
Hadley	203	−19
Haldane	231	−53
Hale	36	−36
Halley	59	−49
Hartwig	16	−39
Heaviside	95	−71
Helmholtz	21	−46
Henry	336	+11
Herschel	230	−14
Hipparchus	151	−44
Holden	34	−26
Holmes	292	−75
Hooke	44	−45
Huggins	204	−49
Hussey	127	−54
Hutton	255	−72
Huxley	259	−63
Huygens	304	−14
Janssen	322	+3
Jarry-Desloges	276	−9
Jeans	206	−70
Joly	42	−75
Jones	20	−19
Kaiser	340	−46
Von Kármán	59	−64
Keeler	152	−61
Kepler	219	−47
Knobel	226	−6
Korolev	196	+73
Kuiper	157	−57
Kunowsky	9	+57
Lambert	335	−20
Lamont	114	−59
Lampland	79	−36
Lassell	63	−21
Lau	107	−74
Le Verrier	343	−38
Liais	253	−75
Li Fan	153	−47
Liu Hsin	172	−53
Lockyer	199	+28
Lomonosov	8	+65
Lowell	81	−52
Lyell	15	−70
Lyot	331	+50
McLaughlin	22	+22
Mädler	357	−11

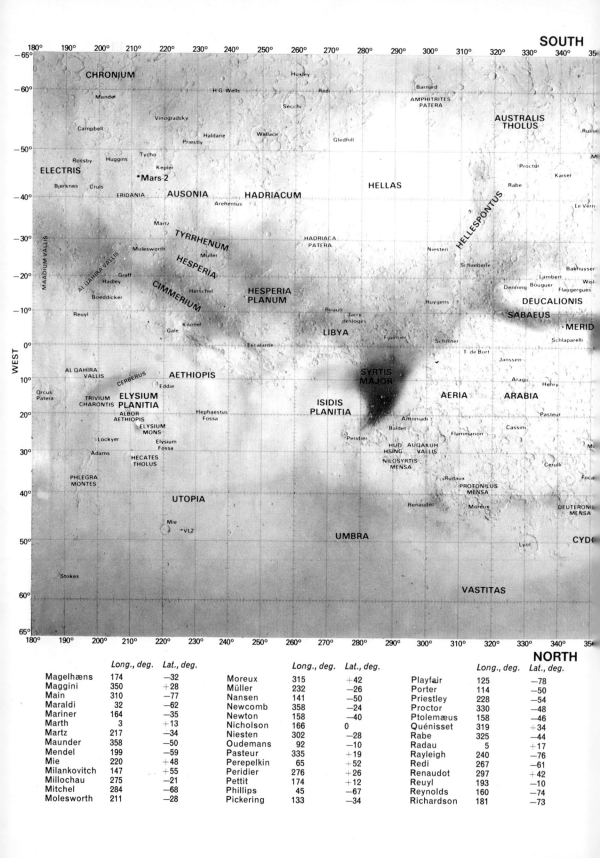

	Long., deg.	Lat., deg.		Long., deg.	Lat., deg.		Long., deg.	Lat., deg.
Magelhæns	174	−32	Moreux	315	+42	Playfair	125	−78
Maggini	350	+28	Müller	232	−26	Porter	114	−50
Main	310	−77	Nansen	141	−50	Priestley	228	−54
Maraldi	32	−62	Newcomb	358	−24	Proctor	330	−48
Mariner	164	−35	Newton	158	−40	Ptolemæus	158	−46
Marth	3	+13	Nicholson	166	0	Quénisset	319	+34
Martz	217	−34	Niesten	302	−28	Rabe	325	−44
Maunder	358	−50	Oudemans	92	−10	Radau	5	+17
Mendel	199	−59	Pasteur	335	+19	Rayleigh	240	−76
Mie	220	+48	Perepelkin	65	+52	Redi	267	−61
Milankovitch	147	+55	Peridier	276	+26	Renaudot	297	+42
Millochau	275	−21	Pettit	174	+12	Reuyl	193	−10
Mitchel	284	−68	Phillips	45	−67	Reynolds	160	−74
Molesworth	211	−28	Pickering	133	−34	Richardson	181	−73

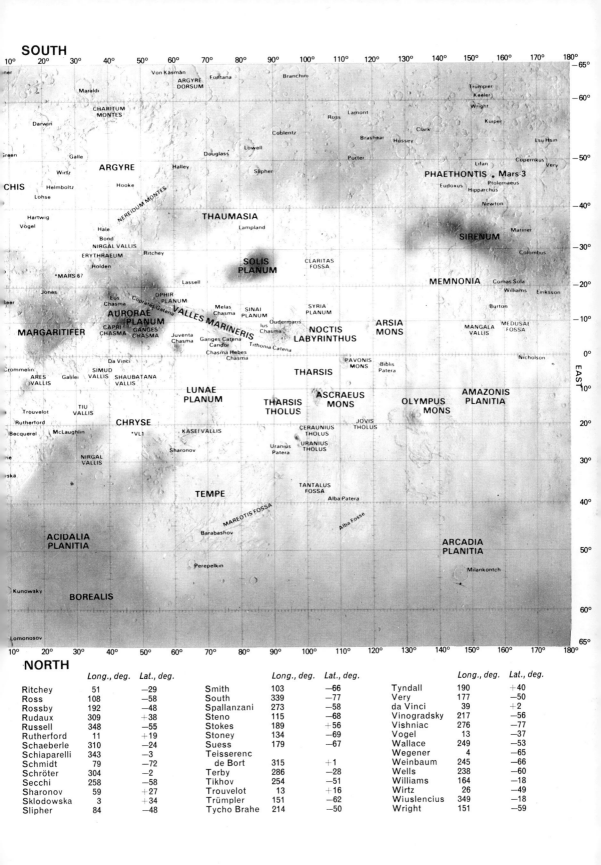

SOUTH

			Long., deg.	Lat., deg.			Long., deg.	Lat., deg.			Long., deg.	Lat., deg.
	Ritchey		51	−29		Smith	103	−66		Tyndall	190	+40
	Ross		108	−58		South	339	−77		Very	177	−50
	Rossby		192	−48		Spallanzani	273	−58		da Vinci	39	+2
	Rudaux		309	+38		Steno	115	−68		Vinogradsky	217	−56
	Russell		348	−55		Stokes	189	+56		Vishniac	276	−77
	Rutherford		11	+19		Stoney	134	−69		Vogel	13	−37
	Schaeberle		310	−24		Suess	179	−67		Wallace	249	−53
	Schiaparelli		343	−3		Teisserenc				Wegener	4	−65
	Schmidt		79	−72		de Bort	315	+1		Weinbaum	245	−66
	Schröter		304	−2		Terby	286	−28		Wells	238	−60
	Secchi		258	−58		Tikhov	254	−51		Williams	164	−18
	Sharonov		59	+27		Trouvelot	13	+16		Wirtz	26	−49
	Sklodowska		3	+34		Trümpler	151	−62		Wiuslencius	349	−18
	Slipher		84	−48		Tycho Brahe	214	−50		Wright	151	−59

NORTH

THE SATELLITES OF MARS

DATA

	Phobos	Deimos
Mean distance from centre of Mars: km	9270	23 400
astronomical units	0·000 062 5	0·000 157 0
Mean angular distance from Mars, at mean opposition:	24″·6	1′01″·8
Mean sidereal period, days:	0·3189	1·2624
Mean synodic period:	7h 39m 26s·6	1d 6h 21m 15s·7
Orbital inclination to Martian equator:	1°·1	1°·8
Orbital eccentricity	0·0210	0·0028
Diameter, kilometres:	20 × 23 × 28	10 × 12 × 16
Escape velocity, km/s:	0·016	0·008
Magnitude at mean opposition:	11·6	12·8
Maximum apparent magnitude as seen from Mars:	−3·9	−0·1
Apparent diameter as seen from Mars:		
maximum	12′·3	2′
minimum	8′	1′·7

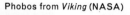

Phobos from *Viking* (NASA)

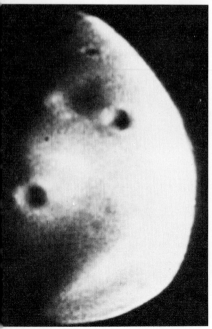

Deimos from *Viking* (NASA)

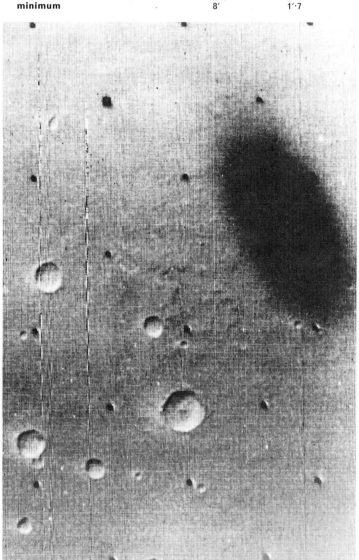

The shadow of Phobos on Mars measured 50 by 110 km when *Mariner 9* took this picture on 4 February 1972

The first mention of possible Martian satellites was fictional – by Jonathan Swift, in Gulliver's *Voyage to Laputa* (1727). Swift described two satellites, one of which had a revolution period shorter than the rotation period of its primary; but there was no scientific basis at that time for such an object. Two satellites were also described in another novel, Voltaire's *Micromégas* (1750). The reasoning was, apparently, that since the Earth had one satellite and Jupiter was known to have four, Mars could not possibly manage with less than two!

The first systematic search for Martian satellites was made by W. Herschel in 1783. The result was negative. H. D'Arrest at Copenhagen, in 1862 and 1864, was similarly unsuccessful.

The first satellite to be discovered was Deimos, by Asaph Hall at Washington on 10 August 1877. Hall discovered Phobos on 16 August 1877. The names Phobos and Deimos were suggested to Hall by Mr Madan of Eton.

The first colour estimates of the satellites were made by E. M. Antoniadi, who in 1930 reported that Phobos was white and Deimos bluish. These results were certainly spurious, even though Antoniadi was using the 83 cm Meudon refractor. It is now thought that the satellites are dust-covered, with very low albedoes.

The first serious suggestion that the satellites might be artificial was made by the Russian astrophysicist I. Shklovsky in 1959. This was based on the assumption that Phobos was spiralling downwards toward Mars, being braked by the atmospheric resistance – though this would involve virtually nil mass, and hence a hollow body! The results have now been shown to be spurious; Phobos is in a stable orbit.

The first proof that Phobos is irregular in shape was obtained in 1969, when Mariner 7 photographed the shadow of Phobos cast on to Mars and revealed that it was elliptical. The first accurate size measurements were obtained by Mariner 9 in 1971–2. Both satellites are irregular, so that diameter values are bound to be somewhat arbitrary.

The first detection of craters on the satellites was also due to Mariner 9, in 1971–2.

The largest crater on Phobos is Stickney, 5 km in diameter. (Stickney was the maiden name of Asaph Hall's wife.) The principal craters on Deimos have been named Swift and Voltaire.

The Martian satellites are curious bodies, and it may well be that they are captured asteroids rather than bona-fide satellites, though there are admittedly difficulties in the way of this hypothesis. Phobos would be invisible to a Martian observer above latitude 69 degrees; the limiting latitude for Deimos would be 82 degrees. To a Martian observer, Phobos would have less than half the apparent diameter of the Moon as seen from Earth, and would give only about as much light as Venus does to us; from Phobos, Mars would subtend an angle of 42 degrees! The apparent diameter of Deimos as seen from Mars would be only about twice the maximum apparent diameter of Venus as seen from Earth, and with the naked eye the phases would be almost imperceptible. Deimos would remain above the Martian horizon for $2\frac{1}{2}$ sols consecutively; Phobos would cross the sky in only $4\frac{1}{2}$ hours, moving from west to east, and the interval between successive risings would be only just over 11 hours. This curious behaviour is due to the fact that Phobos is the only known natural satellite to have a revolution period less than the rotation period of its primary. Total solar eclipses could never occur; Phobos would transit the Sun 1300 times a year, taking 19 seconds to cross the disk, while Deimos would show an average of 130 transits, each taking 1 minute 48 seconds. To a Martian observer, eclipses of the satellites by the shadow of Mars would be very frequent.

The rotation periods of the satellites are synchronous: that is to say they would keep the same faces turned permanently toward the planet, though there would be marked 'libration' effects similar to those of the Moon.

Craters on Phobos

D'Arrest	Roche	Todd
Hall	Sharpless	Wendell
	Stickney	

(There is also one named ridge: Kepler Dorsum.)

Craters on Deimos

Swift	Voltaire

THE MINOR PLANETS

The Minor Planets are also known as asteroids or planetoids. Approximately 2000 have had their orbits computed, and it is estimated that the total membership of the swarm may exceed 40 000.

THE FIRST KNOWN MINOR PLANETS

In the following table:

q = perihelion distance, in astronomical units.
P = revolution period, in years.
i = orbital inclination in degrees.
D = diameter in kilometres

a = aphelion distance, in astronomical units.
e = orbital eccentricity.
M = opposition magnitude.
A = albedo.

(Asteroid diameters have recently been re-measured. Asteroids whose diameter remains very uncertain are indicated by an asterisk.)

No.	Name	Discoverer	q	a	P	e	i	M	D	A
1	Ceres	Piazzi, 1 January 1801	2·55	2·94	4·061	0·079	10·60	7·4	1003	0·054
2	Pallas	Olbers, 28 March 1802	2·11	3·42	4·607	0·237	34·85	8·0	608	0·074
3	Juno	Harding, 1 September 1804	1·98	3·35	4·358	0·257	13·00	8·7	250	0·151
4	Vesta	Olbers, 29 March 1807	2·15	2·57	3·630	0·089	7·14	6·5	538	0·229
5	Astræa	Hencke, 8 December 1845	2·10	3·06	4·139	0·187	5·34	9·8	117	0·140
6	Hebe	Hencke, 1 July 1847	1·93	2·92	3·778	0·203	14·77	8·3	195	0·164
7	Iris	Hind, 13 August 1847	1·84	2·94	3·686	0·230	5·50	7·8	209	0·154
8	Flora	Hind, 18 October 1847	1·86	2·55	3·267	0·156	5·89	8·7	151	0·144
9	Metis	Graham, 26 April 1848	2·09	2·68	3·684	0·123	5·58	9·1	151	0·139
10	Hygeia	De Gasparis, 12 April 1849	2·84	3·46	5·593	0·100	3·81	10·2	450	0·041
11	Parthenope	De Gasparis, 11 May 1850	2·20	2·70	3·840	0·102	4·62	9·9	150	0·126
12	Victoria	Hind, 13 September 1850	1·82	2·85	3·568	0·218	8·37	9·9	126	0·114
13	Egeria	De Gasparis, 2 November 1850	2·36	2·80	4·135	0·085	16·53	10·8	224	0·041
14	Irene	Hind, 19 May 1851	2·16	3·01	4·163	0·164	9·13	9·6	158	0·162
15	Eunomia	De Gasparis, 29 July 1851	2·15	3·14	4·300	0·188	11·73	8·5	272	0·155
16	Psyche	De Gasparis, 17 March 1851	2·53	3·32	5·000	0·135	3·09	9·9	250	0·093
17	Thetis	Luther, 17 April 1852	2·13	3·52	3·880	0·138	5·59	10·7	109	0·103
18	Melpomene	Hind, 24 June 1852	1·80	2·80	3·480	0·218	10·14	8·9	150	0·144
19	Fortuna	Hind, 22 August 1852	2·06	2·83	3·816	0·158	1·56	10·1	215	0·032
20	Massalia	De Gasparis, 19 September 1852	2·06	2·76	3·737	0·145	0·70	9·2	131	0·164
21	Lutetia	Goldschmidt, 15 November 1852	2·04	2·83	3·800	0·162	3·08	10·5	115	0·093
22	Calliope	Hind, 16 November 1852	2·61	3·21	4·962	0·104	13·72	10·6	177	0·130
23	Thalia	Hind, 15 December 1852	2·01	3·24	4·252	0·236	10·17	10·1	111	0·164
24	Themis	De Gasparis, 5 April 1853	2·76	3·52	5·559	0·121	0·77	11·8	234	0·030
25	Phocæa	Chacornac, 6 April 1853	1·79	3·01	3·720	0·256	21·57	10·3	72	0·184
26	Proserpina	Luther, 5 May 1853	2·42	2·89	4·328	0·089	3·56	11·3	80*	?
27	Euterpe	Hind, 8 November 1853	1·94	2·75	3·600	0·172	1·59	9·9	108	0·147
28	Bellona	Luther, 1 March 1854	2·35	3·20	4·625	0·153	9·41	10·8	126	0·132
29	Amphitrite	Marth, 1 March 1854	2·37	2·72	4·083	0·073	6·09	10·0	195	0·140
30	Urania	Hind, 22 July 1854	2·07	2·67	3·637	0·127	2·10	10·6	91	0·144
31	Euphrosyne	Ferguson, 1 September 1854	2·45	3·86	5·606	0·223	26·29	11·0	370	0·030
32	Pomona	Goldschmidt, 26 October 1854	2·38	2·80	4·163	0·082	5·50	11·3	93	0·140
33	Polyhymnia	Chacornac, 28 October 1854	1·89	3·84	4·842	0·340	1·90	11·0	45*	?
34	Circe	Chacornac, 6 April 1855	2·40	3·00	4·404	0·108	5·51	12·3	111	0·039
35	Leucothea	Luther, 19 April 1855	2·32	3·67	5·177	0·225	8·04	12·4	45*	?
36	Atalantë	Goldschmidt, 5 October 1855	1·92	3·58	4·557	0·301	18·45	11·6	118	0·024
37	Fides	Luther, 5 October 1855	2·18	3·10	4·296	0·175	3·08	10·7	95	0·186
38	Leda	Chacornac, 12 January 1856	2·32	3·17	4·535	0·155	6·97	12·2	45*	?
39	Lætitia	Chacornac, 8 February 1856	2·46	3·08	4·608	0·112	10·37	10·3	163	0·169
40	Harmonia	Goldschmidt, 31 March 1856	2·16	2·37	3·413	0·047	4·26	10·5	100	0·123
41	Daphne	Goldschmidt, 22 May 1856	2·02	3·51	4·591	0·270	15·87	10·3	204	0·056
42	Isis	Pogson, 23 May 1856	1·89	2·99	3·812	0·226	8·53	10·2	97	0·125
43	Ariadne	Pogson, 15 April 1857	1·83	2·57	3·270	0·168	3·47	10·2	85	0·113
44	Nysa	Goldschmidt, 27 May 1857	2·05	2·79	3·769	0·151	3·71	9·7	82	0·377
45	Eugenia	Goldschmidt, 27 June 1857	2·50	2·94	4·489	0·081	6·60	11·5	226	0·030
46	Hestia	Pogson, 16 August 1857	2·09	3·00	4·011	0·170	2·32	11·1	133	0·028
47	Aglaia	Luther, 15 September 1857	2·49	3·27	4·881	0·135	4·99	12·1	158	0·027
48	Doris	Goldschmidt, 19 September 1857	2·93	3·30	5·496	0·060	6·55	11·9	250*	0·03?
49	Pales	Goldschmidt, 19 September 1857	2·40	3·79	5·455	0·224	3·16	11·4	70*	?
50	Virginia	Ferguson, 4 October 1857	1·89	3·41	4·313	0·287	2·82	11·6	35*	?
51	Nemausa	Laurent, 22 January 1858	2·21	2·52	3·639	0·066	9·95	10·8	151	0·050
52	Europa	Goldschmidt, 4 February 1858	2·75	3·43	5·447	0·111	7·46	11·1	289	0·035
53	Calypso	Luther, 4 April 1858	2·08	3·16	4·232	0·206	5·15	11·7	45*	?
54	Alexandra	Goldschmidt, 10 September 1858	2·17	3·25	4·458	0·200	11·82	10·9	180	0·030
55	Pandora	Searle, 10 September 1858	2·37	3·15	4·585	0·142	7·20	11·6	65*	?

56	Melete	Goldschmidt, 9 September 1859	1·99	3·21	4·187	0·235	8·07	11·5	146	0·026
57	Mnemosyne	Luther, 22 September 1859	2·84	3·48	5·611	0·100	15·17	12·2	109	0·140
58	Concordia	Luther, 24 March 1860	2·58	2·82	4·435	0·045	5·05	13·1	110	0·030
59	Elpis	Chacornac, 12 September 1860	2·38	3·03	4·471	0·117	8·63	11·4	75*	?
60	Echo	Ferguson, 14 September 1860	1·95	2·83	3·703	0·183	3·59	11·5	51	0·154
61	Danaë	Goldschmidt, 9 September 1860	2·51	3·47	5·168	0·161	18·19	11·7	70*	?
62	Erato	Förster, 14 September 1860	2·60	3·67	5·550	0·169	2·22	13·0	45*	?
63	Ausonia	De Gasparis, 10 February 1861	2·09	2·70	3·706	0·127	5·78	10·2	91	0·128
64	Angelina	Tempel, 4 March 1861	2·35	3·02	4·391	0·124	1·31	11·3	56	0·342
65	Cybele	Tempel, 8 March 1861	3·01	3·83	6·333	0·121	3·54	11·8	309	0·022
66	Maia	Tuttle, 9 April 1861	2·19	3·10	4·306	0·172	3·05	12·5	85	0·029
67	Asia	Pogson, 17 April 1861	1·97	2·87	3·768	0·184	6·01	11·6	58	0·157
68	Leto	Luther, 29 April 1861	2·27	3·30	4·645	0·184	7·97	10·8	126	0·126
69	Hesperia	Schiaparelli, 29 April 1861	2·49	3·47	5·142	0·166	8·55	11·3	90*	?
70	Panopæa	Goldschmidt, 5 May 1861	2·14	3·09	4·227	0·181	11·62	11·5	151	0·039
71	Niobe	Luther, 13 August 1861	2·28	3·23	4·575	0·173	23·27	11·3	115	0·140
72	Feronia	Safford, 29 May 1861	1·99	2·54	3·411	0·120	5·41	12·0	96	0·032
73	Clytia	Tuttle, 7 April 1862	2·55	2·78	4·352	0·048	2·38	13·4	35*	?
74	Galatea	Tempel, 29 August 1862	2·12	3·44	4·632	0·238	4·01	12·0	40*	?
75	Eurydice	Peters, 22 September 1862	1·85	3·49	4·367	0·306	4·99	11·1	40*	?
76	Freia	D'Arrest, 21 October 1862	2·73	4·03	6·213	0·193	2·08	12·3	65*	?
77	Frigga	Peters, 12 November 1862	2·32	3·02	4·362	0·131	2·43	12·1	67	0·113
78	Diana	Luther, 15 March 1863	2·07	3·17	4·238	0·209	8·70	10·9	60*	?
79	Eurynome	Watson, 14 September 1863	1·97	2·92	3·822	0·194	4·62	10·8	76	0·137
80	Sappho	Pogson, 3 May 1864	1·84	2·76	3·481	0·200	8·65	12·8	83	0·113
81	Terpsichore	Tempel, 30 September 1864	2·25	3·46	4·820	0·212	7·89	12·1	45*	?
82	Alcmene	Luther, 27 November 1864	2·15	3·37	4·593	0·220	2·84	11·5	65	0·138
83	Beatrix	De Gasparis, 26 April 1865	2·22	2·64	3·790	0·085	5·00	12·0	123	0·030
84	Clio	Luther, 25 August 1865	1·80	2·92	3·631	0·236	9·34	11·2	90	0·037
85	Io	Peters, 19 September 1865	2·15	3·16	4·325	0·192	11·95	11·2	147	0·042
86	Semele	Tietjen, 4 January 1866	2·42	3·78	5·454	0·220	4·81	12·6	45*	?
87	Sylvia	Pogson, 16 May 1866	3·14	3·82	6·500	0·099	10·85	12·6	90*	?
88	Thisbe	Peters, 15 June 1866	2·32	3·22	4·605	0·162	5·22	10·7	210	0·045
89	Julia	Stéphan, 6 August 1866	2·09	3·01	4·080	0·180	16·08	10·3	155	0·086
90	Antiope	Luther, 1 October 1866	2·59	3·68	5·558	0·174	2·24	12·4	55*	?
91	Ægina	Stéphan, 4 November 1866	2·31	2·87	4·168	0·107	2·11	12·2	104	0·031
92	Undina	Peters, 7 July 1867	2·97	3·43	5·724	0·072	9·93	12·0	250*	0·03?
93	Minerva	Watson, 24 August 1867	2·37	3·14	4·571	0·141	8·57	11·5	168	0·039
94	Aurora	Watson, 6 September 1867	2·83	3·48	5·600	0·102	8·03	12·5	188	0·029
95	Arethusa	Luther 23 November 1867	2·61	3·53	5·377	0·149	13·00	12·1	230	0·019
96	Ægle	Coggia, 17 February 1868	2·65	3·46	5·340	0·134	16·06	12·4	60*	?
97	Clotho	Tempel February 17 1868	1·98	3·36	4·360	0·257	11·79	10·5	95	0·121
98	Ianthe	Peters, 18 April 1868	2·19	3·20	4·417	0·187	15·60	12·6	35*	?
99	Dike	Borrelly, 28 May 1868	2·14	3·19	4·348	0·197	13·88	13·5	25*	?
100	Hecate	Watson, 11 July 1868	2·57	3·61	5·424	0·168	6·42	12·1	60*	?

The names of the next hundred asteroids to be discovered are as follows; they extend to 200 Dynamene, found by C. H. F. Peters on 27 July 1879 (his thirty-fifth discovery).

101	Helena	123	Brunhild	151	Abundantia	173	Ino
102	Miriam	124	Alceste	152	Atala	174	Phædra
103	Hera	125	Liberatrix	153	Hilda	175	Andromache
104	Clymene	126	Velleda	154	Bertha	176	Iduna
105	Artemis	127	Johanna	155	Scylla	177	Irma
106	Dione	128	Nemesis	156	Xanthippe	178	Belisana
107	Camilla	129	Antigone	157	Dejanira	179	Clytæmnestra
108	Hecuba	130	Electra	158	Coronis	180	Garumna
109	Felicitas	131	Vala	159	Æmilia	181	Eucharis
110	Lydia	132	Æthra	160	Una	182	Elsa
111	Ate	133	Cyrene	161	Athor	183	Istria
112	Iphigenia	134	Sophrosyne	162	Laurentia	184	Deiopeia
113	Amalthea	135	Hertha	163	Erigone	185	Eunice
114	Cassandra	136	Austria	164	Eva	186	Celuta
115	Thyra	137	Melibœa	165	Loreley	187	Lamberta
116	Sirona	138	Tolosa	166	Rhodope	188	Menippe
117	Lomia	139	Juewa	167	Urda	189	Phthia
118	Peitho	140	Siwa	168	Sibylla	190	Ismena
119	Althæa	141	Lumen	169	Zelia	191	Kolga
120	Lachesis	142	Polana	170	Maria	192	Nausikaa
121	Hermione	143	Adria	171	Ophelia	193	Ambrosia
122	Gerda	144	Vibilia	172	Baucis	194	Procne
		145	Adeona			195	Eurykleia
		146	Lucina			196	Philomena
		147	Protogenia			197	Arete
		148	Gallia			198	Ampella
		149	Medusa			199	Byblis
		150	Nuwa			200	Dynamene

Ceres, the largest of the minor planets, on 28 November 1975
(R. Arbour)

Some further minor planet data

The following asteroids are of note in some way or other – either for orbital peculiarities, physical characteristics, or because they are the senior members of 'families' – or even because of their unusual names! Diameters are given only in cases where reasonably reliable measurements have been made.

No.	Name	Discoverer	q	a	P	e	i	M	D
103	Hera	Watson, 7 September 1868	2·48	2·92	4·439	0·081	5·42	11·5	96
107	Camilla	Pogson, 17 November 1868	3·25	3·73	6·519	0·070	9·92	12·7	211
110	Lydia	Borrelly, 19 April 1870	2·51	2·95	4·517	0·080	6·00	11·6	
132	Æthra	Watson, 13 June 1873	1·61	3·61	4·222	0·383	25·16	11·9	
137	Melibœa	Palisa, 21 April 1874	2·47	3·78	5·524	C·210	13·44	12·1	150
153	Hilda	Palisa, 2 November 1875	3·37	4·59	7·926	0·153	7·85	13·3	
155	Scylla	Palisa, 8 November 1875	2·00	3·51	4·582	0·273	11·45	13·7	
174	Phædra	Watson, 2 Sept 1877	2·44	3·28	4·830	0·146	12·15	12·6	
175	Andromache	Watson, 1 October 1877	2·55	3·89	5·770	0·208	3·22	13·1	
194	Procne	Peters, 21 March 1879	1·99	3·24	4·232	0·239	18·47	10·6	191
215	Œnone	Knorre, 7 April 1880	2·67	2·86	4·600	0·033	1·70	14·1	
216	Cleopatra	Palisa, 10 April 1880	2·09	3·50	4·668	0·252	13·09	9·9	
232	Russia	Palisa, 31 January 1883	2·11	2·99	4·078	0·173	6·08	13·6	
258	Tyche	Luther, 4 May 1886	2·08	3·15	4·227	0·206	14·28	11·2	
279	Thule	Palisa, 25 October 1888	4·15	4·42	8·864	0·032	2·34	15·4	
298	Baptistina	Charlois, 9 September 1890	2·05	2·48	3·406	0·096	6·29	14·3	
300	Geraldina	Charlois, 3 October 1890	3·15	3·27	5·750	0·020	0·76	14·8	
311	Claudia	Charlois, 11 June 1891	2·89	2·91	4·934	0·003	3·24	14·9	
323	Brucia	Max Wolf, 20 December 1891	1·67	3·10	3·676	0·301	24·20	12·9	
324	Bamberga	Palisa, 25 February 1892	1·78	3·59	4·403	0·336	11·16	9·2	246
344	Desiderata	Charlois, 15 November 1892	1·79	3·40	4·184	0·310	18·49	10·7	
349	Dembowska	Charlois, 9 December 1892	2·66	3·19	5·001	0·090	8·27	10·6	144
360	Carlova	Charlois, 11 March 1893	2·46	3·54	5·197	0·179	11·68	12·7	
386	Siegena	Max Wolf, 1 March 1894	2·40	3·39	4·927	0·170	20·27	11·3	191
399	Persephone	Max Wolf, 23 February 1895	2·82	3·28	5·327	0·074	13·14	14·2	
403	Cyane	Charlois, 18 May 1895	2·55	3·09	4·724	0·096	9·13	13·4	
408	Fama	Max Wolf, 13 October 1895	2·70	3·64	5·631	0·150	9·05	14·2	
434	Hungaria	Max Wolf, 11 September 1898	1·80	2·09	2·711	0·073	22·50	13·5	11
451	Patientia	Charlois, 4 December 1899	2·82	3·30	5·354	0·077	15·23	11·9	276
466	Tisiphone	Max Wolf, 17 January 1901	3·16	3·59	6·201	0·063	19·05	13·5	
471	Papagena	Max Wolf, 7 June 1901	2·21	3·57	4·906	0·024	14·95	9·7	143
475	Ocllo	Stewart, 13 August 1901	1·61	3·60	4·180	0·380	18·80	13·3	
477	Italia	Carnera, 23 August 1901	1·96	2·87	3·754	0·188	5·30	12·7	
499	Venusia	Max Wolf, 24 December 1902	3·08	4·84	7·891	0·222	2·08	14·3	
505	Cava	Frost, 21 August 1902	2·03	3·34	4·403	0·244	9·80	11·8	
511	Davida	Dugan, 30 May 1903	2·66	3·72	5·700	0·166	15·72	10·5	323
694	Ekard	Metcalf, 7 November 1909	1·80	3·54	4·363	0·324	15·79	11·5	101
699	Hela	Hellfrich, 5 June 1901	1·55	3·68	4·225	0·408	15·30	13·7	
704	Interamnia	Cerulli, 2 October 1910	2·58	3·53	5·345	0·155	17·30	11·0	350
724	Hapag	Palisa, 21 October 1911	1·82	3·07	3·836	0·260	11·77	15·9	
747	Winchester	Metcalf, 7 March 1913	1·96	4·02	5·181	0·344	18·14	10·7	205
944	Hidalgo	Baade, 31 October 1920	2·00	9·61	14·041	0·657	42·49	14·9	15
985	Rosina	Reinmuth, 4 October 1922	1·66	2·94	3·489	0·277	4·07	14·6	
1009	Sirene	Reinmuth, 31 October 1923	1·44	3·82	4·260	0·454	15·75	15·9	
1011	Laodamia	Reinmuth, 5 January 1924	1·56	3·23	3·703	0·350	5·47	13·4	7
1019	Strackea	Reinmuth, 3 March 1924	1·77	2·05	2·624	0·072	27·00	15·6	
1038	Tuckia	Max Wolf, 24 November 1924	3·00	4·87	7·808	0·238	9·25	15·6	
1108	Demeter	Reinmuth, 31 May 1929	1·81	3·05	3·784	0·257	24·90	14·2	
1178	Irmela	Max Wolf, 13 March 1931	2·18	3·18	4·385	0·185	6·92	15·1	20
1180	Rita	Reinmuth, 9 April 1931	3·29	4·67	7·943	0·173	7·21	14·7	
1345	Potomac	Metcalf, 4 February 1908	3·27	4·70	7·948	0·179	11·35	15·2	
1355	Magœba	Jackson, 30 April 1935	1·76	1·94	2·523	0·049	22·68	14·8	
1373	Cincinnati	Hubble, 30 August 1935	2·31	4·51	6·300	0·321	38·90	17·6	
1383	Limburgia	Van Gent, 9 September 1934	2·49	3·66	5·394	0·191	0·01	15·8	
1453	Fennia	Väisälä, 8 March 1938	1·84	1·95	2·613	0·028	23·68	15·6	
1486	Marilyn	Delporte, 23 August 1938	1·92	2·47	3·261	0·124	0·08	15·9	
1578	Kirkwood	Cameron, 10 January 1951	3·08	4·84	7·876	0·223	0·82	15·8	
1625	The NORC	Arend, 1 September 1953	2·39	3·95	5·634	0·247	15·48	14·1	
1796	Riga	Chernykh, 16 May 1966	3·14	3·56	6·130	0·062	22·62	15·8	

The Trojans

These asteroids move in virtually the same orbit as Jupiter, but are about 60 degrees from Jupiter, so that there is no fear of collision. They are named after the character of the Trojan War. Of the first fifteen Trojans, ten lie east of Jupiter; the western group includes Patroclus, Priamus, Æneas, Anchises and Troilus – it is unfortunate that the Greeks and the Trojans are mixed up! Co-operative work by T. Gehrels, C. J. van Houten and P. Herget, using plates taken in 1973 by Gehrels with the 48-inch Schmidt telescope, has resulted in the detection of 34 new Trojans. All are, of course, extremely faint. No doubt many more Trojans await discovery.

THE BRIGHTEST TROJANS

No.	Name	Discoverer	q	a	P	e	i	D	M	A
588	Achilles	Max Wolf, 22 February 1906	4·44	5·98	11·896	0·148	10·32		15·3	?
617	Patroclus	Kopff, 17 October 1906	4·48	5·94	11·881	0·140	22·01	147	15·2	0·037
624	Hector	Kopff, 10 February 1907	4·99	5·25	11·589	0·024	18·28	179	15·2	0·038
659	Nestor	Max Wolf, 23 March 1908	4·68	5·84	12·058	0·110	4·51		15·8	
884	Priamus	Max Wolf, 22 September 1917	4·56	5·81	11·809	0·120	8·90		16·1	
911	Agamemnon	Reinmuth, 19 March 1919	4·81	5·50	11·700	0·067	21·90		15·1	
1143	Odysseus	Reinmuth, 28 January 1930	4·73	5·70	11·898	0·093	3·15		15·6	
1172	Æneas	Reinmuth, 17 October 1930	4·64	5·70	11·740	0·102	16·71	130	15·7	0·044
1173	Anchises	Reinmuth, 17 October 1930	4·44	5·89	11·743	0·140	6·95	92	16·2	0·034
1208	Troilus	Reinmuth, 31 December 1931	4·64	5·65	11·741	0·093	33·70		16·0	
1404	Ajax	Reinmuth, 17 August 1936	4·62	5·80	11·888	0·112	18·09		16·6	
1437	Diomedes	Reinmuth, 3 August 1937	4·85	5·32	11·460	0·046	20·61		15·7	
1583	Antilochus	Arend, 20 September 1950	4·99	5·56	12·120	0·054	28·30		16·3	
1647	Menelaus	Nicholson, 23 June 1957	5·08	5·37	11·935	0·028	5·64		18·0	
1749	Telamon	Reinmuth, 23 September 1949	4·68	5·85	12·088	0·111	6·07		17·5	

CLOSE-APPROACH MINOR PLANETS

				Distance from Sun, a.u.					
	No.	Name	Discoverer	Peri-helion	Aph-elion	Period, years	Eccen-tricity	Inclina-tion	Notes
1898 DQ	433	Eros	Witt (Berlin)	1·133	1·783	1·76	0·223	10·83	Close approaches 1931, 1975
1911 MT	719	Albert	Palisa (Vienna)	1·190	2·258	4·42	0·540	10·82	Lost
1918 DB	887	Alinda	Wolf (Heidelberg)	1·148	3·883	3·99	0·544	9·07	Observed in 1969–73
1924 TB	1036	Ganymed	Baade (Bergedorf)	1·220	2·646	4·33	0·542	26·30	Secure
1929 SH	1627	Ivar	Hertzsprung (Johannesburg)	1·125	2·604	2·55	0·397	8·43	Keeps pace with Earth!
1932 EA	1221	Amor	Delporte (Uccle)	1·083	2·758	2·66	0·436	11·95	Seen every 8 years
1932 HA	1862	Apollo	Reinmuth (Heidelberg)	0·647	2·293	1·78	0·566	6·36	Seen in 1973. Close approaches due 1980, 1982
1936 CA	—	Adonis	Delporte (Uccle)	0·441	3·299	2·56	0·764	1·37	Recovered in 1977. No close approach before 2000
1937 UB	—	Hermes	Reinmuth (Heidelberg)	0·617	2·662	2·10	0·624	6·21	Lost
1948 EA	1863	Antinoüs	Wirtanen (Lick)	0·891	3·629	3·40	0·605	18·45	Recovered in 1972
1948 IA	1685	Toro	Wirtanen (Lick)	0·771	1·964	1·60	0·436	9·37	Secure. Will come within 0·15 a.u. in 1980
1949 MA	1566	Icarus	Baade (Palomar)	0·187	1·969	1·12	0·827	22·95	Next close approach (0·15 a.u.) in 1987
1950 DA	—	—	Wirtanen (Lick)	0·839	2·528	2·18	0·501	12·15	Lost
1950 KA	1570	Betulia	Johnson (Johannesburg)	1·119	3·271	3·25	0·490	52·04	Seen in 1976. No close approach before 2000
1950 LA	1980	—	Wilson and Wallenquist (Palomar)	1·049	2·334	2·24	0·365	26·84	Recovered in 1975
1951 RA	1620	Geographos	Wilson and Wallenquist (Palomar)	0·827	1·661	1·39	0·335	13·33	Recovered 1969. Next close approach (0·09 a.u.) in 1983
1953 EA	1915	Quetzalcoatl	Wilson (Palomar)	1·052	3·989	4·50	0·582	20·55	Last seen in 1977
1953 RA	1916	—	Arend (Uccle)	1·250					Recovered in 1974
1960 UA	—	—	Giclas (Flagstaff)	1·049	3·481	3·41	0·537	3·70	Recovered in 1977
1963 UA	—	—	Goethe (Link)	1·240					Recovered in 1976
1968 AA	1917	Cuyo	Cesco and Samuel (Barreal)	1·063	3·234	3·15	0·505	24·01	Recovered 1970. Next close approach (0·12 a.u.) 1989
1971 FA	1864	Dædalus	Gehrels (Palomar)	0·563	2·359	1·77	0·615	22·14	Secure
1971 UA	1865	Cerberus	Kohoutek (Hamburg)	0·576	1·584	1·12	0·467	16·09	Secure. Orbit wholly within that of Mars
1972 RA	—	—	Klemola (Lick)	1·124	3·467	3·48	0·510	8·80	Probably recoverable
1972 RB	—	—	Gehrels (Palomar)	1·101	3·234	3·19	0·492	5·24	Not recoverable until 1988
1972 XA	1866	Sisyphus	Wild (Berne)	0·871	2·916	2·61	0·540	41·12	Secure. Next close approach, 1985
1973 EA	1981	—	Kowal (Palomar)	0·623	2·930	2·37	0·650	39·85	Recovered in 1975. Next close approaches, 1987–1992
1973 EC	1943	—	Gibson (El Leoncite)	1·064	1·798	1·71	0·256	8·70	Recovered in 1975
1973 NA	—	—	Helin (Palomar)	0·880	4·014	3·83	0·641	68·06	No close approach before 1992
1974 MA	—	—	Kowal (Palomar)	0·422	3·083	2·53	0·760	37·65	Probably recoverable
1975 YA	—	—	Kowal (Palomar)	0·905	1·675	1·47	0·298	64·01	Probably recoverable

	Discoverer	Distance from Sun, a.u.: Peri-helion	Aphe-lion	Period, years	Eccen-tricity	Incli-nation	Notes
1976 AA	Helin (Palomar)	0·790	1·143	0·95	0·183	18·93	Discovered 7 January, 1976. First asteroid known to have a period less than 1 year.
1976 UA	Kowal, Sebok and Helin (Palomar)	0·464	1·221	0·76	0·450	5·85	Very faint. Shortest known period. Close approaches due in 1981 and 1983.
1976 WA	Schuster (La Silla)	0·827	3·878	3·73	0·656	24·26	No close approach before 1991.
1977 HA	Helin (Palomar)	0·795	2·398	2·02	0·502	22·98	Probably recoverable.
1977 HB	Kowal (Palomar)	0·701	1·453	1·12	0·349	9·41	Next close approach in 1985. Makes 3 approaches in successive years, then unobservable for 6 years.
1977 RA	Wild (Berne)	1·24					
1977 VA	Helin (Palomar)	1·13					
1978 DA	Schuster (La Silla)	1·025	2·471	3·89	0·585	15·62	Orbit similar to Alinda and Quetzalcoatl.
1978 CA	Giclas (Lowell)	0·883	1·124	1·19	0·214	26·07	
1978 PA	Schuster (La Silla)						
1978 RA	Helin (Palomar)	0·469	0·832	0·8	0·436		Identical with 1975 T.B.

Other possible minor planets of this type are 1959 LM (Hoffmeister, Boyden, Bloem-fontein) and 1947 XC (Giclas, Flagstaff), which were insufficiently observed and are probably lost.

Eros on 21 January 1975 (P. B. Doherty)

RECORD CLOSE APPROACHES

Name	Date	Minimum distance. a.u.
Hermes	1937 October 30	0·006
1976 UA	1976 October 20	0·008
Adonis	1936 February 7	0·015
1977 HA	1977 April 1	0·03
1975 YA	1975 December 26	0·04
Icarus	1967 June 14	0·04
Quetzalcoatl	1953 March 4	0·05
1950 DA	1950 March 12	0·06
Geographos	1969 August 27	0·06
1960 UA	1960 October 7	0·06
Apollo	1932 May 25	0·07
1973 NA	1973 July 2	0·08

All the 'close approach' asteroids are small. Ganymed, with a diameter of about 35 km, is probably the largest of them. Eros is irregular in shape (it has been likened to a cosmical sausage) with a mean diameter of 23 km; Betulia is perhaps 8 km in diameter, Toro 3, Alinda 4 and Icarus possibly no more than 1 km. It is therefore difficult to keep track of them. Apollo (diameter about 3 km) was 'lost' for many years before its recovery in 1973; it passed within 0·01 astronomical units of Venus in 1950 and within 0·07 astronomical units of Venus in 1968; it also by-passed Mars at 0·05 units in 1946. It will approach the Earth to within 0·06 units in November 1980 and again in May 1982.

Toro was within 0·14 units of the Earth on 8 August 1972. It makes periodical fairly close approaches, and there were popular reports that it ranked as a minor Earth satellite. This is, of course, incorrect; it is an ordinary asteroid in orbit round the Sun, though admittedly its mean distance from the Sun is very nearly the same as that of the Earth. (It is not, however, moving in the Earth's orbit, since its path is considerably more eccentric than ours.)

The largest asteroid is Ceres, with a diameter of 1003 km. Ceres is the only asteroid known to have a diameter of over 1000 km (another estimate gives it as 1040 km). The asteroids known to have diameters of 250 km or over are:

1	Ceres	1003 km
2	Pallas	608
4	Vesta	538
10	Hygeia	450
31	Euphrosyne	370
704	Interamnia	350
511	Davida	323
65	Cybele	309
52	Europa	289
451	Patientia	276
15	Eunomia	272
16	Psyche	250
48	Doris	250
92	Undina	250
3	Juno	250

(Most of these are 'carbonaceous', but Pallas is classed as 'peculiar carbon', Hygeia 'peculiar carbon', Vesta 'eucritic', Eunomia 'siliceous' and Psyche 'metallic iron'.)

The smallest known asteroid is probably 1976 UA, with an estimated diameter of about half a kilometre; but this is very uncertain, and no doubt many smaller members of the asteroid swarm exist.

Asteroid diameters are difficult to measure directly, as they are very small; early results were in general too low (thus Ceres was given as only 680 km). A better method, involving infra-red radiation, was used by D. A. Allen and others from 1970. The apparent brightness of a body will depend partly on its size and partly on its albedo or reflecting power; everyone knows that in sunlight a white car will feel cooler than a black one! No object can absorb more than a certain amount of radiation, and the excess is emitted as infra-red. Therefore, the amount of infra-red emission gives a clue to the albedo, and hence to the diameter. A third method has been applied by G. Taylor (Herstmonceux); this involves the occultations of stars by asteroids. The first good results were obtained with 6 Hebe, whose diameter was found to be 195 km with an uncertainty of only 6 km either way. These results are of great value, as they lead to estimates of the densities of the asteroids – and this information is needed for studies of the evolution of the Solar System.

Direct measures of asteroid masses depend upon mutual perturbations. So far measurements have been made of the encounters of 4 Vesta with 197 Arete, and of Ceres with Pallas. The derived masses are: Ceres $(11·7 \pm 0·6) \times 10^{20}$ kg; Pallas, $(2·6 \pm 0·8) \times 10^{20}$ kg; Vesta, $(2·4 \pm 0·2) \times 10^{20}$ kg.

All these are uncertain, particularly in the case of Pallas. Some other derived masses, based on the diameters and assumed albedoes are:

Juno $2·0 \times 10^{19}$ kg;
Hebe $2·0 \times 10^{19}$ kg;
Hygeia $6·0 \times 10^{19}$ kg;
Eunomia $4·0 \times 10^{19}$ kg;
Davida $3·0 \times 10^{19}$ kg;
Icarus $5·0 \times 10^{12}$ kg;
Eros $5·0 \times 10^{15}$ kg;
Adonis $5·0 \times 10^{10}$ kg;

The brightest asteroid is Vesta, which is the only member of the swarm ever visible with the naked eye. There are only 10 asteroids which can exceed the ninth magnitude, and one of these (Eros) does so only during its rare close approaches, as in 1931 and 1975. The list is as follows: (mean opposition magnitudes);

4	Vesta	6·4
1	Ceres	7·3
2	Pallas	7·5
7	Iris	7·8
3	Juno	8·1
6	Hebe	8·3
433	Eros	8·3 (maximum at close approach)
15	Eunomia	8·5
8	Flora	8·7
18	Melpomene	8·9

The most reflective asteroid so far measured is 44 Nysa, with an albedo of 0·377. Others are 64 Angelina (0·342) and 434 Hungaria (0·300).

The darkest asteroid so far measured is 95 Arethusa, with an albedo of 0·019 (blacker than a blackboard); it is darker than the previous record-holder, 342 Bamberga (0·032).

The shortest known rotation periods (based upon variations in magnitude) are for 1566 Icarus (2h 16m) and 321 Florentina (2h 52m) (magnitude range 10·6 to 11·4). The slowest known spinner is 532 Herculina (18h 49m). Some other derived rotation periods, in hours, are:

Ceres	9·08	Hebe	7·28	Psyche	4·30
Pallas	10·00	Iris	7·12	Eros	5·27
Juno	7·22	Hygeia	10·00	Davida	5·17
Vesta	10·68	Eunomia	6·08	Geographos	5·23

The greatest magnitude range due to rotation is for 1971 FA, with a range of 0·9 magnitude and a period of 8h 34m.

The only asteroids to have been detected by radar are Eros, Icarus and Toro. Much the best results were obtained from Eros during its 1975 approach; the surface was found to be rough on a scale of centimetres. It has been found that the longest and shortest diameters are about 36 and 15 km.

The first systematic search for a planet between the orbits of Mars and Jupiter was initiated in 1800, by six astronomers meeting at Lilienthal – at Johann Schröter's observatory. They based their search on Bode's Law (actually first described by Titius, but popularized by Bode). This so-called Law may be summed up as follows:

Take the numbers 0, 3, 6, 12, 24, 48, 96, 192 and 384, each of which (apart from the first) is double its predecessor. Add 4 to each. Taking the Earth's distance from the Sun as 10, the remaining figures give the mean distances of the planets with considerable accuracy out to as far as Saturn, the outermost planet known when Bode popularized the relationship in 1772. Uranus, discovered in 1781, fits in well. Neptune and Pluto do not, but of course these planets were not discovered until considerably later.

Planet	Distance by Bode's Law	Actual distance
Mercury	4	3·9
Venus	7	7·2
Earth	10	10
Mars	16	15·2
—	28	—
Jupiter	52	52·0
Saturn	100	95·4
Uranus	196	191·8
Neptune	—	300·7
Pluto	388	394·6

Whether the Law is anything more than a coincidence is not known, but it was enough for the 'Celestial Police' to hunt for the missing planet corresponding to Bode's number 28. Schröter became president of the association; the secretary was the Hungarian astronomer Baron Franz Xavier von Zach. They very logically concentrated their searches in the region of the ecliptic.

The first asteroid to be discovered was Ceres, by G. Piazzi at Palermo on 1 January 1801. (He was working on a star catalogue, and was not a member of the 'Police', though he joined later.) Ceres was found to have a distance of 27·7 on Bode's scale. Four asteroids were known by the time that the 'Police' disbanded in 1815; the fifth, Astræa, was not found until 1845 as a result of painstaking solo searches by a German amateur, Hencke.

The first asteroid to be discovered photographically was 323 Brucia, on 20 December 1891 by Max Wolf.

The astronomer to have discovered the greatest number of asteroids to which permanent numbers have been given is K. Reinmuth, with 246 discoveries; Max Wolf has 232, and J. Palisa 121. By 1900, 452 asteroids were known.

The first asteroid known to come within the orbit of Mars was 433 Eros, discovered by Witt at Berlin in 1898. Its variability was detected by von Oppolzer in 1900; and in 1931 Finsen and van den Bos, at Johannesburg, saw its elongated shape. Eros can come to within 23 000 000 km of the Earth, and at the approach of 1931 was intensively studied in an attempt to improve our knowledge of the length of the astronomical unit – though the final value (derived by H. Spencer Jones) is now known to have been rather too great. The last close approach occurred in 1975.

The closest approach of an asteroid was that of Hermes, on 28 October 1937, when it approached the Earth to 800 000 km. It was then of the eighth magnitude, moving 5 degrees per hour – so that it crossed the sky in 9 days. (It is not included in the list of the brightest asteroids, since it is so exceptional in every way.)

The first (and so far, the only) asteroid found to approach the Sun within the orbit of Mercury is 1566 Icarus, discovered by W. Baade in 1949. It has a period of 409 days, and can approach the Sun to a distance of 28 000 000 km, so that its surface must then be at a temperature of approximately 500 °C.

The first known asteroid to have a revolution period of less than a year is 1976 AA, discovered by Mrs Helin in January 1976. The period is 346·8 days. It and 1976 UA are the only known asteroids with their aphelion distances less than 1 astronomical unit.

The asteroid with the smallest orbit is 1976 UA, discovered by Mrs Helin on 25 October 1976. The period is 283 days. The distance from the Sun ranges between 69 000 000 and 126 000 000 km, so that the orbit lies well within that of the Earth.

The greatest perihelion distance for a numbered asteroid (excluding Chiron) is that of 588 Achilles; 5·98 astronomical units.

The greatest aphelion distance for a named asteroid (again excluding Chiron) is that of 944 Hidalgo (9·61 astronomical units, slightly greater than the mean distance between the Sun and Saturn). Hidalgo also has the **largest orbit,** and the **longest revolution period** (over 14 years). The magnitude ranges between 10 and 19. This range is greater than for any other asteroid. The greatest aphelion distance for a 'normal' asteroid is that of 1038 Tuckia (4·87 a.u.).

Another exceptional asteroid is 1973 SC2, which was identified in 1978 on plates which had been taken by T. Gehrels with the Palomar 48-in. Schmidt telescope in 1973. (These were the same plates which resulted in the identification of 34 new Trojans.) 1973 SC2 has an orbit intermediate between those of 944 Hidalgo and 2060 Chiron. The revolution period is 20 years, the orbital eccentricity is 0·55, and the aphelion distance is 11·4 a.u., well outside the orbit of Saturn. Unfortunately, 1973 SC2 has been lost, as it was photographed on only four nights (29 Sept. –

Eros, close to Procyon, on 9 February 1975 (P. B. Doherty)

5 Oct. 1973) and its orbit is only approximately known. Its rediscovery will be largely a matter of luck!

The greatest orbital eccentricity is for 1566 Icarus (0·83). Then follow Adonis (0·76) and Hidalgo (0·66).

The least orbital eccentricity is for 311 Claudia (0·0031). Then follow 903 Nealley (0·0039) and 1300 Marcelle (0·0050).

The greatest orbital inclination is for 1973 NA (66·8 degrees). The greatest inclination for a named asteroid is for 1580 Betulia (52·0 degrees).

The least orbital inclination is for 1383 Limburgia (0·014 degrees). Then follows 1486 Marilyn (0·084 degrees).

The 'gaps' in the asteroid zone were predicted by D. Kirkwood in 1857, and confirmed in 1866. **The existence of asteroid 'families'** was demonstrated by K. Hirayama in 1917; among the main families are those of Flora, Themis and Hungaria. These gaps and families are caused by the perturbing influence of Jupiter. For instance, asteroids of the Hilda group have periods approximately 2/3 that of Jupiter.

The first asteroid known to have a satellite is 532 Herculina. On 7 June 1978 Herculina occulted a star (SAO 120 774) and a second occultation showed the existence of a 50-km satellite. Herculina itself is 217 km in diameter (± 3 km). The satellite is assumed to be 3 mag fainter than Herculina, so that its magnitude is about 13; the angular separation is 0″·9. The observers were Baron, A'Hearn, McMahon and Horne (USA). On 24 January 1975 Eros occulted Kappa Geminorum. The fact that Pioneers 10 and 11 passed unharmed through the asteroid zone en route for Jupiter may indicate that there are fewer very small asteroids than had been anticipated.

The first asteroids were given mythological names (Ceres, for instance, is the patron of Sicily, and it was from Palermo in Sicily that Piazzi made the discovery). Later, the mythological names began to become exhausted; one of the first names of more 'modern' type was that of 125 Liberatrix, which seems to have been given either in honour of Joan of Arc or else of Adolphe Thiers, first President of the French Republic. Asteroids may be named after countries (136 Austria, 1125 China, 1279 Uganda, 1197 Rhodesia), people (1123 Shapleya, in honour of Dr. Harlow Shapley; 1462 Zamenhof, after the inventor of Esperanto; 1486 Marilyn after the daughter of Paul Herget, director of Cincinnati Observatory).

The first masculine name was that of 433 Eros.

The only asteroid to be named after an electric calculator is 1625 The NORC – the Naval Ordnance Research Calculator at Dahlgren, Virginia. Plants are represented (978 Petunia, 973 Aralia); universities (694 Ekard – 'Drake' spelled backwards; the orbit was computed by students at Drake University), musical plays (1047 Geisha), foods (518 Halawe – the discoverer, R. S. Dugan, was particularly fond of the Arabian sweet *halawe*), social clubs (274 Philagoria – a club in Vienna), and even shipping lines (724 Hapag – Hamburg Amerika Linie).

The only name to be auctioned was that of 250 Bettina, whose discoverer, Palisa, offered to sell his right of naming for £50. The offer was accepted by Baron Albert von Rothschild, who named the asteroid after his wife. (The Rothschilds are well represented; 719 is named Albert, and 977 Philippa after Philip von Rothschild.) 443 Photographica commemorates the new method of asteroid discovery, and 1581 Abanderada – someone who carries a banner – is said to be associated with Eva Perón, wife of the former ruler of Argentina! Finally, 1000 is, appropriately enough, named Piazzia.

CHIRON

In 1977 C. Kowal, using the Schmidt telescope at Palomar, discovered an extraordinary object which has been named Chiron. It has been given an asteroidal designation (1977 UB) and a number (2060) but it is far from being an ordinary asteroid, since its path lies mainly between the orbits of Saturn and Uranus! At perihelion it comes to within 1278 million km of the Sun, and about one-sixth of its orbit lies within that of Saturn; but at aphelion it recedes to 2827 million km. greater than the minimum distance between the Sun and Uranus. The orbital inclination of approximately 7 degrees prevents any encounter with Uranus at the present epoch, but it is worth noting that in 1664 BC Chiron approached Saturn to within about 16 000 000 km. This is not so very much greater than the distance between Saturn and its outermost satellite, Phœbe.

The orbital elements of Chiron are as follows:
Orbital period: 50·68 years
Perihelion distance: 8·5 a.u. (1 278 000 000 km).
Aphelion distance: 18·5 a.u. (2 827 000 000 km).
Orbital inclination: 6·923 degrees.
Orbital eccentricity: 0·3786.

The last two perihelia occurred in 1895 and 1945; the next will be in 1996. At discovery the magnitude was 18, but at perihelion it will rise to about 15. The diameter may be as much as 650 km, though this is very uncertain. G. Taylor has found that no occultations of catalogue stars are due between now and 1981, but eventually an occultation must occur, and this could be the best way of measuring the diameter of Chiron.

Kowal's discovery was made on 1 November 1977, from plates exposed on 18 and 19 October 1977. It has since been traced on photographs taken as far back as 1895.

The nature of Chiron remains problematical. Suggestions that it could be a huge comet have been discounted. It could therefore be: (a) an exceptional asteroid (b) a planetesimal – a surviving remnant of the 'building-blocks' which condensed to form the outer planets, (c) the brightest member of a trans-Saturnian asteroid group, or even (d) an escaped satellite of Saturn. For the moment it seems best to treat it as asteroidal in nature, but Kowal summed the situation up neatly when he commented that Chiron was – well, 'just Chiron'!

JUPITER

DATA

Mean distance from the Sun: 778·34 million km (5·203 a.u.)
Maximum distance from the Sun: 815·7 million km (5·455 a.u.)
Minimum distance from the Sun: 740·9 million km (4·951 a.u.)
Sidereal period: 11·86 years = 4332·59 days
Synodic period: 398·9 days
Rotation period: System I (equatorial) 9h 50m 30s; System II
(rest of planet) 9h 55m 41s; System III (radio methods)
9h 55m 29s
Mean orbital velocity: 13·06 km/s
Axial inclination: 3° 04′
Orbital inclination: 1° 18′ 15″·8
Orbital eccentricity: 0·048
Diameter: equatorial 142,200 km: polar 134,700 km
Apparent/diameter from Earth: max 50″·1, min 30″·4
Reciprocal mass, Sun = 1 : 1047·4
Density, water = 1 : 1·33
Mass, Earth = 1 : 317·89
Volume, Earth = 1 : 1318·7
Escape velocity: 60·22 km/s
Surface gravity, Earth = 1 : 2·64
Mean surface temperature: −150°C.
Oblateness: 0·06
Albedo: 0·43
Maximum magnitude: −2·6
Mean diameter of Sun, seen from Jupiter: 6′ 09″

The largest and most massive planet is Jupiter,
which is fifth in order of distance from the Sun.

Oppositions of Jupiter, 1977–2000.

Year and date	Diameter seconds of arc	Magnitude
1977 Dec. 23	47·4	−2·3
1979 Jan. 24	45·8	−2·2
1980 Feb. 24	44·7	−2·1
1981 Mar. 26	44·2	−2·0
1982 Apr. 25	44·4	−2·0
1983 May 27	45·5	−2·1
1984 June 29	46·8	−2·2
1985 Aug. 4	48·5	−2·3
1986 Sept. 10	49·6	−2·4
1987 Oct. 18	49·8	−2·5
1988 Nov. 23	48·7	−2·4
1989 Dec. 27	47·2	−2·3
1991 Jan. 28	45·7	−2·1
1992 Feb. 28	44·6	−2·0
1993 Mar. 30	44·2	−2·0
1994 Apr. 30	44·5	−2·0
1995 June 1	45·6	−2·1
1996 July 4	47·0	−2·2
1997 Aug. 9	48·6	−2·4
1998 Sept. 16	49·7	−2·5
1999 Oct. 23	49·8	−2·5
2000 Nov. 28	48·5	−2·4

Jupiter is well placed for observation for several
months in every year, and the opposition magnitude
has a range of only about 0m·5. Generally speaking
it 'moves' about one constellation per year; thus
the 1979 opposition occurs in Cancer and the 1980
opposition in Leo.

On average Jupiter is the brightest of the planets
apart from Venus, though for relatively brief
periods during perihelic oppositions Mars may
outshine it. Note that in the period between 1977
and 2000, the only years in which Jupiter does not
come to opposition are 1978 and 1990. This is,
of course, because oppositions occurred in the last
part of December in 1977 and 1989.

The shortest 'day' in the Solar System is
that of Jupiter, insofar as the principal planets are
concerned (some of the asteroids rotate much more
quickly). It is not possible to give an overall value
for the axial rotation period, because Jupiter does
not rotate in the way that a solid body would do.
The equatorial zone has a shorter rotation than the
rest of the planet. Conventionally, System I refers
to the region between the north edge of the South
Equatorial Belt and the south edge of the North
Equatorial Belt; the mean rotation period is
9h 50m 30s, though individual features may have
periods which differ perceptibly from this.
System II, comprising the rest of the planet, has a
period of 9h 55m 41s, though again different
features have their own periods; that of the Great
Red Spot is approximately 9h 55m 37s·6. In
addition there is System III, which relates not to
optical features but the bursts of decametre radio
radiation; the period seems to be 9h 55m 29s·4.

Jupiter's most feature, the Great Red Spot, shown clearly in this photograph taken on 4 June 1972. North is at the top. (Lowell Observatory).

There are no true 'seasons' on Jupiter, because the axial inclination to the perpendicular of the orbital plane is only just over 3 degrees – less than for any other planet.

The latitudes of the various belts and zones are also variable, over a limited range. The following values are therefore approximate only:

N.N. Temperate Belt:	lat.	+36°
N. Temperate Belt:		+27°
N. Equatorial Belt:	N. edge	+20°
	S. edge	+ 7°
S. Equatorial Belt:	N. edge	− 6°
	S. edge	−21°
S. Temperate Belt:		−29°
S.S. Temperate Belt:		−44°

The first man to find that System I rotates faster than System II was G. D. Cassini, in 1690.

The most famous marking on Jupiter is the Great Red Spot. It may be identical with a feature recorded by Robert Hooke in 1664, and it was certainly seen by Cassini in 1665, but it became famous in 1878, when it was described as very prominent and brick-red. Subsequently it has shown variations in both intensity and colour, and at times (as in 1976–7) it has been invisible, but it always returns after a few years, so that during telescopic times it has been to all intents and purposes a permanent feature – unlike any other of the spots. At its greatest extent it may be 40 000 km long and 14 000 km wide, so that its surface area is then greater than that of the Earth. On 5 September 1831 H. Schwabe found that it encroached into the region of the southern part of the South Equatorial Belt, making up what is now always termed the Red Spot Hollow.

Though the latitude of the Red Spot varies little, it drifts about in longitude. Over the past century the total longitude drift amounts to approximately 1200 degrees! The latitude is generally very close to −22 degrees.

It was once thought that the Red Spot might be a solid or semi-solid body floating in Jupiter's outer gas, in which case it would be expected to disappear if its level sank for any reason (possibly because of a change in the surrounding gas). Alternatively, there was a suggestion by R. Hide, in 1963, that the Spot might be the top of a Taylor column – a sort of 'standing wave' in the Jovian gas formed above a mountain or a depression in a solid surface below the gaseous layer. Both these theories have now been discarded. Results from the Pioneer probes indicate that Jupiter is mainly liquid, which rules out anything in the nature of a Taylor column; and it is believed that the Red Spot is a phenomenon of Jovian meteorology – possibly a cyclonic disturbance of hurricane type. It is certainly in rotation. The prominent colour remains puzzling, but it is now known that there have been shorter-lived 'Little Red Spots' which were also pinkish in hue. It is thought that the Great Red Spot is exceptionally long-lived merely because it is exceptionally large. though its origin is unknown. Neither can we tell whether it will eventually decay.

The longest-observed 'disturbance' on Jupiter was the South Tropical Disturbance, discovered by P. Molesworth on 28 February 1901 and last indicated by many observers during the apparition of 1939–40. It took the form of a shaded zone between white spots. The rotation period of the Disturbance was shorter than that of the Red Spot, so that periodically the Disturbance caught the Spot up and passed it; the two were at the same latitude, and the interactions were of great interest. Nine conjuctions were observed, and possibly the beginning of a tenth in 1939–40, though by then the Disturbance had practically vanished. It has not reappeared, and there is no real reason to suggest that it will do so in the future. Its average rotation period was 9h 55m 27s·6.

The most prominent belt on Jupiter is generally the North Equatorial, though this applies only to the period between 1900 and the present time. The South Equatorial Belt is often double, and may be broad; it never vanishes, but may be obscure at times. The other belts are also subject to marked variations in intensity.

The first proof that Jupiter is not a miniature Sun was given by H. Jeffreys in his classic papers published in 1923 and 1924. Jeffreys proposed a model in which Jupiter would have a rocky core, a mantle composed of solid water and carbon dioxide, and a very deep, tenuous atmosphere.

The first identification of methane and ammonia in Jupiter's atmosphere was due to R. Wildt in 1932; it followed that much of the planet must be composed of hydrogen. In 1934 Wildt proposed a model giving Jupiter a rocky core 60 000 km in diameter, overlaid by an ice shell 27 000 km thick above which lay the hydrogen-rich atmosphere. (This was certainly more plausible than a strange theory proposed by E. Schoenberg in 1943. Schoenberg believed Jupiter to have a solid surface, with volcanic rifts along parallels of latitude; heated gases rising from these rifts would produce the belts!)

The first 'hydrogen models' were proposed independently in 1951 by W. Ramsey in England and W. DeMarcus in America. On Ramsey's theory, the 120 000 km diameter core was composed of hydrogen, so compressed that it assumed the characteristics of a metal. This core was overlaid by an 8000 km deep layer of ordinary solid hydrogen, above which came the atmosphere.

The first suggestion that Jupiter could be mainly liquid seems to have been made by G. W. Hough as long ago as 1771. Hough, incidentally, believed the Red Spot to be a kind of floating island.

The latest models of Jupiter are based mainly upon work by J. D. Anderson and W. B. Hubbard in the United States. There is a relatively small, rocky core made up of iron and silicates, and at a temperature of about 30 000°C. Around this is a thick shell of liquid metallic hydrogen. At about 46 000 km from the centre of the planet there is a sudden transition from liquid metallic hydrogen to liquid molecular hydrogen; in the transition region the temperature is assumed to be about 11 000°C, with a pressure about three million times that of the Earth's air at sea-level. Above the liquid molecular hydrogen comes the gaseous atmosphere, which is about 1000 km deep, and is made up 82 per cent hydrogen, 17 per cent helium and 1 per cent of other elements. It contains water droplets, ice crystals, ammonia crystals and ammonium hydrosulphide crystals.

Gases warmed by the internal heat of the planet rise into the upper atmosphere, and cool, forming clouds of ammonia crystals floating in gaseous hydrogen. The clouds form the bright zones on Jupiter, which are both colder and higher than the dark belts. The colours of the various features have yet to be satisfactorily explained, but are certainly due to characteristics of Jovian chemistry.

Jupiter radiates more heat than it would do if it depended only upon radiation received from the Sun. On the liquid-planet model, this excess heat must be nothing more than the remnant of the heat generated when Jupiter was first formed. It had been suggested that Jupiter might be slowly contracting, and that this contraction was the cause of the excess radiation emitted; but a liquid is to all intents and purposes incompressible.

The first detection of radio radiation from Jupiter was due to B. F. Burke and F. L. Franklin, in the United States, in 1955. (It must be admitted that the discovery was accidental!) The radio emissions are concentrated in wavelengths of tens of metres (decametric emissions) and tenths of metres (decimetric). Efforts to correlate the radio burst with visual features, such as the Red Spot, were not successful, but it seems that the decametric radiation is affected by the orbital position of Io, the innermost of the large satellites of Jupiter. From these early results it was surmised – correctly – that Jupiter has an extremely powerful magnetic field.

The first probe to Jupiter was Pioneer 10. It was launched from Cape Canaveral on 2 March 1972, and by-passed Jupiter on 3 December 1973 at a distance of 131 400 km. Studies of the Jovian magnetosphere and atmosphere were carried out, and more than 300 pictures returned. Pioneer 10 was the first probe to leave the Solar System; it will never return. In case it should ever be found by some alien civilization, it carries a plaque to give a clue as to the planet of its origin – though whether any other beings would be able to decipher the message on the plaque has been questioned!

The second Jupiter probe was Pioneer 11, launched on 5 April 1973 and which by-passed Jupiter at 46 400 km on 2 December 1974. It also studied Jupiter's magnetosphere and atmosphere, and sent back excellent pictures; it was then put into a path en route for a rendezvous with Saturn, after which it too will leave the Solar System permanently.

The magnetic field of Jupiter is stronger than for any other planet. Except near the planet, the field is dipolar, like that of the Earth, but the direction of the field is opposite to ours – so that if a terrestrial compass were taken to Jupiter, it would point south instead of north. The axis of the dipole field is inclined to the axis of rotation by 10·8 degrees. Inside about three Jupiter radii the field is so complex that it cannot be regarded as truly dipolar. The main dipole field has a strength of about 4 gauss (as against 0·3 to 0·8 gauss at the surface of the Earth). Pioneer 10 crossed the shock wave (i.e. the boundary between the Jovian magnetosphere and the general interplanetary field) on 26 November 1973 – the first probe to do so. Energetic particles were detected as far out as 300 Jovian radii from the planet, and during its passage through the radiation belts it received 200 000 rads from electrons and 50 000 rads from energetic protons; the lethal dose for a man if a mere 500 rads. The Jovian radiation belts – that is to say, the regions in which protons and high-energy electrons are trapped by the magnetic field – are about 10 000 times more intense than the Van Allen belts associated with the Earth. The electrical instruments on Pioneer 10 were 95 per cent saturated, and if the probe had approached much closer to Jupiter the mission would have failed. Pioneer 11 was put into a different trajectory, passing quickly over the equatorial zone so as to avoid the worst of the radiation. The Jovian magnetosphere is very extensive, and it has even been found that the magnetic 'tail' extends so far that when suitably placed, Saturn may lie within it.

The Jovian magnetosphere proper is very variable, and matters are complicated because the five inner satellites (the four Galileans, and Amalthea) lie within it. Quite apart from these investigations, the Pioneers provided much additional information; for example, the ratio of Jovian hydrogen to helium appears to be much the same as with the Sun (10 to 1), which is an important conclusion. The polar temperatures at the cloud level seem to be no cooler than at the equator, so that the poles must receive a greater supply of internal heat, and the overall aspect of the clouds is different. Altogether Jupiter is an extraordinary world, and has proved to be much more complex than had been thought before the Pioneer missions. There seems no doubt that the intense radiation zones will permanently prohibit any manned flights close to the planet.

The two Voyager probes launched in 1977 will by-pass Jupiter in 1979.

THE SATELLITES OF JUPITER

Number	Name	Discoverer	Mean distance from Jupiter: thousands of km	a.u.	Mean angular distance from Jupiter, at mean opposition distance ′ ″	Apparent diameter as seen from surface of Jupiter ′ ″
V	Amalthea	Barnard, 1892	181·3	0·0012	0 59·4	7 24
I	Io	Galileo and Marius 1610	421·6	0·0028	2 18·4	35 40
II	Europa		670·9	0 0045	3 40·1	17 30
III	Ganymede		1070	0·0072	5 51·2	18 06
IV	Callisto		1883	0·0126	10 17·6	9 30
XIII	Leda	Kowal, 1974	11 110	0·0687	60 45	0 0·15
VI	Himalia	Perrine, 1904	11 470	0·0767	62 45	0 8·2
X	Lysithea	Nicholson, 1938	11 710	0·0783	64 05	0 0·3
VII	Elara	Perrine, 1905	11 743	0·0785	64 10	0 1·4
XII	Ananke	Nicholson, 1951	20 700	0·142	116	0 0·2
XI	Carme	Nicholson, 1938	22 350	0·151	123	0 0·2
VIII	Pasiphaë	Melotte, 1908	23 300	0·157	129	0 0·2
IX	Sinope	Nicholson, 1914	23 700	0·158	130	0 0·2

Name	Density, water = 1	Magnitude	Name	Orbital inclination, degrees	Orbital eccentricity	Sidereal period, days	Mean synodic period d h m s	Diameter, km	Escape velocity, km/s	Reciprocal mass, Jupiter = 1
Amalthea	3?	14·1	Amalthea	0·4	0·003	0·498	0 11 57 27·6	240	0·16	?
Io	3·5	4·9	Io	0·0	0·000	1·769	1 18 28 35·9	3659	2·56	21300
Europa	3·1	5·3	Europa	0·5	0·0001	3·551	3 13 17 53·7	3050	2·10	39000
Ganymede	1·9	4·6	Ganymede	0·2	0·0014	7·155	7 03 59 35·9	5270	2·78	12700
Callisto	1·7	5·6	Callisto	0·2	0·0074	16·689	16 18 05 06·9	5000	2·43	17800
Leda	3?	20	Leda	26·7	0·1478	238·7	254	8	0·005	?
Himalia	3?	13·5	Himalia	28	0·1580	250·6	266	170	0·11	?
Lysithea	3?	18·4	Lysithea	29	0·1074	259·2	276	19	0·01	?
Elara	3?	15·8	Elara	28	0·2072	259·7	276	80	0·05	?
Ananke	3?	18·6	Ananke	147	0·169	631	551	17	0·01	?
Carme	3?	17·9	Carme	163	0·207	692	597	24	0·02	?
Pasiphaë	3?	18·6	Pasiphaë	148	0·410	744	635	27	0·02	?
Sinope	3?	18·1	Sinope	157	0·275	758	645	21	0·01	?

Jupiter and the four satellites discovered independently by Galileo and Simon Marius in 1610. Europa is on the left of Jupiter; Io, Ganymede and Callisto are on the right, in that order. The photograph was taken by *Voyager 2* at a range of 437 million km and the images have been enhanced by computer processing (NASA)

The four large satellites are known as the Galileans, though they were observed at about the same time independently by Simon Marius. In 1975 C. Kowal, at Palomar, reported a fourteenth satellite of magnitude 21, which has yet to be confirmed. The four outermost satellites (Ananke, Carme, Pasiphaë and Sinope) have retrograde orbits. All the satellites beyond Callisto are so far from Jupiter, and so subject to solar perturbation, that their orbits are not even approximately circular, and orbital elements are subject to change. Indeed, Pasiphaë was 'lost' after its discovery in 1908, found again in 1922, lost once more until 1938, and again between 1941 and 1955!

The first satellites to be discovered (naturally excluding the Moon!) were the four Galileans, in 1610. **The last satellite to be discovered by visual means** was Amalthea, by Barnard in 1892; subsequent discoveries have been photographic. **The last confirmed satellite discovery** was that of Leda, by Charles Kowal in 1974, using the Palomar 48 in Schmidt telescope.

The brightest satellite is Ganymede, which would be an easy naked-eye object if it were not so overpowered by Jupiter. It is one of the largest satellites in the Solar System, and is probably inferior only to Titan in Saturn's system and Triton in Neptune's

The least dense of the large satellites is Callisto, which, according to V. Rakotyan and his colleagues in the USSR, may be covered with a layer of ice, though this remains unconfirmed.

The smallest known body to have an atmosphere is Io. Results from Pioneer 10 showed that there is an ionosphere extending some 700 km above the surface on the sunlight side, and there is a tenuous atmosphere with a surface pressure of about 10^{-9} bar. In 1974 R. Brown, from Harvard, detected the presence of a sodium cloud associated with Io; F. P. Fanale has suggested that Io has a surface layer of salt, and that the sodium cloud is produced by the effects of Jovian radiation upon this salty covering.

Surface details upon the Galilean satellites have been observed from Earth, and in 1961 A. Dollfus and his colleagues in France drew the first maps of the surface features. Ganymede and Europa were shown on the Pioneer pictures, but detailed charts remain to be compiled. In 1979 Voyager 1 is scheduled to study Amalthea, Io, Ganymede and Callisto; the closest approach will be to Io. Voyager 2 will study Callisto, Ganymede, Europa and Amalthea; the closest approach will be to Ganymede.

Since the apparent diameter of the Sun as seen from Jupiter is less than 6 minutes of arc, Amalthea and all four Galileans can produce total eclipses. To a Jovian observer, Io would appear as much the largest of the satellites.

The satellites beyond Callisto may well be in the nature of captured asteroids, and do not in any way resemble the massive Galileans. Their diameters are very uncertain, as are their densities and escape velocities; only Himalia can be as much as 100 kilometres across. It is very likely that more 'asteroidal' type satellites await discovery.

SATURN

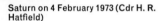

Saturn on 4 February 1973 (Cdr H. R. Hatfield)

DATA

Mean distance from the Sun:
1427·0 million km (9·539 a.u.)
Maximum distance from the Sun:
1507 million km (10·069 a.u.)
Minimum distance from the Sun:
1347 million km (9·008 a.u.)
Sidereal period: 10759·20 days
= 29·46 years
Mean synodic period: 378·1 days
Rotation period (equatorial):
10 hours 14 minutes
Mean orbital velocity: 9·6 km/s
Axial inclination: 26° 44′
Orbital inclination: 2° 29′ 21″·6
Orbital eccentricity: 0·056
Diameter: equatorial 119 300 km; polar
107 700 km
Apparent diameter from Earth: max.
20″·9, min 15″·0
Reciprocal mass, Sun = 1 : 3498·5
Density, water = 1 : 0·71
Mass, Earth = 1 : 95·17
Volume, Earth = 1 : 744
Escape velocity: 32·26 km/s
Surface gravity, Earth = 1 : 1·16
Mean surface temperature: −180°C
Oblateness: 0·1
Albedo: 0·61
Opposition magnitude: max −0·3,
min +0·8
**Mean diameter of Sun, seen from
Saturn:** 3′22″

The most remote planet known in ancient times was Saturn, which is sixth in order of distance from the Sun.

Opposition Dates, 1977–2000:

Date	Magnitude	Date	Magnitude
1977 Feb. 2	0·0	1989 July 2	+0·2
1978 Feb. 16	+0·3	1990 July 14	+0·3
1979 Mar. 1	+0·5	1991 July 26	+0·3
1980 Mar. 13	+0·8	1992 Aug. 7	+0·4
1981 Mar. 27	+0·7	1993 Aug. 19	+0·5
1982 Apr. 8	+0·5	1994 Sept. 1	+0·7
1983 Apr. 21	+0·4	1995 Sept. 14	+0·8
1984 May 3	+0·3	1996 Sept. 26	+0·7
1985 May 15	+0·2	1997 Oct. 10	+0·4
1986 May 27	+0·2	1998 Oct. 23	+0·2
1987 June 9	+0·2	1999 Nov. 6	0·0
1988 June 20	+0·2	2000 Nov. 19	—0·1

The opposition magnitude is affected both by Saturn's varying distance and by the angle of presentation of the rings. The last edgewise presentation was in 1966. The next will be in 1980 and 1995. The intervals between successive edgewise presentations are 13 years 9 months and 15 years 9 months. During the shorter interval, the south pole is sunward, the southern ring-face is seen, and Saturn passes through perihelion.

Ring Diameters

Ring A: outer 272 300 km; inner 239 600 km.
Ring B: outer 234 200 km: inner 181 100 km.
Ring C: inner 149 300 km.

The first observations of Saturn must date back before recorded history, since the planet is a bright naked-eye object (at its most brilliant it surpasses any of the stars apart from Sirius and Canopus).

The first recorded observations of Saturn seem to have been those made in Mesopotamia in the mid-7th century BC. About 650 BC there is a record that Saturn 'entered the Moon', which is presumably a reference to an occultation of the planet.

The first observation of Saturn by Copernicus was made on 26 April 1514, when Saturn lay in a line with the stars 'in the forehead of Scorpio'. Copernicus made three other recorded observations of Saturn: on 5 May 1514, 13 July 1520 and 10 October 1527.

The first observation of Saturn by Tycho Brahe was made on 18 August 1563, when Saturn was in conjunction with Jupiter.

The first telescopic observation of Saturn was made in July 1610 by Galileo, using a magnification of 32 on his largest telescope. He recorded that 'the planet Saturn is not one alone, but is composed of three, which almost touch one another and never move nor change with respect to one another. They are arranged in a line parallel to the Zodiac, and the middle one is about three times the size of the lateral ones.' Galileo's telescope was inadequate for him to see the ring-system in its true guise, and subsequently he lost sight of the 'companions', as the rings were edgewise-on in December 1612; later he recovered them.

The earliest-dated drawing of Saturn to be published was made by Pierre Gassendi on 19 June 1633, though it, together with other drawings made by Gassendi, was not actually published until 1658.

The first true explanation of the nature of the ring system was given by Christiaan Huygens in 1656, from his observations made in 1655. His explanation was challenged, and was not finally accepted by all astronomers until 1665.

The first indications of the Crêpe Ring (Ring C) may have been given by Campani in 1664, though they were not interpreted as being such. Campani also appears to have recorded a belt on the planet's disk.

The discovery of the Cassini Division, separating Rings A and B, was made by G. D. Cassini in 1675 (claims that W. Ball had observed the Division ten years earlier have been discounted). Cassini also recorded the South Equatorial Belt on the planet's disk.

The first suggestion that the rings are not solid was made by J. Cassini in 1705. Theoretical confirmation was provided by James Clerk Maxwell in 1875, who showed that no solid ring could exist; it would be disrupted by the gravitational pull of Saturn.

The discovery of the solar flattening of Saturn was made by W. Herschel in 1789.

Herschel gave the ratio of the equatorial to the polar diameter as 11:10, which is approximately correct.

The first well-defined spots to be detected on Saturn were observed by J. H. Schröter and his assistant, K. Harding, at Schröter's observatory at Lilienthal, near Bremen, in 1796.

The discovery of Encke's Division in Ring A was made by J. F. Encke at Berlin on 28 May 1837.

The discovery of the Crêpe Ring, Ring C, was made by W. Bond at Harvard in November 1850, using the 15 in Merz refractor there. Independent confirmation was provided by Lassell and Dawes in England.

The transparency of Ring C was discovered in 1852, independently by Lassell and by C. Jacob at Madras.

The first explanation of the divisions in the ring system was given by D. Kirkwood in 1866. Kirkwood attributed the Divisions to the perturbing effects of Saturn's inner satellites.

The discovery of dark absorption bands in the spectrum of Saturn was made by A. Secchi in 1863.

The first brilliant white spot to be recorded on Saturn was discovered by A. Hall on 8 December 1876.

The first observation of the globe of Saturn seen through the Cassini Division was made in 1883 by A. Hall and C. A. Young.

The first spectroscopic proof of the meteoritic nature of the rings was given by J. E. Keeler in 1895. Keeler showed that the rings did not rotate as a solid mass would do; the inner sections had the fastest rotation. This was proved by their Doppler shifts.

The first report of a dusky ring exterior to Ring A was made by G. Fournier, using an 11 in refractor, at Mont Revard (France) in 1907. Various other reports have since been made, but the existence of this outermost ring is extremely dubious.

The most famous observation of the passage of a star behind the rings was made by M. A. Ainslie and J. Knight, independently, on 9 February 1917. The dimmimg of the star as it passed behind the rings and divisions provided valuable information about the transparencies of the ring sections. The star concerned was B. D.

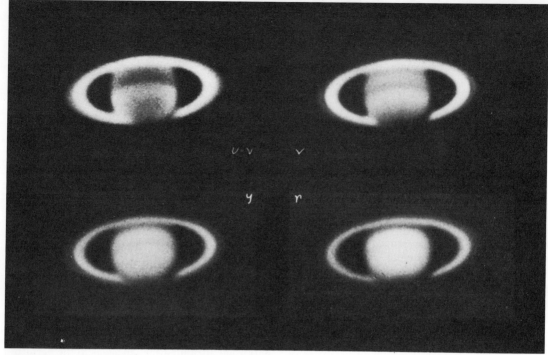

Saturn in ultra-violet, violet, yellow and red light on 13 July 1926

+21°1714. Ainslie used a 9 in refractor at Black-heath; Knight a 5 in refractor at Rye in Sussex.

The first theoretical demonstration that Saturn has a cold surface, and is therefore not a 'miniature sun', was given by H. Jeffreys in 1923.

The spectroscopic detection of methane and ammonia in Saturn's atmosphere was made by T. Dunham at Mount Wilson, California, in 1932. Bands had been seen by V. M. Slipher in 1909, but had not previously been identified.

The most prominent white spot on Saturn to be seen during the present century was discovered by W. T. Hay on 3 August 1933. Hay was using a 6 in refractor, and is the only famous comedian to have discovered a spot on Saturn – he is best remembered as Will Hay! The spot remained identifiable until 13 September 1933. W. H. Wright considered it to have been of and 'eruptive nature'.

The first modern-type model of Saturn was described by R. Wildt in 1938. Wildt believed that there was a rocky core, overlaid by a thick layer of ice and then by the gaseous atmosphere. The model has now been rejected, as has an alternative model proposed in 1951 by W. Ramsey, who

considered that hydrogen was the main constituent but that near the core it acquired metallic characteristics.

The first serious observation of Saturn with the Hale reflector at Palomar was made by G. P. Kuiper in 1954. Kuiper concluded that the Cassini Division was the only genuine gap, Encke's Division being merely a 'ripple' and the other reported minor divisions illusory.

The first observations of infra-red thermal emission from Saturn were made in 1969, from equipment carried in an aircraft.

The first radar reflections from the rings were observed in 1972. It was concluded that the ring particles were icy, with diameters of between 4 and 30 cm.

The first Saturn probe was Pioneer 11, launched on 5 April 1973. It by-passed Jupiter on 2 December 1974, and was then put into on orbit which will take it to a rendezvous with Saturn.

According to the latest model, Saturn has a rocky core 20 000 km in diameter, surrounded by a 5000 km shell of ice and an 8000 km layer of metallic hydrogen. It radiates more heat than it absorbs, indicating that there is an internal heat source. Saturn is the only planet whose mean density is less than that of water.

Name	Discoverer	Mean distance from Saturn thousands of km	a.u.	Mean angular distance from Saturn, at mean opposition distance ′ ″	Apparent diameter as seen from the surface of Saturn ′ ″	Orbital inclination, degrees	Orbital eccentricity
Janus	Dollfus, 1966	169	0·00106	0 25·5	9 30	0·0	0·0
Mimas	Herschel, 1789	186	0·00124	0 30·0	10 54	1·5	0·020
Enceladus	Herschel, 1789	238	0·00159	0 38·4	10 36	0·0	0·005
Tethys	Cassini, 1684	295	0·00197	0 47·6	17 36	1·1	0·000
Dione	Cassini, 1684	378	0·00252	1 01·1	12 24	0·0	0·002
Rhea	Cassini, 1672	528	0·00352	1 25·1	10 42	0·3	0·001
Titan	Huygeus, 1655	1223	0·00817	3 17·3	17 12	0·3	0·029
Hyperion	Bond, 1848	1484	0·00991	3 59·4	0 43	0·6	0·104
Iapetus	Cassini, 1671	3562	0·02380	9 35·	1 48	14·7	0·028
Phœbe	Pickering, 1898	12 960	0·08657	34 51	0 3·2	150	0·163

Name	Sidereal period, days	Synodic period, d	h	m	Diameter, km	Escape velocity, km/s	Reciprocal mass, Saturn = 1	Density, water = 1	Magnitude
Janus	0·815	0	17	59	300?	0·11?	?	?	14
Mimas	0·942	0	22	37	400	0·15	15 000 000	1	12·1
Enceladus	1·370	1	8	53	550	0·20	7 000 000	1	11·9
Tethys	1·887	1	21	18	1200	0·47	910 000	1·1	10·3
Dione	2·737	2	17	42	1150	0·87	490 000	3·2	10·4
Rhea	4·517	4	12	28	1450	0·87	250 000	2	9·8
Titan	15·945	15	23	16	5800	2·47	4 150	1·34	8·3
Hyperion	21·277	21	7	39	300	0·19	5 000 000	3	14·2
Iapetus	79·331	79	22	05	1800	0·67	300 000	3	var
Phœbe	550·337	523	13		200	0·13	?	?	16

The diameter of Iapetus may be considerably greater than this. All satellite diameters are rather uncertain. The value for Titan given here is according to a determination made in 1975 by J. F. Veverka at Cornell University, USA. As the apparent mean diameter of the Sun as seen from Saturn is 3′22″, all the satellites, apart from the three outermost, could produce total eclipses; to a Saturnian Tethys would appear as the largest satellite. Phœbe has a retrograde orbit, and may be classed as 'asteroidal'. There have been suspicions that other faint satellites may exist, but 'Themis', reported by W. H. Pickering in 1904, has been definitely shown to be non-existent.

Mimas, Enceladus, Tethys, Dione, Rhea and Titan photographed by G. P. Kuiper in 1946 with the 82 in McDonald reflector

The largest satellite of Saturn is undoubtedly Titan, which is larger than the planet Mercury or any of the satellites of Jupiter. It may well be the largest satellite in the Solar System; its only rival is Triton, the senior satellite of Neptune, whose diameter is decidedly uncertain.

The smallest satellite is probably Phœbe, with an estimated diameter of 250 km.

The densest satellite is Dione, whose density seems to be very similar to that of our Moon. The inner satellites have densities approximately the same as that of water, so that they may be icy in nature.

The only satellite to move in a retrograde orbit is Phœbe. This, together with its great distance from Saturn and its small size, has led to the suggestion that it may be a captured asteroid rather than a bona-fide satellite.

The first satellite to be discovered was Titan, by Christiaan Huygens on 25 March 1655.

The greatest number of satellite discoveries was made by G. D. Cassini, who found Iapetus in 1671, Rhea in 1672, and Dione and Tethys in 1684. Of the other satellites, Mimas and Enceladus were discovered by W. Herschel in 1789, Hyperion by Bond and Lassell (independently) in 1848, Phœbe by W. H. Pickering in 1898, and Janus by A. Dollfus in 1966.

The first satellite to be discovered photographically was Phœbe. In 1904 W. H. Pickering announced the discovery of another satellite, moving in an orbit between those of Titan and Hyperion; it was said to have a mean distance from Saturn of 908 000 miles, a period of 20·85 days, an orbital eccentricity of 0·23 and an orbital inclination of 39°. It was named Themis. However, it has not been seen since, and all astronomers now agree that it does not exist, so that Pickering must have mistaken a faint star for a satellite.

The last satellite to be discovered was Janus, detected by Dollfus at the Pic du Midi Observatory while the rings were edge-on. Janus is so close to Saturn that it can be observed only during an edgewise presentation of the rings, so that it

will not be recovered until 1980. (The satellite had been recorded several times earlier in 1966 by the present writer, who however did not recognize it as being new, and can therefore claim absolutely no credit!)

The first announcement of the variability of Iapetus was made by G. D. Cassini in 1672. Iapetus is the most variable satellite in the Solar System; it is much brighter when to the west of Saturn than when to the east. Like all fairly large satellites, its axial rotation period is equal to its revolution period (79·3 days in the case of Iapetus), so that clearly the cause of variation lies either in irregularity of shape or the unequal reflecting power of its two hemispheres. If the latter explanation is correct, one hemisphere is as dark as a blackboard and the other as reflective as snow! In 1974 S. Soter (Cornell University) suggested that the leading hemisphere of Iapetus might have been darkened by the accumulation of dust eroded from Phœbe, but this idea will be very difficult to prove or to disprove. Recent estimates indicate that when at maximum brightness Iapetus is not much inferior to Titan, in which case it may be larger than previously thought. At its faintest it fades below magnitude 11.

The first observed eclipse of Rhea by the shadow of Titan was seen on 8 April 1921 by six English observers, independently: A. E. Levin, P. W. Hepburn, L. J. Comrie, E. A. L. Attkins, F. Burnerd and C. J. Spencer. According to the first three observers, Rhea disappeared completely for over half an hour.

The first surface markings to be seen on Titan were observed in the early 1940s by five French observers: Lyot, Bruch, Camichel, Dollfus and Gentili, using the 24in refractor at the Pic du Midi Observatory.

The first suggestion of an atmosphere around Titan was made by the Spanish astronomer J. Comas Solá in 1908. Comas Solá based this upon his observation that the centre of the disk was brighter than the limbs.

The first spectroscopic detection of the atmosphere of Titan was achieved by G. P. Kuiper in 1943–4. Methane (CH_4) is the only gas to have been positively identified, though in 1972 L. Trafton believed

that he had detected molecular hydrogen. It is very likely that the atmosphere is made up chiefly of methane and hydrogen, and that the ground pressure is about 100 millibars – ten times as great as the atmospheric pressure on the surface of Mars. There have been suggestions that the atmosphere may be 'cyclic'; molecules escape, but cannot escape from the pull of Saturn itself, so that they are subsequently re-collected by Titan. However, it has also been suggested that the atmosphere is permanent, and that the ground density is even greater than 100 millibars.

The surface temperature of Titan is about —150°C, and it is probable that clouds form. D. Hunten has estimated the height of the cloud deck as 150 km, though this is naturally very uncertain. It is thought that Titan has a rocky core, surrounded by a wet, rocky mantle with water bound in the rock; outside this is a 'magma' of water containing dissolved ammonia, while the crust may be a mixture of ice and methane. In every way Titan is of exceptional interest, since it is unlike any other body in the Solar System, and it is one of the main objects of the Voyager programme.

Voyager 1 should by-pass Saturn in November 1980. It is scheduled to make a close approach to Titan (as well as less close encounters with Mimas, Enceladus, Dione and Rhea). If this is achieved, Voyager I will merely escape from the Solar System. Voyager 2, which should encounter Saturn in August 1981, will then survey Titan from a much greater distance as well as making observations of Rhea, Tethys, Enceladus, Dione and Mimas; it will then go on to a rendezvous with Uranus. However, if Voyager 1 is not wholly successful, Voyager 2 will be re-routed to rendezvous closely with Titan – in which case it will be unable to go on to Uranus. Astronomers are therefore devoutly hoping that Voyager 1 will not fail!

URANUS

The first planet to be discovered in modern (telescopic) times was Uranus, which comes seventh in order of distance from the Sun.

Opposition Dates, 1977–2000

1977 Apr. 30	1983 May 29	1989 June 24	1995 July 21
1978 May 5	1984 June 1	1990 June 29	1996 July 25
1979 May 10	1985 June 6	1991 July 4	1997 July 29
1980 May 14	1986 June 11	1992 July 7	1998 Aug. 3
1981 May 19	1987 June 16	1993 July 12	1999 Aug. 7
1982 May 24	1988 June 20	1994 July 17	2000 Aug. 11

During this period Uranus passes through several constellations:

1977–1981, Uranus is in Libra.
1982–1984, Uranus is in Scorpio.
1985–1988, Uranus is in Ophiuchus.
1989–1995, Uranus is in Sagittarius.
1996–2000, Uranus is in Capricornus.

In June 1989 Uranus will reach its greatest southerly declination ($-23°.7$). (Greatest northerly declination had been reached in March 1950.)

DATA

Mean distance from the Sun: 2869·6 million km (19·181 a.u.)
Maximum distance from the Sun: 3004 million km (20·088 a.u.)
Minimum distance from the Sun: 2735 million km (18·275 a.u.)
Sidereal period: 84·01 years = 30684·9 days
Synodic period: 369·7 days
Rotation period: $\pm$23 hours
Mean orbital velocity: 6·80 km/s
Orbital eccentricity: 0·047
Diameter: 51 800 km
Apparent diameter from Earth: max 3″·7, min 3″·1
Reciprocal mass, Sun = 1 : 22 800
Density, water = 1 : 1·7
Mass, Earth = 1 : 14·6
Volume, Earth = 1 : 67
Escape velocity: 22·5 km/s
Surface gravity, Earth = 1 : 1·17
Mean surface temperature: $-210°C$
Oblateness: 0·01—0·03
Albedo: 0·35
Maximum magnitude: +5·6
Mean diameter of Sun, seen from Uranus: 1′ 41″

PERIHELIA, APHELIA, DISTANCE FROM EARTH

The most recent aphelion passage was that of 1 April 1925. Uranus was at its maximum distance from the Earth (21·09 a.u.) on 13 March of that year. Aphelion will next be reached on 27 February 2009.

The most recent perihelion passage was that of 20 May 1966. Uranus was at its closest to the Earth (17·29 a.u.) on 9 March of that year. Perihelion will next be reached on 13 August 2050.

The unique axial inclination of Uranus (98°) means that the rotation is technically retrograde, though not generally reckoned as such; the five satellites move virtually in the equatorial plane of Uranus, and move in the same sense as the rotation of the planet. From Earth, the equator of Uranus is regularly presented (as in 1923 and 1966); at other times a pole is presented (the north pole in 1901 and 1985, the south pole in 1946 and 2030).

The first record of Uranus seems to have been an observation by the Astronomer Royal, the Rev. John Flamsteed, on 23 December 1690, when the planet was in Taurus; Flamsteed took it for a star (34 Tauri). Altogether, 22 pre-discovery observations have been listed, as follows:
Flamsteed 1690, 1712, four times in 1715.
J. Bradley, 1748 and 1750.
P. Le Monnier, twice in 1750, 1764, twice in 1768, six times in 1769, 1771.
T. Mayer, 1756.

It is remarkable that Le Monnier failed to identify Uranus because of its movement. He observed it eight times in four weeks (27 December 1768 – 23 January 1769) without realizing that it was anything but a star. He was not blessed with orderly mind, and it is on record that one of his observations of Uranus was later found to be scrawled on a paper bag which had once contained hair perfume!

The discovery of Uranus was made on 13 March 1781 by William Herschel, using a 6·2 in reflector of 7 ft focal length and a magnification of 227. Herschel recognized that the object (in Gemini) was not a star, and he believed it to be a comet. His communication to the Royal Society was indeed entitled 'An Account of a Comet'.

The first recognition of the new body as a planet seems to have been due, independently but at about the same time (May 1781) by the French amateur astronomer J. de Saron – who was guillotined in 1794 during the Revolution – and by the Finnish mathematician Anders Lexell. Lexell calculated an orbit, finding that the distance of the planet from the Sun was 19 a.u. – only slightly too small. He gave a period of between 82 and 83 years, and stated that the apparent diameter was between 3″ and 5″, making the planet larger than any other apart from Jupiter and Saturn.

The first proposal to name the planet 'Uranus' was made in 1781 by the German astronomer J. E. Bode. Other names were proposed – for instance 'Hypercronius' (J. Bernouilli, 1781) and 'the Georgian Planet' (by Herschel himself in 1782, in honour of his patron, King George III). Others called it simply 'Herschel'. Until 1850 the *Nautical Almanac* continued to call it the Georgian Planet, but in that year the famous mathematician John Couch Adams suggested changing over to 'Uranus'. This was done, and the name became universally accepted.

The first attempt to measure the apparent diameter of Uranus was made by Herschel in 1721. His value (4″·18) was rather too great. In 1788 he gave the diameter as 34 217 miles (55 067 km), with a mass 17·7 times that of the Earth; these values were also slightly too high. In 1792–4 Herschel made an attempt to measure the polar flattening, and from his results concluded that Uranus must have a rapid rotation.

The first observation of an occultation of Uranus by the Moon was made on 6 August 1824 by Captain (later Rear-Admiral) Sir John Ross, using a power of 500 on a reflector of 25 ft (7·26 m) focal length. On 4 October 1832 Thomas Henderson observed an occultation of Uranus by the Moon from the Cape of Good Hope.

The first estimate of the axial rotation period was made by J. Houzeau, in France in 1856. Houzeau stated that the rotation period must be between $7\frac{1}{4}$ hours and $12\frac{1}{2}$ hours.

The first reports of markings on the disk of Uranus were made by J. Buffham on 25 January 1870; the telescope used was a 9 in refractor, magnification 320. Two bright round spots were recorded, and on 19 March a light streak was described. It seems rather improbable that these markings were genuine features, though one cannot be sure.

The first observations of dark lines in the spectrum of Uranus were made by A. Secchi in 1869, from Italy. Further lines were detected by the English amateur W. Huggins in 1871, and in 1889 Huggins obtained the first photographs of these lines.

The first really good tables of the motion of Uranus were compiled in America by Simon Newcomb, in 1875.

The first spectroscopic confirmation of the retrograde rotation was obtained by the French astronomer H. Deslandres, in 1902. Final confirmation was obtained by P. Lowell and V. M. Slipher, at the Flagstaff Observatory in Arizona, in 1911. For many years the rotation period was thought to be 10h 49m, but recent work has shown that the true value is almost certainly more like 23 hours (Trafton, 1978).

The first identification of methane in the atmosphere of Uranus was made by R. Mecke, at Heidelberg, in 1933 (it had been suggested, on theoretical grounds, by R. Wildt in 1932). Confirmation was obtained by V. M. Slipher and A. Adel, at Flagstaff, in 1934.

The first widely-accepted model of Uranus was that of R. Wildt, who proposed in 1934 that the planet must have a rocky core, overlaid by a thick layer of ice which was in turn overlaid by the atmosphere. In 1951 W. R. Ramsey, of Manchester University, proposed an alternative model, according to which Uranus was made up largely of methane, ammonia and water. It is still not certain whether Uranus has a hot or cold interior, but very recent results seem to indicate that there is no appreciable heat-source, in which respect Uranus differs from Neptune. According to one model, there is a rocky core 16 000 km in diameter, surrounded by an 8000 km ice layer, with the rest of the planet consisting mainly of molecular hydrogen. The atmosphere itself seems to be remarkably clear, so that the sunlight can penetrate deeply into it. The spectrum shows strong bands of methane, but not ammonia, no doubt because the low temperature means that the ammonia clouds form at too great a depth to be observed. Hydrogen has been detected, and the presence of helium is extremely probable. Ethane, which has been identified in the spectrum of Neptune, has not been found in Uranus.

The first photographs of Uranus from a balloon-borne telescope were obtained in 1970, from a height of 24 km; the balloon was *Stratoscope II*. The resolution was 0″·15, ten times better than can be obtained by a telescope on the ground. No markings were shown. From these results, R. E. Danielson, M. Tomasko and B. Savage, at Princeton Univerrsity, derived a value of 51 800 km for the planet's diameter. More recently, J. Janesick (1977), at Mount Lemmon Observatory, in the United States, has given a value of 55 000 km, though with an uncertainty of 1000 km either way. It must be admitted that the

diameters of both Uranus and Neptune are still uncertain, and possibly the main hopes lie with Taylor's occultation technique; however, it does seem that Uranus may be rather the larger of the two, though less massive – and also less active in its atmosphere.

On Uranus, sunlight would be relatively strong, ranging between 1068 and 1334 times that of full moonlight on Earth. Saturn would be fairly bright when well placed (every 45½ years); Neptune would just be visible with the naked eye when near opposition; Jupiter would have an apparent magnitude of 1·7, but would remain inconveniently close to the Sun in the Uranian sky.

THE SATELLITES OF URANUS

Uranus and its satellites

Name	Discoverer	Mean distance from Uranus: thousands of kilometres	a.u.	Angular distance from Uranus, at mean opposition distance ″	Apparent diameter as seen from surface of Uranus ′ ″
Miranda	Kuiper, 1948	130·5	0·000 87	9·9	17 54
Ariel	Lassell, 1851	191·8	0·001 28	14·5	30 54
Umbriel	Lassell, 1851	267·2	0·001 79	20·3	14 12
Titania	Herschel, 1787	483·4	0·002 93	33·2	15 00
Oberon	Herschel, 1787	586·3	0·003 91	44·5	9 48

Name	Sidereal period, days	Synodic period d h m s	Orbital inclination	Orbital Eccentricity
Miranda	1·4135	1 9 55 31	0·0	0·00
Ariel	2·5204	2 12 29 39·0	0·0	0·003
Umbriel	4·1442	4 3 28 25·8	0·0	0·004
Titania	8·7059	8 17 0 1·2	0·0	0·002
Oberon	13·4633	13 11 15 36·5	0·0	0·001

Name	Diameter, km	Density, water = 1	Escape velocity, km/s	Reciprocal mass Uranus = 1	Mean magnitude at opposition
Miranda	550	5?	0·46	1 000 000	16·5
Ariel	1500	5?	1·25	67 000	14·4
Umbriel	1000	4?	0·75	170 000	15·3
Titania	1800	6?	1·65	20 000	14·0
Oberon	1600	5?	1·34	34 000	14·2

All these figures for the diameters are very uncertain. In 1950 W. H. Steavenson gave the magnitude as: Ariel 13·9, Umbriel 14·8, Titania 14·0 and Oberon 14·1. If this were correct, it would presumably mean that Ariel is the largest of the satellites. As the diameter of the Sun as seen from Uranus, is below 2′, all the satellites could produce total eclipses. To an observer on Uranus, Ariel would appear much the largest of the satellites. As the satellites revolve round Uranus in the same sense as the planet's rotation, they are technically retrograde, though not usually regarded as such.

The first satellites to be discovered were Oberon and Titania, by William Herschel in 1787. Both were seen on 11 January of that year, though Herschel delayed an announcement until he was certain of their nature. Ariel and Umbriel were discovered by the English amateur W. Lassell on 24 October 1851, with his 24-in reflector. (Previous observations made by Lassell in 1847 had been inconclusive.)

The last satellite to be discovered was Miranda, by G. P. Kuiper on 16 February 1948. The discovery was made photographically with the 82-in reflector at the McDonald Observatory in Texas.

The names of the first four satellites were suggested by Sir John Herschel in 1852.

The discovery of four more satellites was announced by William Herschel in 1797, but these were never confirmed, and are certainly non-existent (they cannot be identified with Ariel or Umbriel, which were not then known). In 1894, 1897 and 1899 W. H. Pickering made unsuccessful searches for extra satellites; Kuiper's searches in the late 1940s and early 1950s resulted in the discovery of Miranda. It is quite possible that other small satellites exist.

Nothing is known about the physical details of the satellites, but their escape velocities are much too low for them to retain any appreciable atmospheres.

THE RINGS OF URANUS

The discovery that Uranus has a system of rings came as a great surprise to astronomers. On 10 March 1977 it had been predicted that the planet would occult the star SAO 158687, magnitude 8·9, and the occultation would give a good opportunity of measuring the diameter of Uranus. Calculations made by Gordon Taylor of the Royal Greenwich Observatory indicated that the occultation would be visible from only a restricted area in the southern hemisphere, and observations were made by J. Elliot, T. Dunham and D. Mink, flying at 12·5 km above the southern Indian Ocean in the Gerard P. Kuiper airborne observatory – which is in fact a modified C–141 aircraft carrying a 36 in reflecting telescope. Close watches were also being kept at ground-based observatories at Flagstaff (Arizona), Cape Town and Perth.

Observations from the aircraft were begun well in advance. Thirty-five minutes before the occultation the star 'winked' five times, so that apparently it was being temporarily obscured by something in the vicinity of Uranus. Perth and Cape Town were alerted. The occultation by Uranus occurred at 20.52 hours UT, and lasted for 25 minutes. After emersion, there were more winks, and it was later found that these were symmetrical with the first set, indicating a system of rings. The post-emersion winks were also seen from Cape Town by J. Churms.

It is now thought that there may be at least six rings, denoted by letters of the Greek alphabet. Their distance in kilometres from the centre of the planet are:

Ring α, 44 900
β, 45 900
γ, 47 900
δ, 48 600
ε, 51 700
ζ, 54 300

Like Saturn's rings, the system of Uranus seems to be made up of bands of material separated by gaps. The fact that the star was not totally obscured indicates that the discrete particles cannot be more than about 5 km in diameter. Also like Saturn's

rings, the system lies wholly within the Roche limit.

However, the two systems are by no means identical. Ring ε appears to be the widest in the Uranian system, but spans only about 100 km, while α, β, γ and δ seem to be about 10 km wide (as against 26 500 km for Saturn's Ring B). And moreover the distance across Uranus' rings is only about 7000 km, as against 61 500 km from the inner edge of Saturn's Crêpe Ring to the outer edge of Ring A. Originally it was suggested that the Uranian ring ε might be elliptical rather than practically circular, but it is now thought more likely that two rings, ε and ζ, are involved. (It does not seem probable that one ring in the system would have a shape different from all the rest.) S. F. Dermott and Thomas Gold, of Cornell University in the United States, have put forward the theory that the distribution of the rings and the gaps in Uranus' system may be caused by the gravitational pulls of the largest satellites, Ariel, Titania and Oberon.

Confirmatory observations, again by the occultation method, were obtained on 23 December 1977 and 10 April 1978. On the latter occasion, observations made by E. Persson, using the 2·5-metre reflector at Las Campanas (Chile) indicated that there might be as many as 8 rings, though the original Ring ζ was not confirmed.

Whether the rings will be detected visually from Earth remains to be seen, but they must be at least ten magnitudes fainter than Uranus itself, and are rather close-in. Possibly they will be observable with the proposed space telescope, and further information may come in 1986, when the first Voyager probe makes a rendezvous with the planet.

It is important to note that there is no connection between the ghostly rings of Uranus and the spurious ring reported by Sir William Herschel in 1787. Herschel's 'ring' was undoubtedly due to optical effects caused by the 'front-view' arrangement of the reflector which he was using.

NEPTUNE

The first planet to be discovered by mathematical calculation was Neptune, which comes eighth in order of distance from the Sun.

Because of its great distance from the Sun and from the Earth, Neptune reaches opposition only about two days later every year. The opposition of 1977 occurred on 5 June, when the planet was in the constellation of Ophiuchus.

The first recorded observation of Neptune seems to have been that of the French astronomer J. Lalande, who recorded it as a star on 8 and 10 May 1795.

The first suggestion of an unknown planet beyond Uranus was made by the Rev. T. J. Hussey, Rector of Hayes, in a letter written to the Astronomer Royal (Airy) on 17 November 1834. Uranus had been observed to depart from its predicted position, and in 1832 Airy had himself pointed out that the best available tables of Uranus, compiled by Alexis Bouvard, were in error by half a minute of arc. Various theories were later advanced to explain this (an unknown satellite of Uranus, braking due to 'cosmic fluid', and even a cometary collision!) but Hussey's suggestion was supported in 1835 by J. E. B. Valz, and subsequently by others.

The first plan to search for the trans-Uranian planet was announced by F. W. Bessel in 1840. Bessel hoped to collaborate with his pupil F. W. Flemming, but Flemming died soon after; Bessel himself became ill, and was never able to follow the problem up (he died in 1846).

The first estimated position of the new planet was calculated by John Couch Adams, of Cambridge, in 1845. Adams had made up his mind to tackle the problem while still graduating, in 1841. On 21 October 1845 Adams sought an interview with Airy, who had given him little encouragement, but was unsuccessful. No search was instigated. Meanwhile, U. J. J. Le Verrier, in France, had begun working on the problem (1845), and reached similar conclusions. Le Verrier's memoir reached Airy in December 1845, and finally, on 9 July 1846, Airy requested Challis, professor of astronomy at Cambridge, to begin searching for a new planet on the basis of Adams' calculations. Challis used the 11·75-in 'Northumberland' refractor, but had no proper maps of the area, and his searches were laborious – and, it must be admitted, carried out with a strange lack of energy.

DATA

Mean distance from the Sun: 4496·7 million km (30·058 a.u.)
Maximum distance from the Sun: 4537 million km (30·316 a.u.)
Minimum distance from the Sun: 4456 million km (29·800 a.u.)
Sidereal period: 164·8 years = 60 190·3 days
Synodic period: 367·5 days
Rotation period: ±22 hours
Mean orbital velocity: 5·43 km/s
Axial inclination: 28° 48′
Orbital inclination: 1° 45′ 19″·8
Orbital eccentricity: 0·009
Diameter: 49 500 km
Apparent diameter from Earth: max 2″·2, min 2″·0
Reciprocal mass, Sun = 1 : 19 300
Density, water = 1 : 1·77
Mass, Earth = 1 : 17·2
Volume, Earth = 1 : 57
Escape velocity: 23·9 km/s
Surface gravity, Earth = 1 : 1·2
Mean surface temperature: −220°C
Oblateness: 0·02
Albedo: 0·35
Maximum magnitude: +7·7
Mean diameter of Sun, seen from Neptune: 1′ 04″

The first identification of Neptune was made by J. Galle and H. D'Arrest, at Berlin, on 23 September 1846, on the basis of Le Verrier's calculations. It was found that Le Verrier's predicted position was in error by only 55 minutes of arc. The telescope used by Galle and D'Arrest was a 9 in refractor, which was powerful enough to show the planet as a tiny disk. Subsequently, Challis found that he had seen the planet twice – the second occasion being on 12 August – but had failed to identify it. Following the announcement from Berlin the planet was widely observed; in England J. R. Hind, using a 7 in refractor, saw it on 30 September.

The first suggested name for the new planet was 'Janus', by Galle. Le Verrier suggested 'Neptune', but then changed his mind and proposed to have the planet named after himself; Challis proposed 'Oceanus', but before long the name 'Neptune' became universally accepted.

The first announcement of Adams' independent work (almost as accurate as Le Verrier's) was made by Sir John Herschel on 3 October 1846. The announcement caused deep resentment in France, and led to acrimonious disputes in which however, neither Adams not Le Verrier took much part; they remained on cordial terms for the rest of their lives. Nowadays they are recognized as co-discoverers of Neptune.

(As an aside, I have always wondered why Adams did not make a search personally! He was confident of the rough position of the planet, and it was by no means faint. He had finished his calculations long before Le Verrier, and he would have been quite capable of identifying the planet with modest optical aid if he had tried to do so!)

A ring of Neptune was suspected by the English amateur W. Lassell on 14 October 1846. (Lassell would have taken part in the search for Neptune, on Adams' behalf, had he not had the misfortune of being rendered *hors de combat* with a sprained ankle.) Later observations showed that the Neptunian ring, like Herschel's reported ring round Uranus, does not exist. Lassell did however discover Triton, the senior satellite, shortly after the identification of Neptune itself.

Neptune is a twin of Uranus – but a non-identical twin, since it is more massive, shows more activity in its atmosphere, and seems to have an internal heat-source. There are suggestions that it has a 16 000 km core, overlaid by an 800 km layer of ice and then by the methane-rich atmosphere. Surface details have never been seen.

The rotation period as determined by J. H. Moore and D. H. Menzel at the Lick Observatory was about 15·8 hours, but this seems to be too short; L. Trafton (1978) prefers 22 hours. Neptune does not share Uranus' peculiar axial tilt.

The diameter has been measured and re-measured! On 7 April 1968 the planet occulted a star, and this enabled G. Taylor, at Herstmonceux, to announce a revised diameter value of 50 940 km equatorial and 49 920 km polar. This was slightly larger than the alternative value of 49 500 km. It does seem, however, that Neptune is probably slightly smaller than Uranus.

THE SATELLITES OF NEPTUNE

	Triton	Nereid
Discoverer	Lassell, 1846	Kuiper, 1949
Mean distance from Neptune:	353 000 km	5 560 000 km
	0·0024 a.u.	0·0312 a.u.
Mean angular distance from Neptune, at mean opposition distance:	16"·9	4' 23"·9
Apparent diameter, as seen from surface of Neptune:	1°01'	average 19"
Mean sidereal period, days	5·877	359·881
Mean synodic period	5d 21h 3m 29s·8	362d 1h
Orbital inclination	159°·9	27°·2
Orbital eccentricity	0·000	0·749
Diameter, km	6000?	500?
Reciprocal mass, Neptune = 1	750	?
Density, water = 1	5	?
Escape velocity, km/s	4·9	0·3
Magnitude	13·5	19

As the apparent diameter of the Sun as seen from Neptune is only about 1', Triton can produce a total eclipse. Triton is the only large satellite in the Solar System to have retrograde motion; it was discovered by Lassell on 10 October 1846. In 1944 a methane atmosphere was suspected by Kuiper, but this has not been confirmed. The diameter of Triton is uncertain; if the value of 6000 km is correct, then Triton is the largest satellite in the Solar System, and considerably larger than the planet Pluto. H. L. Alden has given its mass as 0·022 that of the Earth, and it is probably about 1·8 times as massive as the Moon.

Nereid, discovered by Kuiper on 1 May 1949, has the most eccentric orbit of any natural satellite in the Solar System. Its distance from Neptune ranges between 140 000 km and 9 500 000 km, and it has been said that the orbit is more like that of a comet than a satellite!

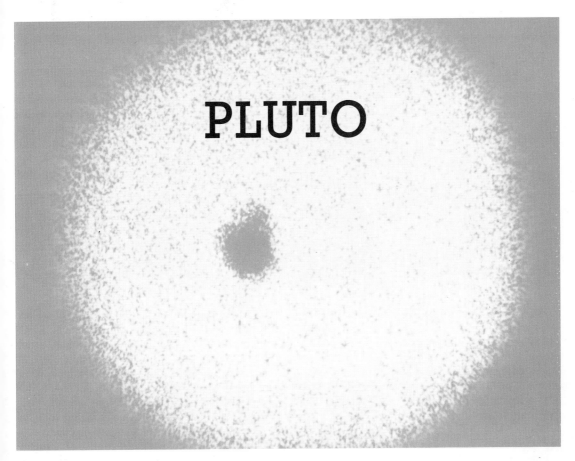

PLUTO

Photograph taken at the US Naval Observatory in 1978 and interpreted as showing that Pluto is a small planet with a relatively very large satellite

DATA

Mean distance from the Sun:
5·900 million km (39·44 a.u.)
Maximum distance from the Sun:
7375 million km (49·19 a.u.)
Minimum distance from the Sun:
4425 million km (29·58 a.u.)
Sidereal period: 247·7 years = 90 465 days
Synodic period: 366·7 days
Rotation period: 6 days 9 hours 17 minutes
Mean orbital velocity: 4·7 km/s
Orbital inclination: 17°·2
Orbital eccentricity: 0·248
Diameter: about 3000 km ?
Apparent diameter from Earth:
below 0·25″
Reciprocal mass, Sun = 1: below 4 000 000
Mean surface temperature: about −230°C
Albedo: about 0·4
Opposition magnitude at perihelion: 14
Mean diameter of Sun, as seen from Pluto: 49″

The most recently discovered planet is Pluto, whose mean distance from the Sun is much greater than for any other planet. As yet our knowledge of it is very far from complete.

Pluto comes to opposition during April at the present epoch; the 1977 opposition occurred on 2 April, and the planet is in Virgo. It is, of course, a very slow mover! The next perihelion is due in 1989, and the next aphelion will occur in 2113.

The magnitude of Pluto is often quoted as 14·9. This, however, is the mean opposition magnitude, and Pluto is at present approaching perihelion. Estimates by T. J. C. A. Moseley and by the present writer make the magnitude exactly 14, so that Pluto is visible in a telescope of 25 cm aperture or more.

The first suggestion of a trans-Neptunian planet seems to have been made by the French astronomer Camille Flammarion in 1879. Flammarion based his suggestion upon the fact that several comets appeared to have their aphelia at approximately the same distance, well beyond the orbit of Neptune. Further comments were made in 1900 by G. Forbes, who assumed a planet larger than Jupiter, moving at a distance of about 100 a.u. in a period of 1000 years; in 1902 by T. Grigull of Münster, whose hypothetical planet was the size of Uranus, moving at 50 a.u. in a period of 360 years (Grigull even gave it a name – 'Hades'); by Gaillot in France – who believed that two planets existed – and by T. J. J. See in America, who increased this number to three.

The first systematic photographic and visual search for the planet was undertaken by Percival Lowell, at Flagstaff, from 1905 to 1907. The results were negative. Lowell's 'Planet X' was believed to have a period of 282 years and a mass 7 times that of the Earth, with a rather eccentric orbit (0·202), and the date of next perihelion was given as 1991. A second search at Flagstaff, carried out by C. O. Lampland in 1914, was equally fruitless, and was given up on Lowell's death in 1916. Lowell's final calculations about his planet were published in 1915.

The second prediction was made by W. H. Pickering in 1909; his planet had a mass twice that of the Earth and a period of 373·5 years. His method – unlike Lowell's – was essentially graphical, but the conclusions were much the same. On the basis of his results, Milton Humason, at Mount Wilson Observatory, undertook a photographic search in 1919, but again the results were negative.

The discovery of Pluto was made at the Flagstaff Observatory by Clyde Tombaugh. Using a 13 in refractor, Tombaugh began work in 1929, again using photographic methods. Pluto was detected upon plates taken on 23 and 29 January, but the announcement was made on 13 March – 149 years after the discovery of Uranus and 78 years after Lowell's birth.

Earlier searches had failed because Pluto was fainter than had been predicted. However, Humason's failure in 1919 was sheer bad luck. When his plates were re-examined, it was found that Pluto had been recorded twice; but once the image fell upon a flaw in the plate, and on the second occasion Pluto was masked by an inconveniently-placed bright star. Moreover, the orbital inclination of 17 degrees was unexpectedly high. At perihelion, Pluto comes 56 000 000 km within the orbit of Neptune, but at the present epoch the distance between the two planets cannot be less than 384 000 000 km, so that there is no fear of collision.

Early diameter estimates indicated that Pluto is small by planetary standards – so small, indeed, that it could produce no measurable perturbations on Uranus or Neptune; yet it was by these alleged perturbations that the planet had been tracked down. In 1936 A. C. D. Crommelin, of Greenwich, suggested the theory of specular reflection. If Pluto were highly reflective, the bright image of the Sun might falsify the diameter estimates, so that Pluto could be much larger – and hence more massive – than it seemed.

The first suggestion that Pluto might be an escaped satellite of Neptune was made by R. A. Lyttleton, of Cambridge, in 1936; it was supported by G. P. Kuiper in 1956. It is true that Pluto is comparable in size with Triton, Neptune's large satellite, and Triton is also the only large satellite with a retrograde orbit, so that the idea of great disturbances in the remote past was logical enough. However, the theory remains highly speculative.

The first direct measurement of Pluto's diameter was made by Kuiper, using the 82 in reflector at the McDonald Observatory in Texas, in 1949. The diameter was given as 10 200 km, with a mass 8/10 of that of the Earth. However, in 1950 Kuiper and Humason made new measurements with the Palomar 200 in reflector, and the diameter of Pluto was reduced to 5800 km – smaller than Mars, and possibly smaller than Triton. **The first partial occultation of star a by Pluto** was observed in 1965. From this, it appeared that the diameter could not be more than 5800 km.

The first measurement of the rotation period of Pluto was made in 1955 by M. Walker and R. Hardie. They found that the brightness of the planet varied by 20 per cent in 6·39 days, and this was presumably due to axial rotation. There are thus over 14 100 Plutonian 'days' in each Plutonian 'year'!

Nothing is known about the surface features of Pluto, but spectroscopic investigations in 1975 indicated that there is a covering of methane ice.

No solution has yet been found to the problem of Pluto's discovery. Sheer chance seems to be most improbable, but it now appears definite that Pluto has insufficient mass to cause any measurable perturbations in the movements of Uranus or Neptune, so that another planet may await detection. This was suggested by Pickering after Pluto had been found, and also in 1950 by K. H. Schütte of Munich on the basis of cometary aphelia. In 1962 A. G. W. Cameron wrote that 'it is difficult to escape the conclusion that there must be a tremendous mass of small solid material on the outskirts of the Solar System', though F. L. Whipple assumed this material to be cometary, left over after the formation of Uranus and Neptune. In 1972 K. A. Brady (U.S.A.) investigated the motion of Halley's Comet, and suggested that there were perturbations due to a planet moving round the Sun in a retrograde orbit at about 8000 million km in a period of 512 years; the mass would be greater than that of Saturn. Subsequent calculations have cast grave doubt upon the validity of this conclusion, though D. Rawlins has calculated that massive planets could exist at reasonably great distances without measurably affecting the motion of Neptune.

Whether a new planet will be found – if it exists – remains to be seen, but there is no theoretical objection to it. Meanwhile, Pluto's orbit marks the boundary of the known planetary system.

The first report of a satellite of Pluto was made on 22 June 1978 by James W. Christy, at the U.S. Naval Observatory at Flagstaff, Arizona. Pluto had been repeatedly photographed with the 1·55-metre reflector with a view to predicting possible occultations of stars by the planet. Pictures taken in April and May showed that Pluto was apparently elongated. Plates taken earlier (1965, 1970, 1971) were then checked, and the same effects were noted. Intensive studies were then carried out at Flagstaff, and confirmation was obtained on 6 July 1978 by J. Graham, using the 401-cm reflector at Cerro Tololo Observatory. The effect was interpreted as being due to the presence of a satellite. It was provisionally designated 1978 P-1, though Christy suggested for it the name Charon (not to be confused with Chiron, the remarkable object discovered by Kowal in 1977).

Current information indicates that the satellite's revolution period is the same as the rotation period of Pluto (6·3867 days). The orbit is almost circular, but highly inclined, and the mean distance from Pluto is of the order of 17,000 km, giving a maximum angular separation of approximately 0″.8.

The magnitude is given as 15 to 16. This is only about two magnitudes fainter than Pluto itself, in which case the system must be in the nature of a binary planet. The movements of the two bodies indicate that they have a combined mass of 1/200 000 000 that of the Sun (= 0·0017 that of the Earth or 0·014 that of the Moon), considerably less than for any other planet.

If Pluto has a diameter of approximately 5800 km, that of the satellite will be about 1200 km if it has the same albedo; but the density of Pluto will be very low – less than that of water. However, Pluto may be much less than 5800 km in diameter. When D. Cruikshank, C. Pilcher and D. Morrison, from Hawaii in 1976, detected methane frost in the spectrum of Pluto, they deduced a higher albedo (0·4 to 0·6, as opposed to the value of 0·1 given by Kuiper), and hence a smaller diameter of between 2800 and 3300 km, smaller than for any of the Galilean satellites of Jupiter, or for Titan or Triton. They also maintained that if Pluto were made up of frozen volatiles with a mean density of $1 - 2$ kg/m^3, the mass would be only a few thousandths of that of the Earth – which now appears to be the case. Pluto was presumably formed out of a low-temperature condensate of the primæval solar nebula.

Mutual eclipses, transits and occultations of Pluto and its satellite presumably occur twice in every 248 years, when the Earth passes through the orbital plane of the system. R. Harrington (USNO) suggests that this may happen in 1983–7, though it is hardly likely that any marked magnitude fluctuations will be observed.

If the mass of Pluto has to be revised downward as the result of this unexpected discovery, the presence of a more remote planet, perturbing Uranus and Neptune, becomes more plausible.

COMETS

Head of Halley's Comet in 1910 (Mount Wilson Observatory)

Comets are members of the Solar System, but are made up of small particles together with extremely tenuous gas, so that they are very different from the planets and are of slight mass. Generally their orbits are highly eccentric.

For all practical purposes, comets may be divided into two classes: (1) Periodical comets, which are seen regularly and which move around the Sun in periods ranging from 3·3 years to over 150 years; and (2) Non-periodical comets, which cannot be predicted because their periods are too long. (There are also comets which may well travel in parabolic orbits, in which case they will never return to the neighbourhood of the Sun; but it is almost impossible to distinguish between a section of a parabola and of an extremely eccentric ellipse.)

The brightest periodical comet is Halley's, with a period of 76 years. Most comets are named in honour of their discoverers, but this is not the case for Halley's Comet, which has been recorded regularly since well before the time of Christ. In 1682 it was observed by Edmond Halley (the actual discovery was made on 15 August of that year by G. Dorffel), and subsequently Halley decided that it must be identical with comets previously seen in 1607 and in 1531. He predicted a return for 1758. On Christmas Night 1758 the comet was duly found, by a German amateur, J. Palitzsch, and it passed perihelion in 1759. This

Halley's Comet in the Bayeux Tapestry

was **the first predicted cometary return,** and it was appropriate that the comet concerned should be known as Halley's.

The first definitely recorded return of Halley's Comet is that of 240 BC, though it is possible that a comet described in 467 BC may also have been Halley's. The return of 163 BC is not on record, but since then the comet has been seen at every perihelion:

83 BC	530	1223
11 BC	607–8	1301
	684	1378
AD 68	760	1456
141	837	1531
218	912	1607
295	989	1682
373	1066	1759
451	1145	1835
		1910

The only doubt refers to 607–608, as two bright comets were recorded at that time, and we cannot be sure which was Halley's.

The only comet to have been referred to by a Roman emperor was Halley's. Comets were once nicknamed 'hairy stars' and were regarded as evil. In AD 79 Vespasian commented that the comet 'menaced rather the King of the Parthians; for he is hairy, while I am bald' – however, Vespasian died in the same year. Another Roman emperor, Macrinus, died at the time of the comet's return in 218.

The first surviving drawing of Halley's Comet refers to the return of 684. It comes from the *Nürnberg Chronicle*.

The only comet recorded in the Bayeux Tapestry is Halley's. The comet was seen in 1066, as the Normans were preparing to invade England, and in the Tapestry it appears clearly, with King Harold toppling on his throne and the Saxon courtiers showing signs of great alarm. (Some scholars consider that the Tapestry was woven by William I's wife.)

The only comet to have been excommunicated by a Pope is again Halley's. This was at the 1456 return, when Pope Calixtus III excommunicated it as an agent of the Devil. (Some scholars doubt this story, but it may well be true!)

The first observer to see Halley's Comet in 1835 was Dumouchel, from Rome; the last to see it in 1836 (May) was Sir John Herschel, from the Cape. The comet was recovered on 12 September 1909 by Max Wolf in Germany, and remained under observation until June 1911, when its distance from the Sun was more than 800 000 000 km. Aphelion was reached in 1948, and the next return is due in 1986, so that it should be recovered some time either in late 1984 or early 1985. Perihelion is due on 9 February 1986.

The first attempt to observe a cometary transit across the face of the Sun was made on 18–19 May 1910, with Halley's Comet. The American astronomer F. Ellerman went to Hawaii to observe under the best possible conditions, but could see no trace of the comet.

The first short-period comet to be predicted was Encke's, again now named in honour of the mathematician who computed the orbit. The comet was first seen on 17 January 1786 by P. Méchain, from France. It was again recorded on 7 November 1795 by Caroline Herschel; the next returns to be seen were those of 1805 (discovered by Thulis at Marseilles, 19 October) and 1818 (Jean Pons, 26 November). Encke, at Berlin, decided that these comets must be identical, and predicted a return for 1822. He was correct, and since then the comet has been seen at every return except that of 1944 (when it was badly placed, and most astronomers were otherwise engaged).

The comet with the shortest known period is also Encke's: 3·3 years. (Comet Wilson-Harrington, discovered in 1949, had a calculated period of only 2·3 years, but has never been seen again.) The period of Encke's comet has shortened slightly since its discovery; the decrease amounted to over a day between 1822 and 1858. Evaporation processes taking place inside the comet are responsible – not, as Encke himself supposed, some resisting medium in the inner part of the Solar System.

The comet which has presented most observed returns is Encke's. The 1977 return was the 51st to be seen. Encke's Comet is therefore far ahead of its nearest rivals: Halley (27 returns, last perihelion passage 1910) and Pons-Winnecke (18 returns, last perihelion passage 1976). Modern instruments are now capable of recording Encke's Comet when it is near aphelion.

The following data relate to comets with periods of between 3·3 and 20 years, and which have been observed at more than one return.

Greatest orbital eccentricity: Encke (0·85)
Smallest orbital eccentricity: Schwassmann-Wachmann 1 (0·11)
Greatest orbital inclination: Giacobini-Zinner (31°·7)
Smallest orbital inclination: Du Toit-Neujmin-Delporte (2°·4)
Greatest perihelion distance: Schwassmann-Wachmann 1 (5·45 a.u.)
Smallest perihelion distance: Encke (0·34 a.u.)
Greatest aphelion distance: Neujmin 1 (12·16 a.u.)
Smallest aphelion distance: Arend (4·00 a.u.)

A few comets, with orbits of exceptionally low eccentricity, are visible throughout their orbits. The first-discovered of these was Schwassmann-Wachmann 1, first seen in 1925. Its orbit lies entirely between those of Jupiter and Saturn. Normally it is a very faint object, but it can show sudden, unpredictable outbursts which bring it within the range of small telescopes. The last occasion was in the autumn of 1976, when the magnitude rose to above 12. Other comets which may be kept in view throughout their orbits are

Gunn and Smirnova-Chernykh. Oterma's Comet used to have a very low eccentricity and a period of 7·9 years, but a close approach to Jupiter in 1963 altered the period to 19·3 years.

The most numerous comet 'family' is Jupiter's. The mean distance of Jupiter from the Sun is 5·2 a.u., and a glance at the table on page 124 shows that many short-period comets have their aphelia at about this distance. 'Comet families' of Saturn and Uranus are very questionable. However, the Neptune 'family' is rather more definite. The mean distance of Neptune from the Sun is approximately 30 a.u., which corresponds to the aphelia of at least four comets, including Halley's. It must however be added that some authorities are highly sceptical about the significance of these alleged comet families, and attempts to predict the existence of a still more remote planet by means of cometary aphelia are very speculative.

The comet with the longest confirmed period is Grigg-Mellish, which has been observed at the returns of 1742 and 1907; the period is 164 years. Second in order comes Herschel-Rigollet (1788 and 1939; 156 years). These are the only two comets which have been seen at more than one return and whose periods are longer than that of Halley's Comet.

The most famous 'lost' periodical comet is Biela's. At the 1846 return it separated into two; the twins returned in 1852, but have never been seen since, though the comet's orbit is associated with a meteor shower. Westphal's Comet was seen in 1852 and again in 1913, but it evidently failed to survive the perihelion passage of 1913, and did not appear on schedule in 1976. Brorsen's Comet was seen at five returns between 1846 and 1879, but has not appeared since, and has evidently disintegrated. However, one must beware of jumping to conclusions. Comet Di Vico-Swift was 'lost' for 38 years after 1897, but was recovered in 1965; Holmes's 'Comet' 'went missing' for 58 years prior to its recovery in 1964. Both these latter successes were due to the mathematical work of the British astronomer B. Marsden.

The closest cometary approach to the Earth ever recorded was that of Lexell's Comet of 1770 (discovered by C. Messier on 14 June of that year; A. Lexell of St. Petersburg computed the orbit). The distance was reduced to 1·2 million km, and the comet was visible with the naked eye. It has never been seen again, since an approach to Jupiter in 1779 has altered the orbit, and the comet now comes nowhere near the Earth.

Brooks' Comet on 21 October 1911 (Lick Observatory)

The closest known cometary approach to Jupiter was that of Comet Brooks 2, in March 1886 (the approach was not, of course, actually observed). Evidently the planet passed within the orbit of Io, the closest to Jupiter of the four large satellites. The comet suffered partial disruption, and at the 1889 return was seen to be attended by four minor companions which were classified as 'splinters', but which did not last for long. The comet has returned regularly since then, the last appearance being that of 1973.

The largest comet ever recorded seems to have been that of 1811; the diameter of the coma was approximately 2 000 000 km, larger than the Sun. The maximum length of the tail was over 160 000 000 km.

The longest tail ever recorded was that of the Great Comet of 1843: 330 000 000 km – consider-

ably greater than the distance between the Sun and the orbit of Mars.

The brightest comet of modern times was probably that of 1843. According to the famous astronomer Sir Thomas Maclear, it was much more brilliant than the comet of 1811; and Maclear saw both.

The most beautiful comet of modern times is said to have been Donati's of 1858, with its wonderfully curved main tail and its two shorter ones. It was discovered by G. Donati, from Florence, on 2 June 1858, and was last seen on 4 March 1859. During October 1838 the length of the tail was as much as 80 000 000 km. The period is unknown; 2000 years has been suggested, but with no certainty.

The most-tailed comet on record was that of 1744, discovered on 9 December 1743 by Klinkenberg in Holland (and independently on 13 December by De Chéseaux in Switzerland; rather unfairly, it is usually called De Chéseaux' Comet). There were at least six bright, broad tails, but records are sparse, as the comet remained brilliant for only a few nights in March 1744.

The comet with the longest computed period is Delavan's Comet of 1914: 24 000 000 years. This is naturally very arbitrary, and all we can really say is that the period is very long indeed.

The only comet to be associated with port wine was that of 1811. In that year the vintage in Portugal was unusually good; for years afterwards 'Comet Wine' appeared in the price-lists of wine-merchants! It would be interesting to know whether any of it remains to be drunk.

The first time that the Earth was known to pass through the tail of a comet was in late June 1861. The comet concerned – discovered by J. Tebbutt from Australia – was brilliant, but the nucleus never came within 1·8 million km of the Earth. No effects of the comet's tail were noticed with certainty.

The first comet to be discovered during a total eclipse of the Sun was Tewfik's Comet of 1882; the eclipse was photographed from Egypt, and the comet was named after the then ruler of Egypt. It was not seen again, so that this our only record of it. The bright comet of 1948 was also discovered during a total solar eclipse.

The first known attempt to photograph a comet was made by the British pioneer W. de la

Rue, with the Great Comet of 1861. However, he was unsuccessful.

The first really good comet photograph was taken in 1882 at the Cape by Sir David Gill. The comet was very brilliant, and was well shown. Many stars were also shown, and it was this picture which made Gill appreciate the endless potentialities of stellar photography.

The first cometary spectrum was studied in 1864 by G. Donati. The comet concerned was reasonably bright, and Donati saw that the spectrum was not merely a reflected solar spectrum; there were lines which could be due only to materials in the comet itself.

The brightest comet so far seen during the 20th century is probably the so-called Daylight Comet of 1910, discovered by some diamond miners in the Transvaal on 12 January. It was certainly brighter than Halley's, which was on view later in the year; and people who claim to remember Halley's Comet are warned that the comet they saw was probably the Daylight! The orbit was virtually parabolic.

The greenest comet of modern times was Jurlov-Achmarov-Hassell (1939). The green colour was evident even in a small telescope.

The most famous 'spiked' comet was Arend-Roland, which was easily visible with the naked eye in April 1957. There was an apparent tail directed sunward – though the appearance was actually due to thinly-spread matter in the comet's orbit, catching the sunlight at a suitable angle. (The Skylab astronauts also reported a 'spike' with Kohoutek's Comet of 1973.)

The first discovery of a hydrogen cloud associated with a comet was made in 1969, when Comet Tago-Sato-Kosaka was studied from the space-vehicle OAO2 (Orbiting Astronomical Observatory 2) and was found to be surrounded by a hydrogen cloud approximately 1·6 million km in diameter. Similar clouds were associated later with Bennett's Comet of 1970 and Kohoutek's of 1973. Kohoutek's Comet was also found to contain large quantities of the ionized water molecule, H_2O^+.

The greatest cometary disappointment of modern times was Kohoutek's Comet of 1973, which was found on 7 March by L. Kohoutek at Hamburg when it was still almost 700 000 000 km from the Sun. Few comets are detectable as far

away as this, and the comet was expected to become a magnificent object in the winter of 1973–4, but it failed to come up to expectations even though it was visible with the naked eye. It was, however, scientifically interesting, and was carefully studied by the astronauts than aboard the US space-station Skylab (Carr, Gibson and Pogue). Perhaps the comet will do better when it next returns to the Sun, in approximately 75 000 years' time!

The last brilliant comet (to the summer of 1978) was West's, first recorded on a photograph taken at La Silla, in Chile, on 10 August 1975. During March 1976 it was a prominent naked-eye object with a major tail, but after perihelion, and as it receded, it showed signs of disintegration. The estimated period is 300 000 years.

The most distant comet ever observed was Stearns' Comet of 1927, which was followed out to a distance of more than 1 500 000 000 km – so that it was then between the orbits of Saturn and Uranus. It was under observation for more than four years, which is a record for a non-periodical comet. Had it come closer it would have been brilliant, but its perihelion distance was well over 450 000 000 km.

The greatest perihelion distance for any comet is that of Schuster's Comet of 1976, discovered by H. Schuster from Chile on 25 February. The perihelion distance was just over one million km, not far from midway between the orbits of Jupiter and Saturn. (The previous record-holder had been Van den Bergh's Comet of 1974: 900 000 000 km).

The record for comet discoveries is held by J. L. Pons, who discovered a total of 37. C. Messier discovered 13 – and in the process compiled a catalogue of star-clusters and nebulæ, so that he could avoid confusing them with comets. Today, it is by his catalogue that Messier is best remembered.

For many centuries comets were regarded as atmospheric phenomena – though Anaxagoras, around 500 BC, attributed them to clusters of faint stars. The first man to prove that they were extra-terrestrial was Tycho Brahe, who found that the comet of 1577 showed no diurnal parallax, and must therefore be at least six times as remote as the Moon (actually, of course, it was much further away than that). There have been occasional panics due to comets – as in France in 1773, when a mathematical paper by J. J. de Lalande was misinterpreted, and people believed that a comet would collide with the Earth on 20 or 21 May. (Seats in Paradise were sold by members of the clergy at high prices.) Another alarm occurred in 1832, when it was suggested – wrongly – that there might be a near-encounter between the Earth and the now-lost Biela's Comet. In 1910, when Halley's Comet was in view, a manufacturer in America made a large sum of money by selling anti-comet pills, and in 1970 some Arabs mistook Bennett's Comet for an Israeli war weapon. Since 1950 there have appeared several books by an eccentric psychiatrist, Dr Immanuel Velikovsky, who confuses planets with comets, and believes Venus to have been a comet only a few thousands of years ago!

According to a theory developed in America by F. L. Whipple, a comet may be described as a 'dirty ice-ball'. The nucleus is made up of a conglomerate of rocky fragments held together with ices such as frozen methane, ammonia, carbon dioxide and water. When the comet approaches perihelion, these substances evaporate, so that a tail develops; the effect of the solar wind means that the tail always points more or less away from the Sun, so that when a comet has passed perihelion, and is receding, it actually moves tail-first. Tails may be of either gas or dust; some comets have tails of both kinds, while many fainter comets never develop tails at all.

The origin of comets is still a matter for debate, but at least we may be sure that they are bona-fide members of the Solar System. According to the Dutch astronomer J. H. Oort, there is a reservoir or 'cloud' of comets orbiting the Sun at a distance of a light-year or so, and when a comet is perturbed for some reason it may enter an orbit which swings it sunward, with the inevitable risk of being perturbed by a planet (usually Jupiter) and forced into a smaller orbit. It seems that short-period comets have limited lifetimes, since they lose some material by evaporation at every perihelion passage so that there must be, presumably, some source of replenishment to maintain the numbers of short-period comets.

Our knowledge of comets is still incomplete, but will learn more when the first comet probe is sent up. There are suggestions of sending a probe to Halley's Comet during the 1986 return, or even a two-comet probe involving both Halley and Giacobini-Zinner. This project will certainly be difficult, but no doubt a successful rendezvous with a comet will be achieved in the foreseeable future.

Comet 1948 I (RAS)

Periodical comets which have been observed at more than one return.

COMET NOMENCLATURE

Normally, a comet is named after its discoverer (e.g. Comet Lovas of 1974), after independent discoverers (Arend-Roland of 1957) or after the mathematicians who computed the orbits (Lexell, 1770). A comet is given a letter to correspond with its discovery; thus the first comet to be discovered in 1977 was 1977a, the next 1977b and so on. This is a provisional designation, and a permanent designation depends upon the time of perihelion; thus the first comet to reach perihelion in 1977 was 1977 I, the next 1977 II, and so on. These two systems may disagree with respect to the year; thus Comet 1978m may well become 1979 III!

The periodical comets are identified by P/ (e.g. P/Halley, P/Encke). They may be named according to their discoverers at different returns (P/Pons-Winnecke, P/Di Vico-Swift). Occasionally a name is altered; thus the comet formerly called Pons-Coggia-Winnecke Forbes is now known as P/Crommelin, since the identity of the four observed comets was established by A. C. D. Crommelin of Greenwich Observatory.

Comet	Period years	Distance from Sun, a.u. Peri-helion	Aphe-lion	Eccen-tricity	Incli-nation	Number of returns	Last return
Encke	3·3	0·34	4·09	0·85	12·0	51	1977
Grigg-Skjellerup	5·1	1·00	4·94	0·66	21·1	13	1977
Tempel 2	5·3	1·36	4·68	0·55	12·5	16	1977
Honda-Mrkós-Pajdusáková	5·3	0·58	5·49	0·58	13·1	5	1974
Neujmin 2	5·4	1·34	4·84	0·57	10·6	2	1927
Tempel 1	5·5	1·50	4·73	0·52	10·5	6	1977
Tuttle-Giacobini-Kresák	5·6	1·15	5·13	0·63	13·6	5	1973
Tempel-Swift	5·7	1·15	5·22	0·64	5·4	4	1908
Wirtanen	5·9	1·26	5·16	0·61	12·3	5	1974
D'Arrest	6·2	1·17	5·61	0·66	16·7	13	1976
Du Toit-Neujmin-Delporte	6·3	1·68	5·15	0·51	2·9	2	1970
Di Vico-Swift	6·3	1·62	5·21	0·52	3·6	3	1965
Pons-Winnecke	6·3	1·25	5·61	0·64	22·3	18	1976
Forbes	6·4	1·53	5·36	0·56	4·6	5	1974
Kopff	6·4	1·57	5·34	0·55	4·7	11	1976
Schwassmann-Wachmann 2	6·5	2·14	4·83	0·39	3·7	8	1974
Giacobini-Zinner	6·5	0·99	5·98	0·71	31·7	10	1978
Churyumov-Gerasimenko	6·6	1·30	3·51	0·63	7·12	2	1976
Wolf-Harrington	6·6	1·62	5·38	0·54	18·4	6	1977
Tsuchinshan 1	6·6	1·49	5·57	0·58	10·5	3	1977
Perrine-Mrkós	6·7	1·27	5·85	0·64	17·8	6	1974
Reinmuth 2	6·7	1·94	5·19	0·45	7·0	5	1974
Borrelly	6·8	1·32	5·84	0·63	30·2	9	1974
Johnson	6·8	2·20	4·96	0·39	13·9	5	1976
Tsuchinshan 2	6·8	1·78	5·40	0·51	6·7	2	1971
Harrington	6·8	1·58	5·60	0·56	8·7	2	1960
Gunn	6·8	2·45	4·74	0·32	10·4	—	constant
Arend-Rigaux	6·8	1·44	5·76	0·60	17·8	5	1977
Brooks 2	6·9	1·84	5·39	0·49	5·6	11	1974
Finlay	6·9	1·10	6·19	0·70	3·6	9	1974
Taylor	7·0	1·95	3·64	0·47	20·6	2	1977
Holmes	7·0	2·16	5·20	0·41	19·2	5	1972
Daniel	7·1	1·66	5·72	0·55	20·1	5	1964
Harrington-Abell	7·2	1·77	5·68	0·52	16·8	4	1976
Shajn-Schaldach	7·3	2·23	5·28	0·41	6·2	2	1971
Faye	7·4	1·62	5·98	0·58	9·1	17	1976
Ashbrook-Jackson	7·4	2·29	5·33	0·40	12·5	5	1977
Whipple	7·5	2·48	5·16	0·35	10·2	7	1977
Reinmuth 1	7·6	2·00	5·76	0·49	8·3	6	1973
Arend	7·9	1·84	4·00	0·54	20·0	4	1975
Oterma	7·9	3·39	4·53	0·14	4·0	3	1958
Schaumasse	8·2	1·20	6·92	0·70	12·0	7	1976
Jackson-Neujmin	8·4	1·43	6·83	0·65	14·1	2	1970
Wolf	8·4	2·52	5·78	0·40	27·3	12	1975
Smirnova-Chernykh	8·5	3·60	4·20	0·15	6·6	—	constant
Comas Solá	8·6	1·77	6·59	0·58	13·4	7	1977
Kwerns-Kwee	9·0	2·23	6·43	0·49	9·0	2	1972
Denning-Fujikawa	9·0	0·78	4·33	0·82	8·9	2	1978
Swift-Gehrels	9·3	1·36	4·40	0·69	9·3	2	1972
Neujmin 3	10·6	1·98	7·66	0·59	3·9	3	1972
Gale	11·0	1·18	8·70	0·76	11·7	2	1938
Klemola	11·0	1·76	4·94	0·64	10·6	2	1976
Väisälä 1	11·3	1·87	8·19	0·63	11·5	4	1971
Slaughter-Burnham	11·6	2·54	7·72	0·50	8·2	2	1970
Van Biesbroeck	12·4	2·40	5·35	0·55	6·62	2	1966
Wild	13·3	1·98	9·24	0·65	19·9	2	1973
Tuttle	13·8	1·02	10·46	0·82	54·4	9	1967
Du Toit 1	15·0	1·29	10·85	0·79	18·7	2	1974
Schwassmann-Wachmann 1	15·0	5·45	6·73	0·11	9·7	—	constant
Neujmin 1	17·9	1·54	12·16	0·78	15·0	4	1966
Crommelin	27·9	0·74	17·65	0·92	28·9	4	1956
Tempel-Tuttle	32·9	0·98	19·56	0·90	162·7	4	1965
Stephan-Oterma	38·8	1·60	21·34	0·86	17·9	2	1942
Olbers	69·5	1·18	32·62	0·93	44·6	3	1956
Pons-Brooks	71·0	0·77	33·51	0·96	74·2	3	1954
Brorsen-Metcalf	71·9	0·49	34·11	0·97	19·2	2	1919
Halley	76·1	0·59	35·33	0·97	162·2	27	1910
Herschel-Rigollet	156·0	0·75	56·94	0·97	64·2	2	1939
Grigg-Mellish	164·3	0·92	58	0·97	109·8	2	1907

Drawing of Coggia's Comet as it appeared over Paris in 1874

It must be remembered that these elements alter considerably for each revolution. The values given here apply to the beginning of 1977. Du Toit 1 (Comet 1944 III) was reported again in 1974, but the observations cannot be regarded as definite.

Orbits with inclinations over 90° indicate retrograde motion. There are three comets in the list with retrograde orbits: Tempel-Tuttle, Halley and Grigg-Mellish.

LOST PERIODICAL COMETS

Comet	Period, years	Distance from Sun, a.u. Peri-helion	Aphe-lion	Eccen-tricity	Incli-nation	Number of returns	Last return
Brorsen	5·5	0·59	5·61	0·81	29·4	5	1879
Biela	6·6	0·86	6·19	0·76	12·6	6	1852
Westphal	61·9	1·25	30·03	0·92	40·9	2	1913

It seems certain that these comets have disintegrated.

Periodical comets seen at only one return

There are various reasons for 'losing' comets. A few, listed separately, seem definitely to have disintegrated. In other cases the orbit has been so violently perturbed that we have lost contact; the classic case is that of Lexell's Comet of 1770. The list also includes several short-period comets which have been discovered recently, and which will no doubt return on schedule; these are identified by an asterisk.

Comet	Period in years	Last seen	Comet	Period in years	Last seen
Wilson-Harrington	2·3	1949	Schorr	6·7	1918
Helfenzrieder	4·5	1766	Longmore*	7·1	1974
Blanpain	5·1	1819	Swift 2	7·2	1895
Du Toit 2	5·3	1945	Tritton*	7·3	1978
La Hire	5·4	1678	Denning 2	7·4	1894
Barnard 1	5·4	1884	Metcalf	7·9	1906
Schwassmann-Wachmann 3	5·4	1930	Gehrels 2*	7·9	1973
Grischow	5·4	1743	Gehrels 3*	8·4	1975
Haneda-Campos*	5·4	1978	Swift 1	8·9	1889
Clark*	5·6	1973	Boethin*	11·0	1975
Brooks 1	5·6	1886	Peters	13·4	1846
Lexell	5·6	1770	Gehrels 1*	14·6	1973
Kulin	5·6	1939	Kowal*	15·1	1977
Pigott	5·9	1783	Perrine	16·4	1916
West-Kohoutek-Ikemura*	6·1	1975	Pons-Gambart	63·9	1827
Wild 2*	6·2	1978	Ross	64·6	1883
Kohoutek*	6·2	1975	Dubiago	67·0	1921
Spitaler	6·4	1890	Di Vico	75·7	1846
Harrington-Wilson	6·4	1951	Väisälä 2	85·5	1942
Barnard 3	6·6	1892	Swift-Tuttle	119·6	1862
Giacobini	6·7	1896	Barnard 2	128·3	1889
			Mellish	145·3	1917

Brilliant Comets seen since 1700

1744 De Chéseaux' six-tailed comet (actually discovered by Klinkenberg).

1811 Great Comet: discovered by H. Flaugergues.

1843 Great Comet, superior to that of 1811.

1858 Donati's Comet, with its curved tails – possibly the most beautiful comet ever seen.

1861 Brilliant comet, discovered by Tebbutt.

1862 Bright comet, discovered by Swift.

1874 Coggia's Comet; bright naked-eye object.

1882 Great Comet, photographed by Gill from South Africa.

1901 Bright southern comet, discovered by Paysandu.

1910 Daylight Comet; brightest of the century (so far).

1927 Skjellerup's Comet. Bright for a brief period.

1947 Bright southern comet; brief visibility.

1965 Comet Ikeya-Seki; brilliant as seen from the southern hemisphere for a brief period (also well seen from the USA).

1970 Bennett's Comet; fairly bright.

1976 West's Comet; bright naked-eye object during early mornings in March.

To this list must be added Halley's Comet at its returns of 1759, 1835 and 1910.

Comet Seki-Lines 1962 taken on 6 July 1962 at 19 45 UT (E. Jönsson)

METEORS

METEORS

Shower	Begins	Maximum	Ends	Max. ZHR	Position of radiant: R.A. h	m	Dec. o	Associated Comet
Quadrantids	1 January	3 January	6 January	110	15	28	+50	
Corona Australids	14 March	16 March	18 March	5	16	20	−48	
Lyrids	19 April	22 April	24 April	12	18	08	+32	1861 I (Thatcher)
Eta Aquarids	2 May	4 May	7 May	20	22	24	00	P/Halley?
June Lyrids	10 June	15 June	21 June	8	18	32	+35	
Ophiuchids	17 June	20 June	26 June		17	20	−20	
Capricornids	10 July	26 July	15 August	6	21	00	−15	
Delta Aquarids	15 July	28 July	15 August	35	22	36	−10	
Piscis Australids	15 July	30 July	20 August	8	22	40	−30	
Alpha Capricornids	15 July	2 August	25 August	8	20	36	−10	
Iota Aquarids	15 July	6 August	25 August	6	22	15	− 9	
Perseids	25 July	12 August	18 August	68	03	04	+58	P/Swift-Tuttle
Kappa Cygnids	19 August	20 August	22 August	4	19	20	+55	
Draconids	10 October	10 October	10 October	var.	18	00	+54	P/Giacobini-Zinner
Orionids	16 October	21 October	26 October	30	06	24	+15	P/Halley?
Taurids	20 October	4 November	25 November	12	03	44	+22	P/Encke
Cepheids	7 November	9 November	11 November	8?	23	30	+63	
Leonids	15 November	17 November	19 November	var.	10	08	+22	P/Tempel-Tuttle
Andromedids	15 November	20 November	6 December	low	00	50	+55	P/Biela
Phœnicids	4 December	4/5 December	5 December	5	01	00	−55	
Geminids	7 December	14 December	15 December	58	07	28	+32	
Ursids	17 December	22 December	24 December	12	14	28	+76	P/Tuttle

Permanent daytime streams include the Arietids (29 March—June 17), the Xi Perseids (1 to 15 June) and the Beta Taurids (23 June to 7 July). The Beta Taurids seem to be associated with Encke's Comet.

Meteors may be regarded as the débris of the Solar System. They are small and friable, and cannot reach ground level intact. Sporadic meteors may appear from any direction at any moment, but many meteors are members of definite showers.

THE PRINCIPAL METEOR SHOWERS

The showers listed in the table occur annually, though not all of them are consistent: the Perseids are the most reliable, whereas the Leonids produce major showers only occasionally (the last being in 1966). The ZHR or zenithal hourly rate is the probable hourly rate of meteors for an observer who has the radiant at his zenith, and is observing under ideal conditions. In practice, of course, these conditions are never fulfilled, so that the ZHR is the theoretical maximum only.

The only meteor shower named after a discarded constellation is that of the Quadrantids. The old constellation, Quadrans Muralis, has been long since rejected.

The maximum velocity of meteors entering the atmosphere is around 72 km/s. Most meteors vaporize above altitudes of 80 km. Micrometeorites, about 0·1 mm in diameter, are too small to cause luminous effects.

The total number of meteors entering the atmosphere daily has been estimated at 75 000 000 for meteors of magnitude 5 or brighter. An observer under ideal conditions could normally be expected to see about 10 naked-eye meteors per hour (except during a shower, when the number would naturally be higher).

Meteor showers were first postulated by Olmsted and Twining in 1834. H. Olbers suggested a period of 33 years for the Leonid shower. H. A. Newton came independently to the same conclusion, and traced the Leonids back to AD 902; he predicted another major shower for 1866, and it duly occurred.

The Perseid shower was identified by Quetelet in 1836. The shower has been nicknamed 'the tears of St. Lawrence'.

The richest showers on record have probably been the Leonids, which on 17 November 1966 reached a rate of 60 000 per hour over a period of 40 minutes. There were major showers also in 1799, 1833 and 1866, but not in 1899 or 1933, because of perturbations by Jupiter and Saturn.

The first radiant point to be measured was that of the Leonids, on 12–13 November 1833.

The first suggestion of an association between meteors and comets was made by D. Kirkwood in 1861; he believed meteors to be the débris of disintegrated comets. In 1862 G. V. Schiaparelli showed the association of the Perseids with comet P/Swift-Tuttle. The November Andromedids were subsequently shown to be associated with the defunct comet P/Biela. On 27 November 1872 the hourly rate of the Andromedids (sometimes termed Bieliids) reached 100 per hour, but in recent years the shower has become very feeble.

The greatest 'cometary' shower of meteors in the present century was that of 9 October 1933, associated with Comet P/Giacobini-Zinner. The rate reached 350 meteors per minute for a brief period.

The first measures of meteor heights by the triangulation method (i.e. observing the same meteors from widely separated points – at least two observers, at two different stations) were made by two German students, Brandes and Benzenberg, in 1798. They concluded that most meteors penetrate down to an altitude of 80 km above the ground.

The earliest meteor photograph (of an Andromedid) was taken on 27 November 1885 by L. Weinek, at Prague.

The first radar measurements of meteors were made during the last war; pioneer work was done by the British radio astronomers Hey and Stewart, who in 1945 measured the Delta Aquarids by radar. Nowadays, radar has largely superseded the old method of visual observation.

The most reliable annual shower is that of the August Perseids, which never fail to produce a spectacular display. It is noteworthy that so far as meteor observation is concerned, the northern hemisphere has the best of matters!

Noctilucent clouds, which occur at heights of from 80 to 100 km, are thought to be due to ice crystals forming along meteor trails, though upon this point there is still no universal agreement among astronomers and meteorologists.

METEORITES

Meteor Crater or Coon Butte, Coconino County, Arizona. The crater is 1200 m in diameter, 183 m deep, and is surrounded by a wall 30 to 45 m high. It is estimated to be 50 000 years old (Patrick Moore)

Meteorites reach ground level without being destroyed, and are not simply large meteors. They do not belong to showers, and any association with comets is dubious. Meteorites may in fact be more closely associated with asteroids, and it has been claimed that there is no distinction between a small asteroid and a large meteorite!

One important sub-group consists of the carbonaceous chondrites, which contain carbon compounds and organic matter (such as hydrocarbons). One famous example is the Orgueil Meteorite, which fell in France on 14 May 1864. In 1964 B. Nagy, at Columbia University together with his colleague Claus, suggested that this meteorite contained 'organized elements' – leading to the suggestion that there might be evidence of past living material. However, others maintain that these substances are due to terrestrial contamination. (In 1908 Svante Arrhenius suggested that life was brought to Earth by a meteorite; this 'panspermia' theory never met with much support, but has recently been revived in modified form, by F. Hoyle and C. Wickramasinghe.)

The earliest reports of meteoritic phenomena are recorded on Egyptian papyrus, around 2000 BC. Early meteorite falls are, naturally, poorly documented, but it has been suggested that a meteorite fell in Crete in 1478 BC; stones near Orchomenos in Boetia in 1200 BC, and an iron on Mount Ida in Crete in 1168 BC. According to Livy, 'stones' fell on Alban Hill in 634 BC, and there is evidence that in 416 BC a meteorite fell at Ægospotamos in Greece. The Sacred Stone in Mecca is certainly a meteorite. **The oldest meteorite which can be positively dated** is the Ensisheim Meteorite, which fell on 16 November 1492 and is now on show in Ensisheim Church (Switzerland).

In India, it is said that the Emperor Jahangir ordered two sword-blades, a dagger and knife to be made from the Jalandhar Meteorite of 10 April 1621. A sword was made from a meteorite which fell in Mongolia in 1670, and in the 19th century part of a South African meteorite was used to make a sword for the Emperor Alexander of Russia. **Meteorite ages** are usually estimated at between 4 and 4·6 æons.

The most famous terrestrial meteorite crater is in Arizona (lat 35°·3 N., long. 111°·2 W.), 1265 metres in diameter and 175 metres deep. It is well-preserved, and there is no reasonable doubt about its origin – even though G. K. Gilbert, who was strongly of the opinion that the main lunar craters were due to impact, regarded the Arizona crater as volcanic! The crater is certainly more than 10,000 years old, and is a well-known tourist attraction. Other craters which are probably meteoritic include those at Wolf Creek (Australia) and Waqar (Arabia).

Name	Diameter, metres (largest crater)	Date of discovery	Notes
Meteor Crater, Arizona	1265	1891	Many metallic meteorites found.
Wolf Creek, Australia	850	1947	
Henbury, Australia	200 × 110	1931	Thirteen craters.
Boxhole Australia	175	1937	
Odessa, Texas	170	1921	
Waqar, Arabia	100	1932	Two craters.
Oesel, Estonia	100	1927	Six craters.
Campo del Cielo, Argentina	75	1933	Many craters.
Dalgaranga, Australia	70	1928	
Sikhote-Alin, Siberia	28	1947	Total of 106 craters. (Fall observed.)

Nickel iron meteorites (Patrick Moore)

Various other craters have been attributed to meteoric impact. In some cases this may be true, but there are serious doubts about, for instance, the New Quebec crater in Canada (discovered 1950; diameter 3400 metres) and others in the Canadian shield, which appear to be associated with geological features. The Vredefort Ring near Pretoria, in South Africa, is often listed as an impact crater, but geologists who have studied it are unanimous in stating that it is of volcanic origin.

Tektites are small glassy objects, found only in a few regions. They seem to have been heated twice, and are usually aerodynamically shaped. Four main groups are known:

Region	Name	Geological age
Australasia	Australites	Middle/Late Pleistocene
Ivory Coast	Ivory Coast tektites	Lower Pleistocene
Czechoslovakia	Moldavites	Miocene
United States	Bediasites (Texas)	
	Georgiaites (Georgia)	Oligocene

The largest tektite known was found in 1932 at Muong Nong, Laos; it weighs 3·2 kg. Whether or not tektites are extraterrestrial is still unknown. In 1897 R. O. M. Verbeek even suggested that they might come from the Moon, but this theory has now been rejected, and we have to confess that tektites remain a complete enigma.

The first suggestion that objects could 'fall from the sky' was made by E. F. Chladni in 1794, but his idea was met with considerable scepticism. (As recently as 1807 Thomas Jefferson, President of the United States, is quoted as saying 'I could more easily believe that two Yankee professors would lie than that stones would fall from heaven.')

The first scientific proof of meteorite falls was obtained by the French scientist J. B. Biot, who investigated the meteoritic shower at L'Aigle on 26 April 1803.

The classification of meteorites is, broadly speaking, into three types: stones (aerolites), stony-irons (siderolites) and irons (siderites).

Irons (Siderites).

Hexahedrites contain between 4 and 6 per cent of nickel, and consist of the cubic mineral kamacite. With increased nickel another cubic mineral, taenite, is formed. These two minerals orient themselves parallel to octahedral planes, and when the surface is etched with acid after polishing the characteristic Widmanstätten patterns are revealed (**octahedrites**). The **nickel-rich ataxites**, with more than 12 per cent nickel, do not show the Widmanstätten patterns, and consist of plessite.

Stony irons (Siderolites).

Pallasites consist of a network of nickel-iron enclosing crystals of olivine. **Mesosiderites** are

Meteorite	Weight, tonnes
Hoba West, Grootfontein, S.W. Africa	60
Ahnighito (The Tent), Cape York, W. Greenland:	30·4
Bacuberito, Mexico	27
Mbosi, Tanganyika	26
Agpalik, Cape York, W. Greenland	20·1
Armanty, Outer Mongolia	20 (estimated)
Willamette, Oregon, U.S.A.	14
Chupaderos, Mexico	14
Campo del Cielo, Argentina	13
Mundrabilla, Western Australia	12
Morito, Mexico	11

heterogeneous aggregates of silicate minerals and a nickel-iron alloy. There are two further groups, **siderophytes** and **lodranites,** but only one example of each is known.

Stones (Aerolites).

Chondrites contain spherical particles called chondrules. These chondrules are fragments of minerals, and show a radiating structure. Chrondrites account for 85 per cent of known specimens. In **achrondrites,** chondrules are absent; the achrondrites are of coarser structure, with little free iron. They may be calcium-rich or calcium-poor.

The early systems of classification were due to G. Rose (1863), G. Tschermak (1883) and A. Brezina (1904); they were extended by G. Prior (1920) and latterly by G. J. H. McCall (1973). Stones are more commonly found than irons in the ratio of 96 per cent to 4 per cent, but this is misleading, as irons are much more durable and are more likely to survive. Over 2000 meteorites have now been located, though few have actually been seen to fall. Among famous falls which have resulted in meteorite recovery are those of the Přibram fireball (Czechoslovakia, which was recorded as magnitude —19!) on 7 April 1959, the Lost City meteorite (Oklahoma) in 1970, the Barwell Meteorite (Leicestershire) of 24 December 1965, and the Sikhote-Alin fall in Siberia on 12 February 1947 – the greatest of the century, apart from that of 1908 – which was an iron, and produced 30 craters, since it broke up during its descent.

There is no record of any death due to a meteorite. Reports that a monk was killed at Cremona in 1511, and another monk at Milan in 1650, are unsubstantiated. However, in 1954 a woman in Alabama, USA, had a narrow escape when a meteorite fell through the roof of her house, and she suffered a minor injury to her arm.

Meteorites are not always easy to identify on sight, but when etched with acid irons will show the characteristic **Widmanstätten patterns**, not found elsewhere.

The largest known meteorite is still lying where it fell, in prehistoric times, at Hoba West, near Grootfontein in South-West Africa; it weighs at least 60 tons. All the known meteorites weighing more than 10 tons are irons. They are:

There is a report of an iron found by W. A. Cassidy in Argentina, in September 1969, which is said to have a weight of about 18 tonnes.

The largest stone meteorite fell in Kirin Province, Manchuria, on 8 March 1976. It weighs 1766 kg.

The largest meteorite in a museum is the Ahnighito, found by Robert Peary in Greenland in 1897; it is now in the Hayden Planetarium, New York, along with two other meteorites found at the same time and on the same site (known as The Woman and The Dog). Apparently the local Eskimos were rather reluctant to let them go! The Willamette meteorite is also in the Hayden Planetarium. (This meteorite was the subject of a lawsuit. It was found in 1902 on property belonging to the Oregon Iron and Steel Company. The discoverer moved it to his own property, and exhibited it; the Company sued him for possession, but the Court ruled in favour of the discoverer.)

The most famous fall of recent times was that of 30 June 1908, in the Tunguska region of Siberia. As seen from Kansk, 600 km away, the descending object was said to outshine the Sun, and detonations were heard 1000 km away; reindeer were killed and pine-trees blown flat over a large area. The first expedition to the site was not dispatched until 1927, led by L. Kulik. No fragments have been found, and it has been suggested that the object was the nucleus of a small comet – which, if icy in nature, would presumably evaporate during the descent and landing. (Inevitably, flying saucer enthusiasts have suggested that it was a space-ship in trouble!)

A second major fall in Siberia occurred on 12 February 1947, in the Sikhote-Alin area. The fall was observed, and more than 100 craters were located, of which thirty were of appreciable size. It is fortunate that both these Siberian falls occurred in uninhabited territory. If a meteorite of this size had hit a city, the death-roll would inevitably have been high.

Twenty-two meteorite falls in the British Isles are known:

1623	January 10	Stretchleigh, Devon. 12 kg.
1628	April 9	Hatford, Berkshire. Three stones; about 33 kg.
1779	—	Pettiswood, West Meath
1795	December 13	Wold Cottage, Yorkshire. 25·4 kg.
1804	April 5	High Possil, Strathclyde (Lanarkshire). 4·5 kg.
1810	August ?	Mooresfort, Tipperary. 3·2 kg.
1813	September 10	Limerick. 48 kg (shower).
1830	February 15	Launton, Oxfordshire. 0·9 kg.
1830	May 17	Perth. About 11 kg.
1835	August 4	Aldsworth, Gloucestershire. Small shower, over 0·5 kg.
1844	April 29	Killeter, Tyrone. Small shower; small amount preserved.
1865	August 12	Dundrum, Tipperary. 1·8 kg.
1876	April 20	Rowton, Shropshire. 3·2 kg. (Iron meteorite.)
1881	March 14	Middlesbrough, Cleveland (Yorkshire). 1·4 kg.
1902	September 13	Crumlin, Antrim. 4·1 kg.
1914	October 13	Appley Bridge, Lancashire. 33 kg.
1917	December 3	Strathmore, Tayside (Perthshire). 13 kg (four stones).
1923	March 9	Ashdon, Essex. 0·9 kg.
1931	April 14	Pontlyfni, Gwynedd (Caernarvon). 120 gr.
1949	September 21	Beddgelert, Gwynedd (Caernarvon). 723 gr.
1965	December 24	Barwell, Leicestershire. 46 kg (total).
1969	April 25	Bovedy, N. Ireland. Main mass presumably fell in the sea.

Both these latter falls were well observed. Many fragments of the Barwell Meteorite were found; one was detected some time later nestling in a vase of artificial flowers on the window sill of a house in Barwell village. Though it broke up during the descent, it is the largest stone known to have fallen over the British Isles.

The largest meteorites to have been found in different regions are:

Region	Meteorite	Weight
Africa	Hoba West, Grootfontein	60 tonnes
U.S.A.	Williamette, Oregon	14 tonnes
Asia	Armanty, Outer Mongolia	20 tonnes
South America	Campo del Cielo, Argentina	13 tonnes
Australia	Mundrabilla	12 tonnes
Europe	Magura, Czechoslovakia	1·5 tonnes
Ireland	Limerick	48 kg
England	Barwell, Leicestershire	± 46 kg (total)
Scotland	Strathmore, Tayside (Perthshire)	10·1 kg
Wales	Beddgelert, Gwynedd (Caernarvon)	723 g

Fragment found by Patrick Moore of the 46 kg Barwell meteorite (Patrick Moore)

The only known case of a meteorite entering the Earth's atmosphere and leaving it again is that of the object of 10 August 1972. It seems to have approached the Earth 'from behind' at a relative velocity of 10 km/s, which increased to 15 km/s as the Earth's gravity accelerated it. The object entered the atmosphere at a slight angle, becoming detectable at a height of 76 km above Utah and reaching its closest point to the ground at 58 km above Montana. It then began to move outward, and became undetectable at just over 100 km above Alberta after a period of visibility of 1 minute 41 seconds; the magnitude was estimated by eye-witnesses to be at least —15, and the diameter of the object may have been as much as 80 m. After emerging from the Earth's atmosphere it re-entered a solar orbit, admittedly somewhat modified by its encounter, and presumably it is still orbiting the Sun.

GLOWS AND ATMOSPHERIC EFFECTS

AURORÆ

Auroræ, or polar lights (Aurora Borealis in the northern hemisphere, Aurora Australis in the southern) must have been observed in ancient times, since they may often become extremely brilliant. There is excellent evidence that the Roman emperor Tiberius, who reigned from AD 14 to 37, once dispatched his fire-fighters to the port of Ostia on account of a red glow in the sky, seen from Rome, which proved to be auroral.

The first use of the term 'aurora' was due to the French astronomer P. Gassendi, following the brilliant display of 12 November 1621. However, **the first really good description of a display of Aurora Borealis** was given by K. Gesner of Zürich for the aurora of 27 December 1560. **The first Antarctic aurora** was described by Captain Cook on 20 February 1773.

The first major book dealing with auroræ was the *Traité physique et historique de l'aurora boréale*, written by J. J. de Mairan in 1731.

The first suggestion of an association between auroræ and electrical discharges may have been due to Halley in 1716, when he linked auroral displays with discharges associated with the Earth's magnetic field. (It must be remembered that Halley saw his first aurora in this year, following the end of the solar Maunder Minimum.)

The first suggestion of a connection between auroræ and magnetic effects was made by Hjorter in 1741.

The first suggestion of a connection between auroræ and sunspot phenomena was made by E. Loomis, of Yale, in 1870. The solar association was defined in more detail by the Italian astronomer G. Donati in 1872, following the great display of 4–5 February. In 1896 Birkeland reproduced 'miniature auroræ' by using a spherical electromagnet and a beam of cathode rays. In 1929 S. Chapman and V. Ferraro suggested that the cause of auroræ was solar plasma. It is now known that auroral displays are indeed due to particles emitted by the Sun, and it is assumed that these electrified particles enter the Van Allen zones round the Earth and 'overload' them, so that particles cascade downwards and produce the auroral glows in the upper atmosphere. **The maximum activity of auroræ** is seen along geomagnetic latitude 68 degrees N or S.

On average, auroræ of some kind or other are seen in 240 nights per year in North Alaska, North Canada, Iceland, North Norway and Novaya Zemlya; 25 nights per year along the Canada/United States border and in Central Scotland; and 1 night per year in Central France. Obviously, however, this varies according to the state of the solar cycle, and in general auroræ are most common 2 years after a spot-maximum.

Exceptionally brilliant auroræ occur now and then. Among many cases are the displays of 24–5 October 1870, 4–5 February 1872 and January 25–26 1938. The latter display was spectacular in Southern England and elsewhere, but was not seen from Tromsø in Norway! In 1909 an aurora was seen from Singapore (latitude N. 1 degree 25 minutes), but auroræ are extremely rare from latitudes as low as this.

The heights of auroræ vary. In general, the sharp lower boundary is at about 98 km above ground level, the maximum region of activity at about 110 km, and the normal upper boundary 300 km. In the cases of sunlit activity the upper limit may reach 700 km, or even 1000 km in extreme cases. Extremely low auroræ have been reported now and then, but with no certainty. There have also been many reports of noise accompanying auroral displays – either hissing or crackling. Many cases were listed by S. Tromholt in Norway in 1885 (reported in *Nature*, vol. 32, p. 499); it is extremely hard to explain the noise, but there does seem to be considerable weight of evidence that it occurs, even though final proof is still lacking.

Auroræ may be seen in various forms. There are glows; arcs, with or without rays; bands, more diffuse and irregular; draperies, or curtains made up of very long rays; individual rays or streamers; coronæ, or radiating systems of rays converging at the zenith; 'surfaces', which are diffuse patches, sometimes pulsating; flaming auroræ, composed of quickly-moving sheets of light; and ghost arcs, which may persist long after the main display has ended.

THE ZODIACAL LIGHT

The Zodiacal Light may be seen as a faint cone of light rising from the horizon after sunset or before sunrise. It extends away from the Sun, and is generally observable for a fairly short period only after the Sun has disappeared or before it rises. On a clear, moonless night, under ideal conditions it contributes about one-third of the total sky light, and may be brighter than the average Milky Way region. It is due to particles scattered in the Solar System along and near the main plane of the system. The diameters of the particles are of the order of 0·1 to 0·2 micron. (One micron is equal to one-millionth of a millimetre.) Since the Zodiacal Light extends along the ecliptic, it is best seen when the ecliptic is most nearly vertical to the horizon – i.e. February to March and again in September to October.

The discovery of the Zodiacal Light was due to G. D. Cassini in 1683. He correctly suggested that it was caused by sunlight reflected by interplanetary 'dust'.

THE GEGENSCHEIN

The Gegenschein or Counterglow is seen as a faint patch of radiance in the position in the sky exactly opposite to the Sun. It is extremely elusive, and is generally visible only under near-perfect conditions. The best opportunities come when the anti-sun position is well away from the Milky Way (i.e. in February/April, and September/November) and when the anti-sun position is high, at local midnight.* It has generally an oval shape, measuring 10 by 20 degrees, so that its maximum diameter is roughly 40 times that of the full moon.

The discovery of the Gegenschein seems to have been due to Esprit Pézénas, who reported it to the Paris Academy in 1731. **It was named** by Humboldt, who saw it on 16 March 1803. It was described in more detail by the Danish astronomer Theodor Brorsen, who saw it in 1854 and wrote about it in 1863, but (contrary to statements in many books) Brorsen did not claim the discovery for himself, as he was familiar with Humboldt's description.

The Zodiacal Band is a very faint, parallel-sided band of radiance which may extend to either side of the Gegenschein, or be prolonged from the apex of the Zodiacal Light cone, or join the Zodiacal Light with the Gegenschein. It may be from 5 to 10 degrees wide, and is extremely faint. Like the Zodiacal Light and the Gegenschein, it is due to sunlight being reflected from interplanetary particles in the main plane of the Solar System.

OTHER GLOWS AND ATMOSPHERIC EFFECTS

The Kordylewski Clouds were reported by the Polish astronomer K. Kordylewski in 1961. He believed them to be due to collections of interplanetary débris lying at the 'Lagrangian points' of the Moon's orbit – that is to say, moving in the same path as the Moon, but at distances of 60 degrees ahead and 60 degrees behind respectively. (The Trojan asteroids behave in this way with respect to Jupiter.) It has been claimed that the clouds have been seen with the naked eye as excessively faint patches of light, but their existence has yet to be properly confirmed.

The Airglow (dayglow and nightglow) was named by O. Struve in 1950, who suggested the name in correspondence with C. T. Elvey. Nightglow prevents the sky from being completely dark at any time, quite apart from the diffusion of starlight. There are various causes for it; chemical reactions of the neutral constituents of the upper atmosphere, with light emission; reactions from

ionized constituents, again producing light emission; the influx of incoming particles from space, and so on. There also seems to be a sort of 'geocorona' in the exosphere (1000 to 10 000 km altitude), due to emissions from hydrogen and helium atoms in this highly rarefied region of the atmosphere.

Diffused starlight is quite appreciable – coming as it does from the thousand million stars above magnitude 20, of which approximately one million are above magnitude 11.2, 4850 above magnitude 6, and 1620 above magnitude 5. Over a full visible hemisphere of the sky (20 626 square degrees) the total starlight is approximately equal to 51 stars the brilliance of Sirius or 4 planets the brilliance of Venus at maximum. It is, however, a tiny fraction of the light sent by the Moon.

The Green Flash (or Green Ray) is an atmospheric effect. As the Sun sinks below the horizon, the last segment of it may flash brilliant green for an instant. There are vague references to this in Egyptian and Celtic folklore, but **the first scientific reference to it** was made by W. Swan, who saw it on 13 September 1865. However, Swan's account was not published until 1883. **The first published reference** to the phenomenon was due to J. P. Joule, in 1869. Much more recently, some superb photographs of it have been taken by D. J. K. O'Connell at the Vatican Observatory; these were published in book form in 1958.

The Green Flash has also been seen with the planet Venus. The first published reference to it seems to be due to Admiral Murray, who saw it from HMS *Cornwall*, off Colombo, at 13 50 (1½ hours after sunset) on 28 November 1939. Venus was setting over a sea horizon; Admiral Murray was using binoculars, and described the flash as 'emerald'.

---

*From England I have seen it only once – in March 1942, when the whole country was blacked out as a precaution against German air-raids.

THE STARS

CLASSIFICATION AND EVOLUTION

The stars are of many different types, and show tremendous range in size and luminosity, though rather less in mass. The Sun is a normal star, and the only one sufficiently close to be studied in great detail.

The first attempt to classify the stars according to their spectra was made by the Italian Jesuit astronomer, Angelo' Secchi, in 1863–7. He divided the stars into four types:

I. White or bluish stars, with broad, dark spectral lines of hydrogen but obscure metallic lines. Example: Sirius.

II. Yellow stars; hydrogen less prominent, metals more so. Examples: Capella, the Sun.

III. Orange stars: complicated banded spectra. Examples: Betelgeux, Mira. The class included many long-period variables.

IV. Red stars, with prominent carbon lines: all below magnitude 5. Example: R Cygni. This class also included many variables.

Secchi's catalogue included over 500 stars. His system was, however, superseded by that worked out at Harvard in the United States.

The Harvard system was introduced by E. C. Pickering in 1890, and was extended by two famous women astronomers, Miss A. Cannon and Mrs W. Fleming. It has been further modified into what is now called the MKK (after Morgan, Keenan and Kellerman of the USA) or Yerkes system. The types are:

Type	Spectrum	Surface Temperature	Examples	Notes
W	Many bright lines: divided into WN (nitrogen sequence) and WC (carbon sequence)	Up to 80 000°C	Rare: about 150 have been found in our galaxy, and 50 in the Large Magellanic Cloud. 16 cases are known as planetary nebulæ in which the central star is a Wolf-Rayet; one of these is NGC 7009, the 'Saturn Nebula' in Aquarius.	Known as Wolf-Rayet stars. Have expanding shells, moving outwards at up to 3000 km/sec. All very remote, and appear faint despite their considerable luminosity
O	Both bright and dark lines	40 000°C— 35 000°C	γ Velorum	Represent a transition between W and B stars, though this does not imply any evolutionary sequence
B	Bluish-white (B0) to white (B9). No emission lines, but dominant absorption lines of hydrogen and (particularly) helium.	Over 25 000°C for B0 12 000°C for B9	Rigel B8 (i.e. 8/10 of way from B0 to A0)	Rigel is particularly luminous, with peculiarities in its spectrum
A	White stars: spectra dominated by hydrogen lines	10 000°C – 8000°C	Sirius, Vega, Altair	
F	Yellowish hue. Calcium very conspicuous, with less prominent hydrogen lines	7500°C – 6000°C	Procyon, Polaris	Yellow hue so elusive that to the naked eye most F-stars will be regarded as white
G	Yellow; weaker hydrogen lines, numerous conspicuous metallic lines	5500°C – 4200°C (giants) 6000°C – 5000°C (dwarfs)	Capella (giant) Sun (dwarf)	Beginning of division into dwarf or Main Sequence stars and the giants
K	Orange; weak hydrogen lines, strong metallic lines	4000°C 3000°C (giants) 5000°C 4000°C (dwarfs)	Arcturus, Aldebaran, Pollux	K-stars are more numerous than any other type
M	Orange-red; very complicated spectra with many bands due to molecules	3400°C (giants) 3000°C (dwarfs)	Mira Ceti (variable), Betelgeux, Antares (giants) Proxima Centauri (dwarf)	Strong differences between giants and Main Sequence dwarfs. Proxima Centauri closest star beyond the Solar System
R	Reddish	2600°C	V Arietis, T Lyræ	Remote, and appear faint
N	Reddish; strong carbon lines	2500°C	R Leporis, V Aquilæ	Remote, with many variable examples
S	Reddish; prominent bands of titanium oxide and zirconium oxide	2600°C	χ Cygni, R Cygni	Examples given are long period variables

Most of the stars lie in the sequence from B to M. The very hot W stars are (as noted) very rare; O stars are also rare. Types R, N and S consist entirely of giants, and are regarded as no 'later' than Type M.

Originally it had been planned to make the sequence alphabetical, beginning with the hottest stars and ending with the coolest; but, as so often happens, major modifications had to be introduced (for instance, types C and D were found to be unnecessary), and the final result was alphabetically chaotic. A good and famous mnemonic is 'Wow! Oh Be A Fine Girl Kiss Me Right Now Sweetie' (or, if you prefer it, 'Smack'!)

In addition, P is used for gaseous nebulæ, and Q for novæ.

The following additional letters are used: e (emission lines present), n (nebulous lines), s (sharp lines), k (interstellar lines), m (metallic lines), v (variable), and p or pec (peculiar).

Stellar Evolution. It was originally believed that a star began its career as a red giant, condensing out of nebular material (type M; giant). shrank and heated up to become a Main Sequence star (type B or A), and then cooled while continuing to shrink, ending its career as a red dwarf of type M. This was – it was thought – shown by the famous Hertzsprung-Russell or H/R diagram.

The H/R diagram was originally due to Ejnar Hertzsprung, a Danish astronomer, in 1911; it was proposed independently in 1913 by H. W. Russell of the United States. In an H/R diagram, a star is plotted according to its luminosity (or equivalent, such as absolute magnitude) against spectral type (or surface temperature). The Main Sequence, running diagonally from hot O and B stars at the top left, down to faint M-type dwarfs at the lower right, is very marked; so are the giant and supergiant areas, to the upper right. White dwarfs (not known in 1913) are to the lower left.

The key to stellar energy was discovered in 1939 by H. Bethe (and, at about the same time, by G. Gamow). Bethe actually worked it out during a train journey from Washington to Cornell University! The stars – or, at least, the normal ones – shine by means of nuclear transformations. Thus inside the Sun, hydrogen is being converted into helium. It takes four hydrogen nuclei to make one helium nucleus; each time this happens, a little mass is lost and a little energy is released. Each second, the Sun converts 600 million tonnes of hydrogen into helium – and loses 4 million tonnes in mass. This may sound a great deal, but

the Sun is certainly older than the Earth, and will continue in virtually its present state for 5000 million years in the future – perhaps rather longer.

The career of a star depends upon its initial mass, and the H/R diagram itself is not indicative of a definite evolutionary sequence. In every case the star begins by condensing out of nebular material; this star formation is certainly going on in M.42, the Orion Nebula. We can also observe fairly small dark 'globules' against some bright nebulæ, known as Bok globules in honour of their discoverer, Bart J. Bok, who came originally from Holland but who settled in the United States. Bok drew attention to the globules in the 1930s. Some of them are over 6 light-years in diameter, though others are much smaller. They may well indicate regions of star formation by gravitational contraction, though it is unlikely that all stars begin as Bok globules. There are many known cases in which many stars of similar type, and presumably similar age, are concentrated in a limited area; these are **stellar associations.**

Moreover, it is logical to assume that the stars in any cluster (such as the Pleiades) had a common origin in the same nebular cloud. Not all open clusters are similar in age: for instance the Pleiades cluster is relatively young, while M.67 in Cancer is comparatively old.

(a) Stars of less than 0·1 solar masses. – Condensation out of nebular material causes a rise in temperature, but the cores never become hot enough for nuclear reactions to begin (about 10 000 000 degrees is the 'starting point'). The star therefore turns into a dim red dwarf of type M, and then simply fades until it has lost all its energy.

(b) Stars of between 0·1 and 1·4 solar masses. – The star begins by condensing gravitationally out of nebular material. The 'protostar' will be large and red, though by no means the same as the red giants such as Betelgeux. As contraction continues, the surface temperature remains the same, so that the star becomes less luminous. On the H/R diagram, it passes along what are known as the Hayashi and Henyey tracks (in honour of the astronomers who first described them) and then joins the Main Sequence. The T Tauri stars, named after the most famous member of the class, are still contracting toward the Main Sequence, and are irregularly variable. Condensation to the

Main Sequence takes several million years, and a young star will also have to 'blow away' a surrounding cocoon of dust. It will then brighten, and cases have been found; thus the star V.1057 Cygni brightened up from magnitude 16 to above magnitude 11, in 1969, over a period of only about 250 days. Not surprisingly, stars of this type are strong emitters of infra-red radiation.

With the Sun and similar stars, the core temperature is about 14 000 000 °C. Hydrogen is being converted into helium, admittedly by a rather roundabout process known as the **proton-proton cycle.** (In hotter stars the so-called **carbon-nitrogen cycle** is dominant; the end result is much the same, with four hydrogen nuclei forming one helium nucleus, but carbon and nitrogen are used as catalysts.) Obviously, the core is being depleted of hydrogen and enriched with helium. When the supply of hydrogen in the core runs low, energy production ceases, and the core contracts under the influence of gravitation, though hydrogen conversion continues in the shell surrounding the core. Eventually the star leaves the Main Sequence, and enters the giant branch of the H/R diagram; the core now has a temperature of at least 100 million degrees, and this is enough to start reactions involving the conversion of helium into carbon (the 'helium flash'). It has been calculated that when the Sun reached this stage in its evolution it will have a diameter of at least 25 000 000 miles, and will be 100 times as luminous as it is now, though the surface temperature will be lower; the Sun will have become a red giant. Life on Earth cannot survive this ordeal, and it is very likely that the Earth itself will be destroyed.

For a star such as the Sun, the red giant stage may last for 100 million years, which is not long on the cosmical scale. A whole series of reactions takes place inside the star, and heavier elements are built up in succession, but at last all the nuclear energy is exhausted. There is now nothing to halt gravitational shrinkage, and the star contracts into what is termed a **white dwarf.** The dim Companion of Sirius is of this type.

In a white dwarf, the atoms are broken up and packed tightly together with little waste space. This leads to amazing density values, up to a million times the density of water in extreme cases; a tablespoonful of white dwarf material might well weight a thousand tonnes. For instance, the Companion of Sirius is as massive as the Sun, but smaller than planets such as Uranus and Neptune; we even know of white dwarfs which are smaller than the Moon. They have been described as 'bankrupt stars', though their surface temperatures are still high (10 000 to 40 000 °C). We must assume that eventually a white dwarf will fade away into a cold, dead 'black dwarf' sending out no energy at all.

The Indian astronomer S. Chandrasekhar has shown that a star more than 1·4 times as massive as the Sun cannot become a white dwarf unless it sheds some of its mass. Some stars appear to do this during the preceding giant stage, when the outer part of the envelope is thrown completely off, producing a **planetary nebula.** This may dispose of 10 to 20 per cent of the star's mass. The shell will not fall back, but will expand and dissipate. Planetary nebulæ, as such, are therefore rather short-lived, and do not last for more than 30 000 to 40 000 years.

(c) Stars with initial mass appreciably more than 1·4 times that of the Sun. Here, everything happens at an accelerated pace, and if the protostar is, say, 10 to 40 times as massive as the Sun it may reach the main sequence in a few thousand years. Things proceed much as before, though more quickly, until the end of helium conversion. The carbon-rich core reaches a temperature of between 600 and 700 million °C, and carbon conversion begins, producing elements such as magnesium, sodium, neon and oxygen. When the carbon is exhausted, the core contracts once more; oxygen nuclei are then fused to produce sulphur and phosphorus, though shells nearer the surface are still using up carbon, helium and hydrogen. The situation is very complicated, and when the central temperature has reached 3000 million °C or so even iron nuclei are being produced.

This, however, is the beginning of the end. When enough iron has accumulated in the core, nuclear reactions stop. There is an 'implosion' (the opposite of an explosion), and the outer shells, where reactions are still going on, are heated to incredible temperatures. In a matter of seconds or minutes, a violent shock wave disrupts the star, producing a supernova outburst. Much of the star's material is hurled into space – and since this material contains many heavy elements, the supernova enriches the interstellar medium. The end product is a cloud of expanding material, while the remnant of the original star becomes a **neutron star** or **pulsar.**

The collapse is so dramatic that the protons (positively charged) and the electrons (negatively charged) are fused into neutrons (no electrical charge).

The concept of neutron stars was proposed in 1932 by the Russian physicist Landau, and again in 1934 by F. Zwicky and W. Baade at Caltech (USA) but the first neutron star was not detected until 1967 – not by visible light, but by its radio emissions. A neutron star is indeed an amazing object. The diameter is of the order of 10 km, and the density perhaps a thousand million million tonnes per cubic metre, or 100 million million times that of water. A pin's head of neutron star material would 'weigh' more than the liner *QE2*.

According to theory (which may or may not be correct!) the outer crust of a neutron star is crystalline and iron-rich, composed of what we may term 'normal' matter. Cracks occur in it, producing 'starquakes' which affect the object's behaviour. Below the crust comes the neutron-rich liquid mantle; below again, a superfluid core composed mainly of neutrons; and at the centre, material made up of 'hyperons', about which we

can only speculate. If one could stand on the surface of a neutron star, one's weight would be 10 000 million times greater than on Earth. There are strong magnetic fields – perhaps a hundred million tesla, as against 30 millionths of a tesla for the Earth.

The first neutron star was discovered in 1967 by Miss Jocelyn Bell (now Dr Jocelyn Bell-Burnell) working with the Cambridge team led by Professor A. Hewish. During radio surveys of the sky, she found a weak source which was fluctuating quickly as though 'ticking'. The radio pulses were very regular and rapid, so that the object was called a **pulsar** (CP 1919). The official announcement was made on 29 February 1968. More pulsars were soon found, and today over 300 are known. Their periods have a limited range; that of CP 1919 is 1·3373 seconds.

The fastest pulsar is NP 0532 (in the Crab Nebula); period 0·330 911 second, so that it pulses over 30 times each second. **The slowest pulsar** is NP 0527; period 3·745 491 second. In

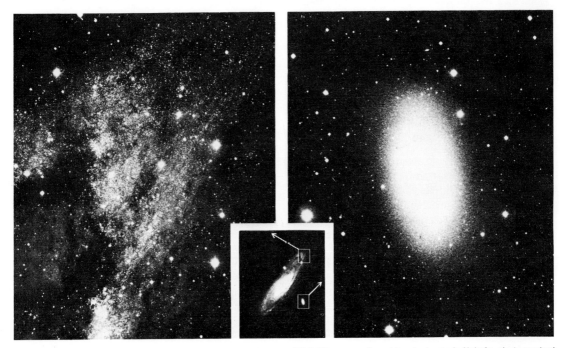

Andromeda Nebula, photographed in blue light showing giant and super-giant stars of Population I in the spiral arms. The hazy patch at upper left is composed of unresolved Population II stars

NGC 205, companion of the Andromeda Nebula, photographed in yellow light shows stars of Population II. The brightest stars are red and a hundred times fainter than the blue giants of Population I

The very bright, uniformly distributed stars are foreground stars belonging in our own Milky Way system

general all pulsars seem to be slowing down by very small amounts; thus the period of CP 1919 is lengthening by a thousand millionth of a second each month, so that in 3000 years' time the period will be 1·3374 second instead of the present-day 1·3373 second. However, some pulsars show irregular changes in period, or **glitches;** thus on 1 March 1969 the pulsar PSR 0833–45 speeded up a full quarter of a millionth of a second!

It was some time before astronomers decided upon the nature of pulsars. There was even the short-lived LGM or 'Little Green Men' theory – that the radio signals were artificial! Pulsating or rotating white dwarfs were also suggested, but it became apparent that a pulsar must be smaller than even a white dwàrf, and the idea of pulsars as neutron stars is now universally accepted. The pulsar is spinning rapidly, and the more normal surface protons and electrons are accelerated outward at the north and south magnetic poles of the star in rather the manner of searchlight beams; every time the Earth passes through the beam, we receive a pulse. Glitches are due to the disturbances in rotational period caused by starquakes.

The first pulsar to be seen visually was NP 0532 in the Crab Nebula. This was achieved in January 1969 by a team at the Steward Observatory in Arizona, using a 36-in reflector. They identified a faint, flashing object whose mean magnitude was about 17, and whose period was the same· as that of the pulsar. Later in 1969 the pulsar was photographed from Kitt Peak (Arizona). It is undoubtedly the remnant of the supernova of 1054. (R. Minkowski had observed it in 1942, but had not realized its significance.) The second visual identification was that of the pulsar PSR 0833–45 in the southern constellation of Vela; the pulsar lies in the Gum Nebula (named after its discoverer, the late Dr C. S. Gum), and it may be assumed that the nebula is a supernova remnant. The Vela pulsar itself was found in 1968 at the Molonglo Radio Observatory in Australia. The period is 0·089 second – the third shortest known. In 1977 a team working with the 3·9 m Anglo-Australian telescope at Siding Spring (Australia) identified the pulsar as a faint flashing object with a mean magnitude of 24·2.

No doubt other pulsars will be optically identified in future years, but their extreme optical faintness makes the searches very difficult.

It seems, then, that a star with initial mass between 1·4 and 3 times that of the Sun will eventually explode as a supernova, and become a neutron star (though the upper limit is somewhat uncertain, and may have to be revised upward, perhaps even to 10).

(d) Stars too massive initially to become supernovæ. Here we are by no means certain of the course of events, but it is suggested that once the final collapse starts nothing can stop it. Eventually the star has become so small and so dense that its escape velocity becomes greater than the velocity of light. Nothing can therefore escape, and to all intents and purposes the collapsed star or **collapsar** is cut off from the rest of the universe.

For obvious reasons, these **black holes** can be detected only by their effects upon objects which emit detectable radiation. One candidate is Cygnus X-1, so called because it is an X-ray source. The system consists of a BO type supergiant, HDE 226868, with about 30 times the mass of the Sun and a diameter 23 times that of the Sun (18 000 000 km), together with an invisible secondary with 14 times the mass of the Sun. The orbital period is 5·6 days, the distance 6500 light-years, and the magnitude of the primary star 9 (it lies at RA 19h 56m 29s, dec. +35° 03'55", near η Cygni). The secondary would certainly be visible if it were a normal star, and it may be a black hole. Another candidate is the secondary component of the eclipsing binary ε Aurigæ.

The size of a black hole will depend upon the mass of the collapsed star. The critical radius of a non-rotating black hole is called the Schwarzschild radius, after the German astronomer K. Schwarzschild, who investigated the problem mathematically in 1916; the boundary around the mass having this radius is the **event horizon**. For a body the mass of the Sun, the Schwarzschild radius would be about 3 km; for the Earth, less than 1 cm.

Conditions inside the event horizon are so different from our everyday experience that we cannot visualize them; it may even be that the old star finally crushes itself out of existence. However, it is important to add that the whole black hole concept is questioned by some astronomers – and we have as yet no decisive proof that they exist.

Note, also, that the basic principle was forecast as long ago as· 1798 by the great French mathematician Laplace. Like Newton, he believed light to consist of a stream of particles, and commented that if a body were sufficiently small and dense it would be invisible, since the light-particles would not be able to travel fast enough to escape from it.

Meanwhile, all we can really say is that black holes are theoretical possibilities, and that the concept has been widely accepted. If valid, then black holes represent the end products of the most massive stars in the universe.

Stellar populations were first noted by W. Baade in the early 1950s. Population I stars are metal-rich, with about 2 per cent of their mass being made up of elements heavier than helium. The most brilliant Population I stars are of types W, O and B. The disk and arms of the Galaxy (and other spirals) are mainly Population I.

Population II stars are metal-poor, and are clearly older, so that their most brilliant members are red giants which have already left the Main Sequence. They are found in the halo and nucleus of the Galaxy (and other galaxies) and in globular clusters. It must, however, be stressed that there is no hard and fast boundary between Population I and Population II regions.

The nearest star (excluding the Sun) is Proxima Centauri, at 4·28 light-years; it is a member of the α Centauri system. The two bright components of α Centauri are only slightly further away. Stars within 13 light-years of the Sun are as follows:

Star	R.A. h m (epoch 1950)	Dec. o	Spec.	Apparent Mag	Absolute Mag +	Parallax "	Radial Velocity, km/s	Distance l/y
Proxima	14 26·3	−62 28	Me	10·7	15·1	0·760		4·3
α Centauri A	14 36·2	−60 38	G4	0·0	4·4	0·760	−25	4·3
α Centauri B	14 36·2	−60 38	K5	1·7	5·8	0·760	−21	4·3
Barnard's Star	17 55·4	+04 33	M5	9·5	13·2	0·552	−108	5·8
Wolf 359	10 54·1	+07 20	M8	13·5	16·7	0·431	+13	7·6
Lalande 21185	11 00·7	+36 18	M2	7·5	10·5	0·402	−87	8·1
Sirius A	06 42·9	−16 39	A0	−1·4	1·4	0·377	−8	8·7
Sirius B	06 42·9	−16 39	dA	8·5	11·4	0·377	−8	8·7
UV Ceti A	01 36·4	−18 13	M6e	12·5	15·3	0·365	+29	8·9
UV Ceti B	01 36·4	−18 13	M6e	13·0v	15·8v	0·365	+29	8·9
Ross 154	18 46·1	−23 53	M6	10·6	13·3	0·345	−4	9·5
Ross 248	23 39·5	+43 56	M6	12·2	14·7	0·317	−81	10·3
ε Eridani	03 30·6	−09 38	K0	3·8	6·2	0·305	+15	10·7
Ross 128	11 43·5	+01 06	M5	11·1	13·5	0·301	−13	10·8
Luyten 789-6	22 35·7	−15 36	M7	11·3	13·3	0·302	−60	10·8
61 Cygni A	21 04·7	+38 30	K5	5·2	7·5	0·292	−64	11·2
61 Cygni B	21 04·7	+38 30	K7	6·0	8·4	0·292	−64	11·2
ε Indi	21 59·6	−57 00	K5	4·7	7·0	0·291	−40	11·2
Procyon A	07 36·7	+05 21	F5	0·4	2·7	0·287	−3	11·4
Procyon B	07 36·7	+05 21	—	10·8	13·1	0·287	−3	11·4
Σ 2398 A	18 42·2	+59 33	M4	8·9	11·1	0·284	+1	11·5
Σ 2398 B	18 42·2	+59 33	M5	9·7	11·9	0·284	+1	11·5
Groombridge 34 A	00 15·5	+43 44	M1	8·1	10·3	0·282	+14	11·6
Groombridge 34 B	00 15·6	+43 44	M6	11·0	13·3	0·282	+21	11·6
Lacaille 9352	23 02·6	−36 09	M2	7·4	9·6	0·279	+10	11·7
τ Ceti	01 41·7	−16 12	K0	3·5	5·7	0·273	−16	11·9
Luyten's Star	07 24·7	+05 29	M5	10·9	13·0	0·266	+26	12·3
Lacaille 8760	21 14·3	−39 04	M1	6·7	8·8	0·260	+23	12·5
Kapteyn's Star	05 09·7	−45 00	M0	8·8	12·7	0·256	+242	12·7
Krüger 60 A	22 26·3	+57 27	M4	9·8	11·9	0·254	−24	12·8
Krüger 60 B	22 26·3	+57 27	M6	11·3	13·3	0·254	−28	12·8

In 1976 O. J. Eggen reported the discovery of a red star, magnitude 10·8, in Sculptor (near τ Sculptoris) which was regarded as very close; later work showed that it is well beyond the 13 light-year limit of the above table.

If placed at the standard distance of 10 parsecs (32·6 light-years) only α Centauri, Sirius, Procyon and τ Ceti would be visible with the naked eye; it is interesting to note that if our Sun could be observed from Sirius it would be a second-magnitude star at R.A., 18h 44m, declination −16°4'. Most of the stars in the table are faint red dwarfs; thus the very feeble Wolf 359 has only 0·000 02 of the luminosity of the Sun. The companions of Sirius and Procyon are white dwarfs. There are four flare stars; UV Ceti B (Luyten L 726–8 B), Proxima, Krüger 60 B and Ross 154. Lalande 21185 and Groombridge 34A are excessively close binaries (too close to be separated visually) and there is a third body, of low mass, in the 61 Cygni system.

Radial velocities (the towards or away motion of the star relative to the Sun) are given as − for a velocity of approach, + for a velocity of recession.

The star with the greatest known proper motion is Barnard's Star, discovered in June 1916 by E. E. Barnard at the Yerkes Observatory. The annual proper motion is 10″·31, so that in 180 years it crosses the sky by a distance equal to the apparent diameter of the Full Moon. It is approaching us at 108 km/s, so that its distance is decreasing at the rate of 0·036 light-years per century. The proper motion is increasing by about 0″·0013 per year, and will reach 25″ by AD 11800, when the star will be at its closest to us – 3·85 light-years, closer than Proxima; the parallax will be 0″·87. The apparent magnitude will than have increased to 8·5. Subsequently the star will begin to recede. Barnard's Star is also known as Munich 15040.

The brightest star within 13 light-years, both apparently and absolutely, is of course Sirius A, with 26 times the luminosity of the Sun. The **least massive star** is UV Ceti B, with a mass of only 0.035 that of the Sun. Apart from the two white dwarfs, all the stars within 13 light-years belong to the Main Sequence; there are no giants.

The first predictions of the positions of invisible stars were made by F. W. Bessel in 1844, when he forecast the positions of the companions to Sirius and Procyon; he based his calculations upon irregularities in the proper motions of the primaries. In 1862 Clark discovered Sirius B in almost the position predicted by Bessel; the orbital period of the system is 50 years. The companion to Procyon was found in 1898 by Schaeberle, again almost in Bessel's position; the orbital period of the system is 39 years.

Efforts have been made to detect planets of other stars. The **most reliable instance** concerns Barnard's Star, which has been closely studied by P. van de Kamp and his colleagues at the Sproul Observatory (Swarthmore, Pennsylvania) ever since 1937, using the 61 cm reflector there. It appears that there may be two planets, one with and orbital period of about 11·7 years and the other of about 18·5 years; the masses are estimated at 0·9 and 0·4 that of Jupiter respectively – too low for stars. (The mass of Barnard's Star itself is 0·14 that of the Sun.) Other 'suspect' cases involve 61 Cygni B, 70 Ophiuchi and ε Eridani. Obviously all extra-solar planets are very faint, and are beyond the range of our present telescopes; the apparent magnitude of the brighter Barnard planet can hardly exceed +30. However, it is possible that visual confirmation will be obtained when telescopes can be set in orbit round the Earth, or on the surface of the Moon.

The first identification of a 'disk-star' which may be in the process of forming its own planets was made in 1977 by R. Thompson, P. Stritmatter, E. Erickson, F. Witteborn and D. Strecker at the Steward Observatory, University of Arizona. Observations were made with the 91 cm infra-red telescope of the Kuiper Airborne Observatory and with the Steward 2·3 m infra-red telescope. (Infra-red observations were essential to see through the dusty veil surrounding the star.) The star is MWC 349, in Cygnus. The surrounding disk of intensely glowing gas seems to have a diameter 20 times that of the central star, and to emit 10 times as much light. At the outer edge, the disk is about as thick as the star's diameter. The star itself is 10 times the diameter and 30 times the mass of the Sun. The luminous disk is believed to be the inner part of a surrounding larger disk of non-luminous gas in which outer planets may already have formed; this disk has a diameter greater than that of the orbit of Pluto, while the luminous disk has a diameter slightly greater than that of the orbit of the Earth. The luminous disk is wedge-shaped in cross-section, and where it joins the star's surface the thickness is 1/40 of the star's diameter.

The disk is estimated to be only 1000 years old, and is fading at about 1 per cent per month, as luminous material from the disk spirals into the central star. Therefore, the luminous disk may have vanished in a century from now.

The evidence rests upon several facts. The star is much brighter in visible light than it should be; it has declined since first identified forty years ago, and the spectrum is not that of a hot star – it is that of a hot, glowing disk. The distance from us is about 10 000 light-years.

The nearest stars which are at all like the Sun are τ Ceti and ε Eridani. From 4 April 1960 to March 1961 astronomers at the Green Bank Radio Astronomy Observatory in West Virginia, led by Dr Frank Drake, carried out a 'listening operation' at a wavelength of 21 cm, using the 80 ft. radio telescope, in the hope of picking up rhythmical signals which could be interpreted as artificial. The project was known officially as Project Ozma, though unofficially as Operation Little Green Men. Not surprisingly, the results were negative, but since 1969 operations of rather similar type have been carried out in the USSR.

The most luminous star known is probably η Carinæ, which is unique in many ways. Telescopically it has been described as looking more like a 'red blob' than a normal star, and it is associated with nebulosity; the distance is estimated at 6400 light-years, and the maximum luminosity may amount to several million times that of the Sun (one estimate gives 4 000 000 Suns). For some years in the 1830s and early 1840s it was the brightest star in the sky apart from Sirius, but for a century now it has been below naked-eye visibility, though it emits strongly in the infra-red and the decline in real luminosity is not so dramatic as might be thought. There have been speculative suggestions that it may suffer a supernova outburst in the foreseeable future (that is to say, within the next few tens of thousands of years); it has also been regarded as an exceptionally slow nova, or as an extraordinary type of variable star. The spectrum can only be classified as 'peculiar'.

Excluding η Carinæ, the most luminous star known is S Doradûs, in the Large Magellanic Cloud; the absolute magnitude is —8·9, so that it is about a million times as powerful as the Sun (it is somewhat variable). From Earth, it cannot be seen with the naked eye.

The most massive star known is Plaskett's Star, HD 47129 Monocerotis, identified by K. Plaskett in 1922, when he discovered the binary nature of the O7-type supergiant. The orbital period is 14 days. The mass of each component is thought to be 55 times that of the Sun, and the primary has a radius of 25 times that of the Sun; the absolute magnitude is —7.

The largest star known may be the invisible secondary of the eclipsing binary ε Aurigæ, whose diameter was estimated to be about 5 700 000 000 km (about the same as that of the orbit of Uranus). However, it is also possible that the secondary is a black hole. The system of α Herculis, made up of a red variable supergiant with a double companion, is enveloped in a huge cloud of gas which could have a diameter of 250 000 000 000 km. Some red supergiants, such as Betelgeux, have diameters exceeding 400 000 000 km. In 1978 it was announced that Betelgeux is surrounded by an extremely tenuous 'shell' of potassium, about 11,000 astronomical units in diameter.

The smallest known star, excluding neutron stars, is the white dwarf LP 327–16, discovered in May 1962 at Minneapolis. The diameter is about 1700 km (half that of the Moon), and the distance from us is 100 light-years. It is possible that another white dwarf, LP 768–500, is even smaller (less than 1600 km).

The first measures of stellar diameters using an intensity interferometer have been made by J. Hanbury Brown, from Australia. Some of the values found are as follows:

Star	Diameter, km	Star	Diameter, km
Canopus	115 000 000	Sirius	4 800 000
β Carinæ	12 400 000	Altair	4 700 000
Vega	7 700 000	Fomalhaut	4 600 000
Regulus	6 600 000	ζ Puppis	4 400 000
α Gruis	6 100 000	Spica	2 200 000
Procyon	5 700 000		

The number of stars visible with the naked eye is approximately 5780, though this value is bound to be arbitrary. This means that it is seldom or never possible to see more than 2500 stars with the naked eye at any one time.

Luminosity and absolute magnitude

Many people prefer to visualize a star's luminosity in terms of the Sun as unity rather than as absolute magnitude. The following conversion table is approximate, but good enough for most purposes.

Absolute Mag.	Luminosity, Sun= 1	Absolute Mag.	Luminosity, Sun = 1
+7·9	0·063	+0·0	95
+7·5	0·087	—0·5	138
+7·0	0·14	—1·0	220
+6·5	0·22	—1·5	350
+6·0	0·35	—2·0	550
+5·5	0·55	—2·5	870
+5·0	0·87	—3·0	1380
+4·8	1·00	—3·5	2200
+4·5	1·38	—4·0	3470
+4·0	2·19	—4·5	5500
+3·5	3·47	—5·0	8700
+3·0	5·5	—5·5	13800
+2·5	8·7	—6·0	21900
+2·0	13·8	—6·5	34600
+1·5	21·9	—7·0	55000
+1·0	34	—7·5	87000
+0·5	55		

DOUBLE STARS

The term 'double star' was first used by Ptolemy, who wrote that η Sagittarii was 'διπλους'. There are, of course, several doubles which may be separated with the naked eye, so that they have presumably been known from antiquity; the most celebrated is ζ Ursæ Majoris (Mizar), which makes a naked-eye pair with 80 Ursæ Majoris (Alcor). The Arabs gave full descriptions of it – though it is true that they apparently regarded Alcor as rather a difficult object. (This is not true today, but it is unlikely that there has been any real change.)

The first double star discovered telescopically was Mizar itself, which is made up of two components 14"·5 apart. The discovery was made by Riccioli in 1651. Alcor is 700" from the main pair, which is rather too wide for a recognized 'double' as entered in the official catalogues. The duplicity of γ Arietis was dis-

covered by Robert Hooke in 1665, while he was searching telescopically for a comet.

The first southern double star to be discovered was α Crucis, by Father Guy Tachard in 1685. Tachard was on his way to Siam, by sea, and stopped off at the Cape of Good Hope, where he was warmly welcomed by the Dutch settlers and set up a temporary observatory, mainly for navigational purposes. He recorded that 'the foot of the Crozier marked in Bayer is a Double Star, that is to say, consisting of two bright stars distant from one another about their own Diameter, only much like to the most northern of the Twins; not to speak of a third much less, which is also to be seen, but further from these two.' The Crozier is the Southern Cross; 'Bayer' refers to J. Bayer's famous star catalogue of 1603, and the 'northern of the Twins' is Castor, already known to be double.

Like most of his contemporaries, Tachard believed the stars to show definite apparent diameters rather than being virtual point sources.

Other doubles discovered at an early stage included α Centauri (1689), Castor, γ Virginis, and the 'Trapezium', θ Orionis, in the Orion Nebula, which is a multiple system.

The first true list of doubles was published in 1871 by C. Mayer of Mannheim. His list included γ Andromedæ, ζ Cancri, α Herculis and β Cygni. Mayer used an 8 ft mural quadrant, with magnifications of 60 and 80.

The first comment upon possible physically-associated or binary pairs was made by the Rev. John Michell in 1767, who wrote: 'It is highly probable in particular, and next to a certainty in general, that such double stars as appear to consist of two or more stars placed very near together, do really consist of stars placed near together, and under the influence of some general law.' Michell repeated this view in 1784. However, in 1782 William Herschel had commented that it was 'much too soon to form any theories of small stars revolving round large ones'.

The first proof of the existence of binary pairs was given by Herschel in 1802. From 1779 he had been attempting to measure the parallaxes of stars, and he had concentrated upon the doubles, since if one member of the pair were more remote than the other it followed that the closer member should show an annual parallax relative to the more distant member. He failed, because his equipment was not sufficiently sensitive, but he made the fortuitous discovery that some of the pairs under study (such as Castor) showed orbital motion, and by 1802 he was confident enough to publish his findings. His classic paper actually appeared in the *Philosophical Transactions* on 9 June 1803.

The first successful measurement of a star-distance was announced in 1838 by F. W. Bessel of Königsberg; the star concerned was 61 Cygni, which Bessel had selected because it had relatively large proper motion and was a wide binary – presumably indicating that by stellar standards it was comparatively close. At about the same time T. Henderson announced a parallax value for α Centauri, based upon measurements which he had made earlier (1831–33) at the Cape, while Struve in Russia gave a rather less accurate value for the parallax of Vega.

Rather surprisingly, binaries are more common than optical pairs, in which the components are not genuinely associated. It has even been estimated that more than 50 per cent of all stars are members of binary systems, with a mean separation from 10 to 20 astronomical units, though this may be too high a ratio. A visual binary will have an orbital period of at least 2 years. It is important to note that both components of a binary move round their common centre of gravity, and the orbits are not in general so dissimilar as might be thought, since in mass there is a far lower spread among the stars than there is in luminosity and in size. Incidentally, there are a few surprising cases – thus γ Arietis, which appears telescopically as a perfect pair of twins, has proved to be an optical double, not a binary. Appearances can be misleading!

The first reliable orbit for a binary pair (ξ Ursæ Majoris) was worked out by the French astronomer Felix Savary in 1830. (The period is 60 years.) Such calculations are of great importance, since they lead to a determination of the combined masses of the components – something which is much more difficult to calculate for a single star. In fact, our knowledge of stellar masses depends very largely upon the orbital movements of binaries.

Many important catalogues of double stars have been published since Mayer's. Among them are the catalogues of F. G. W. Struve (1822, with later additions), E. Dembowski (Naples, 1852; over 20 000 measures), S. W. Burnham (1870, and again in 1906, listing 13 665 pairs – he was personally responsible for the discovery of 1340 of them), R. Aitken (1932; 17 180 pairs) and the Lick Index Catalogue or IDS (1963; 65 000 pairs, of which 40 000 are binaries). Work of the greatest importance was carried out at the Republic Observatory, Johannesburg (formerly the Union Observatory) between 1917 and 1965, under the successive directorships of R. T. A. Innes (1917–27), H. E. Wood (1927–41), W. H. van den Bos (1941–1956) and W. S. Finsen (1957–65). The telescope used for most of the work was the 27 in refractor. It is a matter of great regret that the virtual closing of the Observatory, following the concentration of South African astronomy at Sutherland in Cape Province, has brought this work to an end – temporarily, one must hope.

The first spectroscopic binary (Mizar A) was discovered in 1889 by E. C. Pickering at Harvard; another identification (β Aurigæ) soon followed. Spectroscopic binaries have too small a separation for the components to be seen individually, but the binary nature of the system betrays itself because of the Doppler shifts in the spectra. If both spectra are visible, the absorption lines will be periodically doubled; if one spectrum is too faint to be seen, the lines due to the primary will oscillate about a mean position. There are some 'borderline' cases; thus Capella was long known to be a spectroscopic binary, but the world's largest telescopes can just indicate that it is not a single star. The orbital period is 100 days.

The first astrometric binaries to be studied were Sirius and Procyon, by F. W. Bessel in 1844. In an astrometric binary, the presence of an invisible companion is inferred from slight 'wobblings' of the primary. In both these first cases, the companions were subsequently discovered, but with other astrometric binaries the secondary stars remain unseen.

The first White Dwarf binary companion to be discovered (in fact, the first white dwarf of any kind) was Sirius B, or the Companion of Sirius. It was first seen in 1862 by Clark, at Washington. The orbital period is 50 years, and the maximum separation is 11″·5 (as in 1975). For many years after its discovery the Companion was assumed to be large and red, but in 1915 W. S. Adams, at Mount Wilson, studied its spectrum and found that it was white; the surface temperature was at least 8000 °C. Since the luminosity was only 1/10 000 that of Sirius itself, the Companion had to be very small; smaller, in fact, than Uranus or Neptune, and with a diameter only 3 times that of the Earth. The resulting density turned out to be 70 000 times that of water.

It is worth noting that in ancient times Sirius was described as being red. There is no chance that the primary has changed in colour, and suggestions that the Companion was then going through its red giant stage, before collapsing into the white dwarf condition, seem to be untenable. We must therefore assume that there was some error in observation or interpretation in the old records, though an element of mystery remains.

Binary stars are of many different kinds; sometimes the components are dissimilar (as with Sirius), sometimes they are identical twins – as with γ Virginis, which has a period of 180 years. Several decades ago it was wide and easy, but is now closing, and by 2016 except in giant telescopes. This does not, of course, indicate

any actual closing; everything depends upon the angle from which we see the pair. Another binary, formerly easy but now much less so, is Castor. In fact both components of the bright pair are spectroscopic binaries, and also associated with the system is Castor C or YY Geminorum, made up of two red dwarfs; it is an eclipsing variable. Castor therefore consists of six stars, four luminous and two dim.

The first eclipsing binary to be discovered was Algol, by G. Montanari in 1669 (it is not now thought that the 'winking' was known in ancient times, even though Algol was nicknamed the Demon Star). The cause of the variability was explained by J. Goodricke, the deaf-mute astronomer, in 1782.

The most extreme eclipsing binary is probably ε Aurigæ. The variability was discovered by Fritsch in 1821, but was then thought to be irregular. In fact the system is made up of a powerful supergiant together with an invisible secondary, detectable only in infra-red, which may be either a very young star of immense size or else a black hole. Close to it in the sky is another extreme eclipsing binary ζ Aurigæ, with a K-type supergiant primary and a B-type secondary; when the B-star passes behind the supergiant, its light comes to us for a period via the outer layers of the primary, and the spectral effects are both complicated and highly informative. Another interesting system is β Lyræ. Here, and also with other stars of the same type, the two components are almost in contact; they are tidally distorted into elliptical shapes, and the whole system is enveloped in gas.

The binary with the shortest known period is AM Canum Venaticorum, studied by J. Smak in 1967. The period is only 17 minutes 31 seconds! Needless to say, the components cannot be seen separately.

It is impossible to say which is the binary with the longest known period. With very large separations the period may amount to millions of years, and it is better merely to say that very widely separated components share a common binary motion through space.

The first binary White Dwarf was found by W. Lutyen and P. Higgins in 1973. Its position is R.A. 9h 42m, dec. +23°41′. The separation is 13″, the position angle 052, and the period 12 000 years; at present the two components are 600 astronomical units apart.

The first binary neutron star was discovered in July 1974. The orbital period is 8 hours. The discovery was made by means of radio observations, and the object (pulsar OSR 1913 +16) has not been seen optically, but it is assumed to be a binary made up of two neutron-star components.

The first X-ray binaries were reported following the launch of the first X-ray astronomical satellite, UHURU, in 1970. X-ray binaries may be composed of a B-type star and a neutron star; the neutron star is pulling material off its companion, perhaps at 100 million million million tonnes per second, and X-rays are being emitted. However, studies of this kind are still at a very early stage, and it would be dangerous to be dogmatic.

The origin of binary systems is not known with certainty. The old theory – that a binary formed as a result of the fission of a single star – has been rejected by most astronomers. It has been suggested that binaries may be due to the mutual capture of the components, but this seems unlikely, and it is much more plausible to assume that the components were formed from the same cloud of interstellar material in the same region of space. If the components are identical or near-identical twins, the spectra are generally the same. When there is a definite difference in brightness between the components, the spectra also differ. If both stars belong to the Main Sequence, the primary is usually of earlier type than the secondary, while if the primary is a giant the secondary is either a giant of earlier type or else a dwarf of similar spectral type. Novæ are often found to be binaries (DQ Herculis 1934 is a good example; it is an eclipsing system with a period of 4h 39m made up of the old nova itself, probably so dense that it resembles a white dwarf, together with an invisible companion which is probably a red dwarf of type M). We also have the SS Cygni or U Geminorum stars, all of which seem to be binaries. In a star of this kind the primary is a white dwarf, the secondary a Main Sequence star, presumably a red dwarf. The components are very close together, and the white dwarf is pulling material away from the less massive secondary; gas moves across to the primary in a narrow stream, producing a disk or shell of gas round the white dwarf. Where the stream hits the disk there is a bright spot, and irregularities in the stream make the light we receive flicker rapidly with small amplitude; this flickering is detectable only with very sensitive photoelectric equipment. The larger outbursts in the SS Cygni stars originate in the white dwarf component of the system. It is easy to see why these variables are sometimes termed dwarf novæ.

The most spectacular multiple stars in the sky are probably ε Lyræ and θ Orionis. ε Lyræ has two main components, making up a naked-eye pair; each component is again double, and all four may be distinguished with a fairly small telescope. θ Orionis, in the great nebula M.42, is nicknamed the 'Trapezium' for obvious reasons. It lies on the outskirts of the nebula, and is responsible for making the nebulosity luminous.

The position angle of a visual double star (either a binary or an optical pair) is measured according to the angular direction of the secondary (B) from the primary (A), reckoned from 000 at north round by east (090), south (180) and west (270) back to north. With rapid binaries the separations and position angles alter quickly; a good example is the fine binary ζ Herculis, where the magnitudes are 3 and 5·6, and the period is only 34 years, Some of the published catalogues are already out of date, and need revising. There is scope here for the skilful and well-equipped amateur as well as for the professional, and certainly double-star research is a fascinating branch of astronomy.

VARIABLE STARS AND NOVÆ

Nova HR Delphini on 10 August 1967
(Cdr, H. R. Hatfield)

THE CLASSIFICATION OF VARIABLE STARS

Variable stars are of many kinds. Elaborate systems of classifying them have been proposed, and the notes given below are not intended to be more than a general guide.

Eclipsing Variables.

Better termed 'eclipsing binaries', since there is no intrinsic variation. The light-changes are caused by one component of the binary passing in front of the other.

Algol type (EA). A principal minimum and a very small (almost imperceptible) secondary minimum. The star thus remains at its normal or 'maximum' magnitude for most of the time.

β **Lyræ type** (EB). Components so close that they are distorted into elliptical shapes, and may be almost in contact. Light-variation continuous, with alternate deep and shallow minima.

W Ursæ Majoris type (EW). Dwarf pairs, also ellipsoidal and almost in contact. Periods of less than one day.

Short-period Variables.

Pulsating stars; the variation is intrinsic. The period gives a key to the star's real luminosity, so that these variables act as invaluable aids in distance-estimation.

Classical Cepheids (Cδ). Named after δ Cephei. Periods from 3 to 50 days. F to G supergiants.

W Virginis stars or Type II Cepheids. Also F to G supergiants, but about 1·5 times fainter than classical Cepheids (absolute magnitudes (median) 0 to —3·5, instead of —1·5 to —5).

RR Lyræ stars (formerly termed cluster-Cepheids). A to F giants. Periods from 0·05 day to 1·2 days; amplitudes below 2 magnitudes. All appear to be about equal in luminosity, though the class includes several subdivisions.

β **Cephei (or** β **Canis Majoris) stars.** B-type giants. Periods less than 0·5 day. Very small amplitude (generally less than 0·3 magnitude).

δ **Scuti stars.** F subgiants. Periods less than 1 day. Very small amplitudes (less than 0·3 magnitude). Rare.

Magnetic or spectrum variables. Prototype: α^2CVn. Very small amplitude visually, but with powerful and variable magnetic fields. Rare.

RV Tauri Variables.

Pulsating stars; G to K giants. Alternate deep and shallow minima, with spells of virtual irregularity. About 100 are known, of which the brightest is R Scuti.

Long-period Variables.

Known as Mira stars, after the prototype Mira

The most recent bright nova. V. 1500 Cygni, photographed on 3 September 1975 (P. B. Doherty)

(o Ceti). Late-type giants, with median absolute magnitudes from $+2$ to -2. Periods from about 80 to over 600 days; no Cepheid-type period-luminosity law. Considerable amplitudes (11 magnitudes in the case of χ Cygni). Often intensely red. Periods and amplitudes are not constant from one cycle to another.

Semi-regular Variables.

Late-type giants and supergiants. Periods often so ill-defined as to be virtually unrecognizable. Amplitudes generally much less than for Mira stars. There are various subdivisions of this group.

Irregular Variables.

Generally with small or moderate amplitude. most (not all) are of late spectral type. The group includes the **T Tauri stars** or nebular variables, which are in an early phase of evolution and are associated with nebulosity.

R Coronæ stars.

F to K supergiants. Low in hydrogen abundance, but rich in carbon. Remain 'normally' at maximum, but suffer sudden, unpredictable drops to minimum. Relatively rare.

Eruptive Variables.

These are of various types (R Coronæ stars are sometimes included in the group).

SS Cygni or U Geminorum stars (often called 'dwarf novæ'). Binary systems, generally made up of a Main Sequence dwarf together with a white dwarf. Remain 'normally' at minimum, but show novalike outbursts with amplitudes of a few magnitudes, at mean intervals from about 20 to over 600 days.

Z Camelopardalis stars. Similar to SS Cygni stars except that they undergo 'standstills', which cannot be predicted, and when the light-variations are suspended.

Flare stars (or UV Ceti stars). M-type dwarfs. Sudden flare-ups in light of a few magnitudes, taking place in a few minutes, due to what may presumably be localized flares on the star's surface.

P Cygni stars. Slow, erratic and sometimes novalike fluctuations. Gas-shells ejected.

Z Andromedæ or symbiotic stars. Heterogeneous group, usually with composite spectra.

Recurrent novæ. Stars which have been known to show more than one novalike outburst; the most famous example is T Coronæ. May form a link between true variables and novæ.

Novæ.

Usually O to A subdwarfs; very faint stars which undergo tremendous outbursts, of many magnitudes (almost 20 magnitudes in the case of

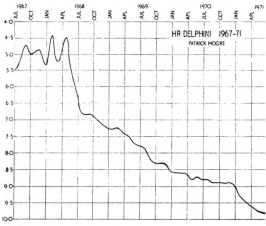

Variation of magnitude of HR Delphini 1967-71 (Patrick Moore)

Nova Cygni 1975). They then decline again. Gas-shells ejected. Probably all novæ are binaries.

Supernovæ.

Tremendous outbursts, when a star virtually destroys itself, ending up as an expanding gas-cloud containing a neutron star (pulsar). Relatively rare. The last to be seen in our Galaxy was Kepler's Star of 1604, though many supernovæ have been observed in external galaxies.

There are a few variables whose type is un-certain (such as ρ Cassiopeiæ), and some which do not seem to fit into any category, such as η Carinæ. In each constellation, variable stars not given a Greek letter are designated by the letters R to Z, then RR to RZ, followed by SS to SZ, TT to TZ etc. After ZZ is reached, we return to AA to AZ, BB to BZ, ending with QZ. This allows for 334 variables in each constellation, after which new variables are designated V.335, V.336 and so on. There have been a few misidentifications; thus the object formerly classed as the variable BW Tauri is now known to be a Seyfert galaxy, while W Comæ has proved to be a quasar.

Secular variables are stars which may have brightened or faded in historical times. Thus Ptolemy and others ranked β Leonis and θ Eridani as being of the first magnitude, whereas today they are respectively of magnitude 2 and below 3. δ Ursæ Majoris was ranked as equal to the other stars of the Plough, but is now at least a magnitude fainter; on the other hand α Ophiuchi was ranked as of magnitude 3, and is now 2. These variations

must be regarded as suspect, since it would be unwise to trust the old estimates too far.

The first variable star to be positively identified as such was Mira (o Ceti), in 1638. It had been previously recorded by Fabricius (1596) and by Bayer (1603) and had even been allotted a Greek letter, so that its surprising that its fluctuations were not tracked down earlier. In the latter part of the 17th century two more variables were identi-fied – Algol, now known to be an eclipsing binary and the long-period χ Cygni.

The following variables were identified between 1638 and 1850 (novæ are not included);

Star	Discoverer of variability	Date
Mira (o Ceti)	Holwarda	1638
Algol (β Persei)	Montanari	1669
χ Cygni	Kirch	1686
R Hydræ	Maraldi	1704
Rasalgethi (α Herculis)	W. Herschel	1759
μ Cephei	W. Herschel	1782
R Leonis	Koch	1782
δ Cephei	Goodricke	1784
β Lyræ	Goodricke	1784
η Aquilæ	Pigott	1784
R Scuti	Pigott	1795
R Coronæ Borealis	Pigott	1795
R Virginis	Harding	1809
R Aquarii	Harding	1811
ε Aurigæ	Fritsch	1821
R Serpentis	Harding	1826
η Carinæ	Burchell	1827
S Serpentis	Harding	1828
U Virginis	Harding	1831
δ Orionis	J. Herschel	1834
S Vulpeculæ	Rogerson	1837
Betelgeux (α Orionis)	J. Herschel	1840
β Pegasi	Schmidt	1847
λ Tauri	Baxendell	1848
R Orionis	Hind	1848
R Pegasi	Hind	1848
R Capricorni	Hind	1848
S Hydræ	Hind	1848
S Cancri	Hind	1848
S Geminorum	Hind	1848
R Geminorum	Hind	1848
T Geminorum	Hind	1848
R Tauri	Hind	1849
T Virginis	Boguslawsky	1849
T Cancri	Hind	1850
R Piscium	Hind	1850

Sir John Herschel regarded Alphard (α Hydræ) as a variable, but this has not been confirmed. It is now known that δ Orionis has a very small range, so that Herschel's discovery may not have been valid.

Several famous stars are included in this list. Betelgeux is the **brightest variable star in the**

sky, excluding those variables with exceedingly small range (such as Rigel) and also excluding η Carinæ, which is unique and which has remained below naked-eye visibility now for something like a hundred years. χ Cygni, with a period of 407 days, has the greatest amplitude of any Mira variable: 11 magnitudes. ε Aurigæ is the remarkable eclipsing system with the remarkably long period of 27 years.

Variable star studies are highly complex, because there are so many types of variables. The eclipsing stars, such as Algol and β Lyræ, are not genuinely variable at all; if for instance Algol were observed from another position in the Galaxy it would be constant. The intrinsic variables are entirely different. There are **rotating variables,** mainly with peculiar A-type spectra, which may have surface characteristics which cause variability as the star spins. **T Tauri variables** are F, G or K giants above the Main Sequence, still surrounded by clouds of dust and gas; mass is being ejected, and as the star must spend several million years in contracting to the Main Sequence it probably loses a significant amount of mass during the T Tauri stage. In **YY Orionis variables** there are red shifts indicating infalling material, so that these stars are extremely young. **RW Aurigæ and T Orionis variables** are also subclasses of T Tauri stars. All these variables are irregular, and may have great chromospheric activity.

Classical Cepheids are yellow supergiants; the period of pulsation is the time required for a vibration to travel from the surface of the star to the centre and back again, so that large stars have longer periods than small ones. Classical Cepheids are massive, young stars, while **W Virginis stars** (Population II Cepheids) are old stars with low masses, and are about 2 magnitudes fainter than Classical Cepheids with equal periods. They have exhausted their available hydrogen, and are now 'burning' helium. In the 1960s one star, RU Camelopardalis, stopped pulsating, and has never really recovered! This shows that the W Virginis stars are evolving rapidly through the Cepheid instability strip of the H-R diagram. **RR Lyræ variables**, formerly termed cluster-Cepheids, are all of approximately the same age, mass and helium content, and all are in the initial stage of core helium-burning; conditions for this kind of variability are critical, which explains why the RR Lyræ stars are so uniform – if they were not, then they would not be RR Lyræ stars. The dwarf Cepheids or **AI Velorum variables** have periods

of from 0·05 to 0·3 day, and ranges of from 0·3 to 0·7 magnitude; it is not known whether they are low-mass stars, younger and rather richer in heavy elements than the RR Lyræ variables, or whether they are young stars of normal mass and therefore close relations of the δ Scuti variables. The **δ Scuti variables** lie near the Main Sequence on the H-R diagram, not far from the Cepheid instability strip; their light curves tend to be irregular. The **β Cephei variables** are hot and massive; the precise nature and cause of their pulsations is unknown as yet. The **RV Tauri variables** lie where the Cepheid instability strip merges with the upper right of the H-R diagram; they are infra-red emitters, so that apparently the pulsation of an RV Tauri star drives off a shell of material. The **Mira variables, the semi-regular variables** and most **irregular variables** are also pulsating. In the **UV Ceti or Flare stars** the flares may be of exaggerated solar type, and are accompanied by radio emission; they are cool, faint Main Sequence stars, and it has been suggested that about 4 per cent of all stars may be of this type, though their faintness means that only the nearer members are detectable. **P Cygni stars** are intensely luminous and produce shells of expanding gas; the brightening of P Cygni itself in the 17th century may have been caused by the expansion of the star's outer layers and the ejection of a gaseous shell. The unique η Carinæ is embedded in a cloud of dust and gas, and is a very strong infra-red source. The **SS Cygni or U Geminorum stars,** also called dwarf novæ, are now believed to be binary systems, made up of a normal cool star and a white dwarf; the white dwarf pulls material away from its companion, forming a ring, so that the result is a layer of hydrogen-rich material. The addition of further hydrogen-rich material creates an unstable situation, and an outburst occurs; B. Warner and his colleagues have found that there is a 'hot spot' on the ring, where the stream of material from the cool star hits it, and there is also very rapid flickering, undetectable visually. The **Z Camelopardalis stars** are basically like SS Cygni stars, but have 'standstills' of unknown cause. **R Coronæ variables** have been described as 'novæ in reverse'. They are supergiants, with surface temperatures similar to Cepheids; their atmospheres are rich in carbon, and very deficient in hydrogen. It is possible that the minima are due to the formation of solid carbon (soot) in the upper atmosphere; the minimum lasts until the outward pressure of radiation blows the soot away. It has been suggested that they are stars which have ejected their outer, hydrogen-rich layers. Alternatively, they could be stars in which the carbon produced near the core has been mixed outward toward the surface by convection.

It is therefore clear that variable stars are not only of different kinds, but very different ages. The very youngest are probably the pre-main sequence stars of which FU Orionis and V.1057 Cygni are typical examples; they brighten when their light can penetrate the surrounding cloud of infalling material. They are infra-red sources,

because of the clouds of warm gas and dust. Both FU Orionis and V.1057 Cygni have brightened up by several magnitudes during the last half-century, and it is not likely that they will decrease again, though they show minor fluctuations.

Novæ are now thought to be binary systems, again with one cool Main Sequence star and a white dwarf. The nuclear reactions caused by the infalling hydrogen-rich material become 'out of control', and there is a violent outburst. There is no similarity between novæ and **supernovæ**, which are not binaries (at least so far as we know). Confusingly, Type I supernovæ are Population II stars of low mass; at maximum the absolute magnitude be as great as —18·6. Type II supernovæ, with maximum absolute magnitude of about —16·6, are produced by massive stars of Population I.

A CATALOGUE OF BRIGHT VARIABLE STARS

The following catalogue includes all variable stars with maximum above magnitude 6·0, apart from those with a range too small to be detected visually.

ECLIPSING STARS

Star		Max.	Min.	Period, d.	Spectrum	Type
R	Aræ	5·9	6·9	4·4	B9	Algol
ε	Aurigæ	3·3	4·2	9899	Fp	
ζ	Aurigæ	4·9	5·5	972	K+B7	
u	Herculis	4·6	5·3	2·1	B	β Lyræ
δ	Libræ	4·8	6·1	2·3	A1	Algol
β	Lyræ	3·4	4·3	12·9	Bp	β Lyræ
U	Ophiuchi	5·8	6·5	1·7	B	Algol
β	Persei	2·1	3·3	2·9	B8	Algol
ζ	Phœnicis	3·6	4·1	1·7	B7	Algol
λ	Tauri	3·3	4·2	4·0	B+A	Algol
V	Puppis	4·3	5·1	1·5	B1+B3	β Lyræ

CEPHEIDS

Star		Max.	Min.	Period, d.	Spectrum	
η	Aquilæ	3·7	4·7	7·2	F—G	
RT	Aurigæ	5·4	6·5	3·7	F—G	
l	Carinæ	3·4	4·8	35·2	F—G	
SU	Cassiopeiæ	5·9	6·3	1·9	F—G	
δ	Cephei	3·5	4·4	5·4	F—G	
β	Doradûs	4·5	5·7	9·8	F—G	
ζ	Geminorum	3·7	4·3	10·2	F—G	
T	Monocerotis	5·8	6·8	27·0	F—G	
χ	Pavonis	4·0	5·5	9·1	F—G	(W Vir)
X	Sagittarii	4·1	5·1	7·0	F—G	
W	Sagittarii	4·0	5·2	7·6	F—G	
AH	Velorum	5·8	6·4	4·2	F—G	

MIRA STARS

Star		Max.	Min.	Period, d	Spectrum	
R	Andromedæ	5·9	14·9	409	S	
R	Aquarii	5·8	11·5	387	M+pec	
R	Aquilæ	5·7	12·0	300	M	
R	Carinæ	3·9	10·0	381	M	
S	Carinæ	4·5	9·9	150	M	
R	Cassiopeiæ	5·5	13·0	431	M	
R	Centauri	5·4	11·8	547	M	
T	Cephei	5·4	11·0	389	M	
o	Ceti	1·7	10·1	332	M	Mira
χ	Cygni	3·3	14·2	407	S	
R	Horologii	4·7	14·3	403	M	
R	Hydræ	4·0	10·0	386	M	
R	Leonis	5·4	10·5	313	M	
R	Leporis	5·9	10·5	432	N	
X	Ophiuchi	5·9	9·2	334	M	
U	Orionis	5·3	12·6	372	M	
R	Serpentis	5·7	14·4	357	M	
R	Trianguli	5·4	12·0	266	M	

SEMI-REGULAR VARIABLES

Star		Max.	Min.	Mean period.*	Spectrum	
U	Antliæ	5·7	6·8	365	N	
UU	Aurigæ	5·1	6·8	235	M	
VZ	Camelopardalis	4·8	5·2	24?	M	
X	Cancri	5·9	7·3	170?	N	
Y	Canum					
	Venaticorum	5·2	6·6	158	N	
T	Centauri	5·5	9·0	91	M	
W	Cygni	5·0	7·6	130	M	
η	Geminorum	3·1	3·9	233	M	
g	Herculis	4·6	6·0	70	M	
α	Herculis	3·0	4·0	100	M	Rasalgethi
R	Lyræ	4·0	5·0	47	M	
W	Orionis	5·9	7·7	212	M	
α	Orionis	0·1	0·9	2070	M	Betelgeux
Y	Pavonis	5·7	8·5	233	N	
β	Pegasi	2·4	2·8	36	M	Scheat
ρ	Persei	3·2	4·2	33—55	M	
TV	Piscium	4·8	5·2	49	M	
L²	Puppis	3·4	6·2	141	M	
R	Sculptoris	5·8	7·7	363	N	
CE	Tauri	4·3	4·7	165	M	
RR	Ursæ Minoris	4·6	5·1	40	M	

*Often so ill-defined as to be unrecognizable, though with some stars, such as R Lyræ, definite periodicity can be established. Some catalogues class U Antliæ as irregular; certainly the period of Betelgeux is very ill-defined.

IRREGULAR VARIABLES

Star		Max.	Min.	Spectrum	
AF	Aurigæ	5·4	6·1	O9	
η	Carinæ	−0·8	7·9	Pec	Unique object. Now below mag. 7.
ρ	Cassiopeiæ	4·1	6·2	F	Occasional 'fades'.
γ	Cassiopeiæ	1·6	3·2	Bp	Generally around mag. 2·3.
μ	Cephei	3·6	5·1	M	Herschel's 'Garnet Star'.
R	Coronæ Borealis	5·8	15	Gp	Prototype RCrB star.
T	Coronæ Borealis	2·0	10·8	Q+M	'Blaze Star'. Recurrent nova
P	Cygni	3	6	Pec	'Nova' 1600. Now about mag. 5.
T	Cygni	5·0	5·5	K	
U	Delphini	5·6	7·5	M	
δ²	Lyræ	4·3	4·7	M	Type unknown.
χ	Ophiuchi	4·1	5·0	B	Novalike, but with small range.
RS	Ophiuchi	5·2	12·2	Op	Recurrent nova.
TX	Piscium	4·3	5·1	N	
BU	Tauri	5·0	5·5	Bp	Pleione (Pleiades). Shell star.

RV TAURI STAR

Star		Max.	Min.	Period, d.	Spectrum
R	Scuti	5·7	8·6	144	G+K

There are thus 78 variables whose range is sufficient to be detected by visual observation, and which are at maximum visible with the naked eye. We may add the flare star UV Ceti, which was once observed to rise briefly to magnitude 5·9.

ALGOL TYPE (EA)

Star		Max.	Primary Min.	Period, d	Spectrum
β	Persei (Algol)	2·1	3·3	2·867	B8
R	Aræ	5·9	6·9	4·425	B9
U	Cephei	6·6	9·8	2·493	B8 + G2
VV	Cephei	6·7	7·5	7430	M2 + B0
δ	Libræ	4·8	6·1	2·327	A1
U	Ophiuchi	5·8	6·5	1·677	B5
U	Sagittæ	6·4	9·0	3·381	B9 + G2
λ	Tauri	3·3	4·2	3·953	B3 + A

β LYRÆ TYPE (EB)

Star		Max.	Primary Min.	Secondary Min.	Period, d.	Spectrum
β	Lyræ	3·4	4·3	3·8	12·91	B8p
S	Antliæ	6·4	6·8	6·8	0·65	A8 + A8
YY	Canis Minoris	8·5	9·1	8·9	1·07	F5
RR	Centauri	7·6	8·1	7·9	0·61	F2
V	Crateris	9·5	10·2	9·9	0·70	A6
u	Herculis	4·6	5·3	4·8	2·05	B3 + B3
V	Puppis	4·3	5·1	5·0	1·45	B1 + B3

W URSÆ MAJORIS TYPE

Star		Max.	Primary Min.	Secondary Min.	Period, d.	Spectrum
W	Ursæ Majoris	8·3	9·1	9·0	0·33	F9 + F8
AB	Andromedæ	10·0	10·9	10·7	0·33	G5 + G6
SW	Lacertæ	9·2	10·0	9·9	0·32	G3 + G3

SS CYGNI (OR U GEMINORUM) TYPE

Star		Max.	Min.	Mean period, days
UU	Aquilæ	11·0	16·8	56
SS	Aurigæ	10·5	14·5	54
SS	Cygni	8·1	12·1	50
U	Geminorum	8·8	14·4	103
VW	Hydri	8·4	13·4	35
X	Leonis	12·0	15·2	22
RU	Pegasi	10·0	13·1	70
SW	Ursæ Majoris	10·8	16·2	Very long (±1000?)

Z CAMELOPARDALIS TYPE

Star		Max.	Min.	Mean period, days
RX	Andromedæ	10·3	13·6	14
Z	Camelopardalis	10·2	13·4	20
AH	Herculis	10·8	13·9	20
CN	Orionis	11·8	14·7	19
TZ	Persei	12·1	15·4	17

RECURRENT NOVÆ

Star		Max.	Min.	Years of observed outbursts
T	Coronæ Borealis	2·0	10·8	1866, 1946
RS	Ophiuchi	5·1	11·7	1901, 1933, 1958, 1967
T	Pyxidis	7·0	14·0	1890, 1902, 1920, 1945, 1965
WZ	Sagittæ	7·0	±16	1913, 1946, 1979

FLARE (UV CETI) TYPE

Star		Max.	Min.	Spectrum
YZ	Canis Minoris	10·3	11·6	M
DO	Cephei	9·9	11·4	M
UV	Ceti	5·9	12·9	M
EV	Lacertæ	8·2	10·2	M
AD	Leonis	9·0	9·5	M

R CORONÆ BOREALIS TYPE

Star		Max.	Min.	Spectrum
S	Apodis	9·6	15·2	R
UW	Centauri	9·6	16·0	K
R	Coronæ Borealis	5·5	15	Gp
RY	Sagittarii	6·0	14·0	Gp
SU	Tauri	9·5	16	G
RS	Telescopii	9·3	14·6	R

RV TAURI TYPE

Star		Max.	Min.	Period, d.	Spectrum
IW	Centauri	7·9	9·6	68	G + A
SX	Centauri	9·9	12·4	33	F5
SS	Geminorum	9·2	10·7	89	F8—G0
AC	Herculis	7·1	9·4	76	Fp
U	Monocerotis	6·8	8·5	92	G2—G8
R	Sagittæ	8·9	11·5	71	G0—K0
R	Scuti	5·7	8·6	144	G + K
RN	Tauri	9·8	13·3	79	G2—M2
V	Vulpeculæ	9·0	11·1	76	G2—K0

NEBULAR TYPE

Star		Max.	Min.	Spectrum
RW	Aurigæ	9·0	12·0	G5
AN	Orionis	10·7	11·7	K
T	Tauri	9·5	13·0	G
RY	Tauri	8·6	11·0	G

RR LYRÆ TYPE

Star		Max.	Min.	Period, d.	Spectrum
SW	Andromedæ	9·3	10·3	0·4423	A3—F8
RS	Boötis	9·7	11·2	0·3774	B8—F0
UY	Cygni	10·4	11·4	0·5607	A5—F0
RR	Lyræ	7·1	8·0	0·5668	A2—F0

Examples of typical Mira stars, Cepheids and semi-regular variables are, of course, included in the list of bright variables (page 156).

The following list includes all novæ since 1600 to have attained magnitude 6·0 or brighter. The list does not include supernovæ, recurrent novæ, or η Carinæ, which is sometimes to be found in lists of novæ even though it is unique.

Nova	Year	Constellation	Max. Mag.	Discoverer
CK Vul	1670	Vulpecula	3	Anthelm
WY Sge	1783	Sagitta	6	D'Agelet
V.841 Oph	1848	Ophiuchus	4	Hind
Q Cyg	1876	Cygnus	3	Schmidt
T Aur	1891	Auriga	4·2	Anderson
V.1059 Sgr	1898	Sagittarius	4·9	Fleming
GK Per	1901	Perseus	0·0	Anderson
DM Gem	1903	Gemini	5·0	Turner
OY Ara	1910	Ara	6·0	Fleming
DI Lac	1910	Lacerta	4·6	Espin
DN Gem	1912	Gemini	3·3	Enebo
V.603 Aql	1918	Aquila	−1·1	Bower
GI Mon	1918	Monoceros	5·7	Wolf
V.476 Cyg	1920	Cygnus	2·0	Denning
RR Pic	1925	Pictor	1·1	Watson
XX Tau	1927	Taurus	6·0	Schwassman & Wachmann
DQ Her	1934	Hercules	1·2	Prentice
CP Lac	1936	Lacerta	1·9	Gomi
V.630 Sgr	1936	Sagittarius	4·5	Okabayasi
BT Mon	1939	Monoceros	4·3	Whipple & Wachmann
CP Pup	1942	Puppis	0·4	Dawson
DK Lac	1950	Lacerta	6·0	Bertaud
V.446 Her	1960	Hercules	5·0	Hassell
V.533 Her	1963	Hercules	3·2	Dahlgren & Peltier
HR Del	1967	Delphinus	3·7	Alcock
LV Vul	1968	Vulpecula	4·9	Alcock
FH Ser	1970	Serpens	4·4	Honda
V.1500 Cyg	1975	Cygnus	1·8	Honda

GALACTIC SUPERNOVÆ

Four definite galactic supernovæ have been observed in historical times: the stars of 1006, 1054, 1572 and 1604. Several other 'possibilities' have been listed by R. Stephenson (Newcastle University). The list is as follows:

1006	Supernova in Lupus. Max. mag. about		−9·5
1054	,,	Taurus. ,,	−4
1572	,,	Cassiopeia. ,,	−4
1604	,,	Ophiuchus. ,,	−3

185 December   Discovered 7 December, near α and β Centauri; Chinese sources. Maximum magnitude about −8. Possible identification with radio source G.315.4–2.3 and optical object RCW 86 (also an X-ray emitter). Probably observed by the Chinese for 20 months.

386   In Sagittarius. Chinese sources; visibility extended over 3 months. Several radio sources lie in the general area (near λ Sagittarii) but positive identification is impossible. This star may or may not have been a supernova.

393 February/March   In Scorpio, near the tail (λ Scorpionis). Visible for about 8 months according to Chinese sources. Maximum magnitude about 0. Possible identification with radio source

	G.348.5+0.1 or G.348.7+0.3; distance ±20 000 light-years. Probably a supernova.
1006 April	In Lupus, near β. Maximum magnitude —9 to —10; according to Chinese and other sources, visible for 2 years. Definitely a supernova; probable identity with radio source G.327.6+14.5. Distance not over 3000 light-years.
1054 July	In Taurus, near ζ. Chinese and Japanese sources. Maximum magnitude about the same as that of Venus (—4 to —4·5). Seen for 23 days in daylight and 21 months at night. Definite identification with the Crab Nebula; distance 6000 light-years.
1181 August	In Cassiopeia, between ε and ι. Chinese and Japanese sources. Possible identification with radio source G.130.7+3.1. Distance about 25 000 light-years? Possible supernova; evidence by no means conclusive.
1572 November	In Cassiopeia, near κ. Tycho's Star. Seen by W. Schüler of Wittenberg on 6 November, but may have been seen on 3 November (Tycho saw it on 11 November). Maximum magnitude slightly brighter than —4. Visible for 18 months. Identity with radio source G.120.1+1.4. The remnant is optically visible, and is an X-ray emitter. Distance about 20 000 light-years.
1604 October	In Ophiuchus, near λ. Kepler's Star, discovered on 9 October; Kepler saw it on 11 October. (There may have been a first sighting on 7 October.) Maximum magnitude —2·5. Identified with radio source G.4.5+6.8.

The brightest of these galactic supernovæ was certainly the star of 1006. There has been another supernova about 1700 which has left the remnant which we know as the radio source Cassiopeia A, but the supernova was not observed, as it was too heavily obscured by interstellar material near the plane of the Galaxy.

Many supernovæ have been seen in external galaxies; the only one to have reached the fringe of naked-eye visibility was S Andromedæ (1885) in Messier 31. Only one supernova in an external galaxy has been discovered visually. This stands to the credit of the South African amateur Jack Bennett, of Pretoria, who was actually hunting for comets when he made the discovery.

STAR CLUSTERS AND NEBULÆ

Clusters and nebulæ are among the most striking of stellar objects. Several are easily visible with the naked eye. Few people can fail to recognize the lovely star-cluster of the Pleiades or Seven Sisters, which has been known since prehistoric times and about which there are many old legends. The nebula in the Sword of Orion, the Sword-Handle in Persèus, and Præsepe in Cancer are other objects visible without optical aid; keen-sighted people have little difficulty in locating the great Andromeda Spiral and the globular cluster in Hercules, while in the far south there are the two Clouds of Magellan, which cannot possibly be overlooked, as well as the bright globulars ω Centauri and 47 Tucanæ.

The most famous of all catalogues of nebulous objects was compiled by the French astronomer Charles Messier, and published in 1781. Ironically, Messier was not interested in the objects he listed; he was a comet-hunter, and merely wanted a quick means of identifying misty patches which were non-cometary in nature! In 1888 J. L. E. Dreyer, Danish by birth (though he spent much of his life in Ireland, and ended it in England), published his New General Catalogue (N.G.C.), augmented in 1895 and again in 1908 by his Index Catalogue (I. or I.C.).

The brightest nebular object is the Large Cloud of Magellan, which may be seen even in moonlight. Next comes the Small Cloud of Magellan. The brightest cluster visible from England is the cluster of the Pleiades. (This claim may be disputed; the Hyades are also bright, but are somewhat drowned by the brilliant orange-red light of Aldebaran, which is not a bona-fide member of the cluster, but merely happens to lie roughly half-way between the Hyades and ourselves.)

Nebular objects may be divided into various types, all of which are represented in Messier's Catalogue and the N.G.C.:

Open or Loose Clusters. Aggregations of stars, with no particular shape. The Pleiades, the Hyades and Præsepe are good examples. These three (and many others!) are contained in our Galaxy, but other galaxies also include open clusters.

Globular Clusters. Symmetrical systems, which may contain up to a million stars, strongly concentrated toward the centre. In our Galaxy, the globulars lie around the edge of the main system, so that all are very remote. Leading examples are ω Centauri, 47 Tucanæ and M.13 Herculis. Again, other galaxies have globular clusters of their own.

Gaseous nebulæ, also termed **galactic nebulæ.** These may be described as 'stellar birth-places', since fresh stars are condensing out of the nebular gas and dust. The best known example is M.42, the Sword of Orion. They are immensely rarefied; D. A. Allen has pointed out that if it were possible to take a 1-in core sample right through the Orion Nebula, which is of the order of 30 light-years in diameter, the total 'weight' of material collected would weigh no more than one new British penny. Some nebulæ (such as that in the Pleiades) shine by reflecting the light of stars in or very near them; others (such as the Orion Nebula) shine also because the very hot stars embedded in them ionize the hydrogen, and cause it to emit light on its own account. For this reason, bright emission nebulæ are termed H.II regions. If a nebula has no suitable stars in or near it to cause luminosity, it remains dark, and is detectable only because it blots out the light of stars beyond. The best-known dark nebula is the Coal Sack in Crux. The first exhaustive catalogue of dark nebulæ was published by E. E. Barnard in 1919. There is no basic difference between a bright nebula and a dark one; and indeed a nebula which appears bright to us could well seem dark if viewed from elsewhere in the Galaxy. We cannot see the objects deep inside nebulæ, but they can be detected by their infra-red radiations; thus the Orion Nebula contains the so-called Becklin's Object, which may well be a remarkably powerful star permanently concealed from us.

Supernova remnants are also known. The classic example is M.1 Tauri (the Crab Nebula), identified with the supernova of 1054. This was first seen by Messier in August 1758; he had not known of its existence, though it had been recorded by J. Bevis in 1731 – and the observation led directly on to Messier's resolve to draw up his famous Catalogue.

Planetary nebulæ are neither planets nor nebulæ, but small, hot stars which are at an advanced stage of evolution, and are surrounded by shells of tenuous gas. The most famous example (though not actually the brightest) is M.57, the Ring Nebula in Lyra. The brightest planetary is NGC 7293, in Aquarius, which was not included in Messier's list. The faintest planetary in Messier's

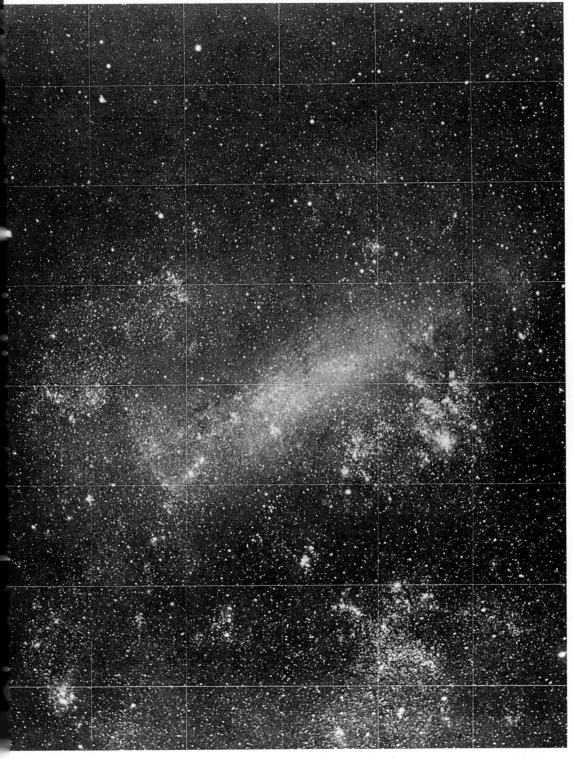

The Large Cloud of Magellan is the brightest nebular object. At only 160 light-years, it is a very near member of the Local Group

M.74, a spiral galaxy in Pisces, photographed by E. Hubble with the 60 in reflector at Mount Wilson. This is one of the fainter Messier objects, and it is not too easy to find with a small telescope. It is of type Sc

catalogue is M.76 Persei, magnitude 12·2. (This is in fact the faintest of all Messier objects.)

Galaxies are external systems, beyond the limits of the Galaxy in which we live. Our Galaxy is often termed the Milky Way, though this name is better restricted to denoting the luminous band seen stretching across the night sky. Galaxies are of various types. In 1925 the American astronomer E. E. Hubble drew up a simple system of classification by shape, which is still widely used even though many more complex classifications have been proposed. Hubble's classes are as follows:

Spirals, resembling Catherine-wheels. Sa: conspicuous, often tightly-wound arms issuing from a well-defined nucleus. Sb: arms looser, nucleus less condensed. Sc: inconspicuous nucleus, loose arms. Our Galaxy belongs to type Sb.

Barred spirals, in which the spiral arms issue from the ends of a kind of 'bar' through the nucleus. They are divided into types SBa, SBb and SBc in the same way as for the normal spirals.

Ellipticals, with no indications of spirality; they range from E7 (highly flattened) down to E0 (virtually spherical, looking superficially like globular clusters even though there is a great difference in mass).

Irregular galaxies, with no definite form.

It is thought that among the galaxies approximately 30 per cent are spiral, 60 per cent elliptical and 10 per cent irregular, though these values are very approximate.

The first spiral form was identified in 1845 by the third Earl of Rosse. Using his newly-completed 72-in. reflector at Birr Castle in Ireland, he saw the spiral shape of M.51 in Canes Venatici, now often nicknamed the Whirlpool. Others were subsequently found, though for some years the 72-in. was the only telescope of aperture large enough to show the spiral forms clearly. Our Galaxy is an Sb spiral. It has an overall diameter of about 100 000 light-years (though some astronomers believe this to be an over-estimate), and contains approximately 100 000 million stars. The central bulge surrounding the galactic nucleus has a thickness of about 20 000 light-years. The Sun lies about 32 000 light-years from the centre of the Galaxy, on the inner edge of one of the spiral arms (known as the Carina-Cygnus arm). The Galaxy is in rotation; the Sun takes about 225 000 000 years to complete one circuit – a period sometimes termed the 'cosmic year', moving at about 250 km/s.

Other terms are also in use. **Seyfert galaxies,** first identified by Carl Seyfert in 1943, have bright, almost stellar nuclei which show variations in brightness; a good example is M.77 Ceti, the most massive galaxy in Messier's list. Many Seyferts are thought to be highly active. Some emit radio waves, infra-red and even X-rays which are detectable from Earth. The so-called N-galaxies also have bright nuclei, and may be closely associated with the Seyferts. We also know of **exploding galaxies,** such as M.82 in Ursa Major, which is a strong radio source. Hydrogen structures in M.82 appear to be moving at over 900 km/s, so that a vast outburst presumably occurred there in the remote past – about 1·5 million years before our view of it, though M.82 is approximately 8·5 million light-years away.

It is not now generally thought that spiral galaxies turn into ellipticals or vice versa, so that Hubble's classification does not indicate an evolutionary sequence. Ellipticals consist mainly of Population II objects, so that the most brilliant stars are old red giants, while Population I ob-

jects are dominant in the spirals, though the populations are always to some extent mixed, and no hard and fast boundaries can be drawn. Certainly the ellipticals contain relatively little interstellar matter – whereas there is plenty of it in the spirals, where star formation is still going on.

The first positive proof of interstellar matter in our Galaxy was obtained by Hartmann in 1904, when he was studying the spectrum of the star δ Orionis. δ Orionis is a spectroscopic binary, so that its lines show Doppler shifts corresponding to the orbital motion. However, some of the lines remained stationary, so that clearly they were associated not with the star itself, but with material between δ Orionis and ourselves. Further indications were obtained by J. Trümpler in 1930, because some of the Milky Way clusters appeared fainter than logically they should have done, and were presumably being dimmed. The gas between the stars is mainly made up of hydrogen and helium, with smaller amounts of carbon, nitrogen,

Lord Rosse's drawing of the Whirlpool Galaxy made in 1845, the first recognition of the spiral structure of a galaxy. The photograph reproduced in the frontispiece demonstrates the accuracy of Rosse's observation

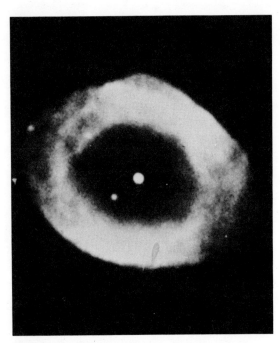

The Ring Nebula (M. 57), which is not actually a nebula but a small, hot star surrounded by a shell of tenuous gas

However, the average density of interstellar gas in our Galaxy is very low – about 500 000 hydrogen atoms per cubic metre, which is much lower than any laboratory vacuum. There are also dust grains, about 0·1 micron (10^{-7} metres) in radius, which are chiefly responsible for the reddening and dimming of distant objects. We cannot see through to the galactic centre, though as long ago as 1927 J. H. Oort established that the nucleus lies beyond the glorious star-clouds in Sagittarius. Dust is thought to make up about 1 per cent of the total mass of the Galaxy, though this figure is very arbitrary.

More molecules are being detected yearly. Also, in 1974 M. Cohen, who was engaged in studying the T Tauri variable HL Tauri, detected associated ice grains. HL Tauri is thought to be a very young star – perhaps no more than 300 000 years old.

As yet we have little information about intergalactic matter, but presumably it must exist, and certainly there seem to be streamers of tenuous gas linking the Magellanic Clouds with our Galaxy and with each other.

The first proof that the 'resolvable nebulæ' are independent galaxies was obtained by Hubble in 1923, when he was able to measure the periods of Cepheids in the Andromeda Spiral, M.31, and hence to find their distances. The idea was not new. It had been suggested as early as 1755 by Immanuel Kant, and had at one time been supported by Sir William Herschel. However, it had been very much of an open question, and was the subject of a great debate in 1920 between two American astronomers, Harlow Shapley and Heber D. Curtis. By studying short-period variables in the globular clusters, Shapley, in 1918, had been able to give the first reasonably accurate measurement of the size of our Galaxy, but he regarded the spirals as being contained in the Galaxy, whereas Curtis believed them to be external. Hubble's work, carried out with the Mount Wilson 100-in reflector (then the only telescope in the world of sufficient power) proved that Curtis had been right. It was thought that the Andromeda Spiral must be 900 000 light-years away, a figure subsequently modified to 750 000 light-years. In 1952 W. Baade found that there had been an error in the Cepheid scale; it was found that there were two kinds of Cepheids, one more luminous than the other, and those in the Spiral had been wrongly identified. They were more powerful than had been believed, and hence more remote. The distance of the Andromeda Spiral is

oxygen and neon. In the 1930s astronomers at Mount Wilson found the first indications of interstellar molecules, but firm proof was postponed until 1963, when the hydroxyl radical, OH, was identified.

In 1968 C. Townes and his colleagues at Berkeley detected interstellar ammonia (NH_3) at a wavelength of 1·26 cm. The first organic molecules (that is to say, molecules containing carbon) were identified in 1969 by L. Snyder, D. Buhl, B. Zuckerman and P. Palmer, who detected formaldehyde (H_2CO). Others have since been found, including carbon monoxide, hydrogen cyanide and methanoic acid. Ethyl alcohol (CH_3CH_2OH) has also been detected. One 'cloud' contains enough of it to fill the entire globe of the Earth with alcohol!

Of supreme importance was the detection, in 1951, of cold hydrogen clouds emitting radiation at a wavelength of 21·1 cm. This had been predicted in 1945 by the Dutch astronomer H. C. van de Hulst; the 1951 identification was made by Ewen and Purcell in America. Studies of the 21-cm line have enabled astronomers to map the spiral arms of the Galaxy, since these are the regions in which the hydrogen clouds lie.

Galaxy NGC 4565. This is a spiral system, but almost edge-on to us. The dark line of obscuring matter along the main plane is clearly shown.

now known to be 2 200 000 light-years. It is **the most remote object clearly visible with the naked eye.** It has been reported that the Triangulum Spiral, M.33, has been seen without optical aid, but this needs exceptional eyesight. The only external systems brighter than these two spirals are, of course, the Magellanic Clouds.

The first indication of an expanding universe was given by V. M. Slipher, who from 1912 to 1925 found that almost all the galaxies showed red shifts in their spectra, indicating recession. However, the significance of this was not immediately realized, since when Slipher began his work – and for more than a decade after – it was by no means certain that the 'resolvable nebulæ' were external galaxies, and men of the calibre of Shapley believed otherwise. Subsequently, Hubble showed a relationship between distance and velocity of recession; the further away the galaxy, the greater the recessional velocity. The value of what is termed Hubble's Constant is critical in any discussion of cosmology. One favoured figure is 55 km/s^{-1}M^{-1} (M = megaparsecs; one megaparsec = 1 million parsecs).

Assuming that the spectral red shifts are true Doppler effects, the only galaxies not receding from us are the members of the **Local Group.** This consists of three or four large systems and more than two dozen dwarf galaxies. The whole Local Group is at least 5 000 000 light-years across.

The largest galaxy in the Local Group is M.31, the Andromeda Spiral, which is 1·5 times the size of our Galaxy. It contains over 300 globular clusters, and objects of all kinds (one supernova has been seen – S Andromedæ of 1885 – and numerous normal novæ). Second in size comes our Galaxy – possibly excluding Maffei 1, to be described below. Third is the Triangulum Spiral, M.33, half the size of our Galaxy and with 10 000 million stars. The Large Magellanic Cloud is a quarter the size of our Galaxy, and the Small Cloud one-sixth the size; both are classified as irregular, though it has been suggested that the Large Cloud shows traces of spirality.

In 1968 the Italian astronomer Maffei discovered two galaxies, only about half a degree from the plane of our Galaxy and therefore heavily obscured by interstellar matter. It now seems that the first of these, Maffei 1, is about 3 000 000

light-years away, and is a member of our Local Group; it appears to be a giant elliptical, but since we receive only about one-hundredth of its light our information about it is scanty. Maffei 2 was orginally thought to be in the Local Group, but recent work indicates that it is well beyond, at a distance of about 15 000 000 light-years.

In 1977 a new dwarf elliptical galaxy in the Local Group was found on a plate taken with the 48 in Schmidt telescope at Siding Spring, Australia. The object lies in Carina, and there is a distinct chance that it is associated with the Magellanic Clouds.

The other Local Group members are dwarfs, mainly less than 5000 light-years in diameter. Some of them, such as the Ursa Minor, Draco, Sculptor, Fornax, and Leo I and Leo II systems are well within a million light-years of us, but are so sparse that they are inconspicuous. No doubt many more of these dwarfs exist, concealed by interstellar matter in the region of the galactic plane. **The smallest known galaxy is GR-8**, a dwarf irregular, a mere 1000 light-years in diameter. **The least massive Messier object is** M.32.

It is also worth noting that one globular cluster, NGC 5694, seems to be moving in a way that indicates its escape from our Galaxy. If so, it will become what is termed an 'intergalactic tramp'!

The following are some of the members of the Local Group:

Name	Distance, thousands of light-years
The Galaxy	—
Large Magellanic Cloud	160
Small Magellanic Cloud	190
Ursa Minor	250
Draco	260
Sculptor	280
Fornax	430
Carina	±500
Leo I	750
Leo II	750
NGC 6722	1700
M.31	2200
M.32	2200
NGC 205	2200
M.33	2400
Maffei 1	3000

Other clusters of galaxies are known, some of them much more populous than the Local Group. The most famous is the Virgo Cluster, which lies at a mean distance of 65 000 000 light-years and contains over 1000 members – notably the giant elliptical M.87, which is a strong radio source. According to E. Holmberg, its mass is as great as 790 000 million Suns, and the diameter is almost 40 kiloparsecs. Even more populous is the Coma cluster, 450 000 000 light-years away; the diameter is at least 10 million light-years, and there are two huge elliptical systems considerably larger than our Galaxy. Their combined absolute magnitude is about —24, as against —21 for our Galaxy.

For some years the remote radio galaxy 3C–295, in Boötes, was the most distant object known (5000 million light-years). It has been suggested that the distance of the galaxy 3C–153 is 8000 million light-years. However, this is exceeded by the strange quasars, which have diameters much smaller than those of conventional galaxies, but which appear to be almost incredibly powerful.

The first quasar to be identified was 3C–273, in Virgo. (The prefix 3C indicates the third Cambridge catalogue of radio sources, published in 1962.) 3C–273 was known to be a strong radio emitter, but identifying it with a visual object proved to be difficult. On 5 August 1962, however, radio astronomers in Australia, working at the Parkes observatory, followed an occultation of the radio source by the Moon, and were able to fix its position very exactly. From this result, the source was identified with what seemed to be a faint bluish star of magnitude 12·8. In 1963 M. Schmidt, at Palomar in California, obtained an optical spectrum, and found that the object was not a star at all; its spectrum was quite different, and showed hydrogen lines which were tremendously red-shifted, indicating a high recessional velocity and therefore immense distance and luminosity. Other quasars were quickly found, together with quasar-like objects which were not radio emitters.

Some quasars are much more remote than any known galaxies. In 1973, for instance, E. J. Wampler found that the quasar OQ 172 has a recessional velocity of 91 per cent of the speed of light. (OQ indicates the Ohio quasar survey.) In 1977 it was announced that Australian researchers at the Siding Spring Observatory had detected a quasar receding at 98 per cent of the velocity of light. This would have indicated a distance of 26 000 million light-years. However, it was subsequently found that there had been an error in interpretation.

The brightest quasar is 3C–273. No others exceed magnitude 16.

It is difficult to say which is **the remotest object known,** but we may be sure that the title belongs to a quasar – assuming that the red shifts are true Doppler effects, as is believed by most astronomers (though not all). If Hubble's law of distance and recessional velocity holds good out to these vast distances, it seems that radio observations are leading us out to almost the boundary of the observable universe, since an object receding at the full velocity of light would naturally be beyond our range.

Possibly associated with quasars are the BL Lacertæ objects, named after the brightest member of the class, identified as a radio source by Schmidt at Palomar in 1968. Not many are known, and most of them have featureless optical spectra; among early examples are AP Libræ and W Comæ, though some authorities have re-classified the latter as a quasar (it was originally thought to be a variable star within our own Galaxy!). In 1976 J. Wampler, at the Lick Observatory, was able to find some lines in the spectrum of the BL Lacertæ object PKS 0548–322, and measured the red shift. Subsequently R. Fosbury and M. Disney, using the 3·9 m Anglo-Australian telescope, obtained a spectrum of the 'fuzz' round the object, and found that it resembled the spectrum of an elliptical galaxy. They were able to show that the total luminosity was about the same as that of a faint quasar. However, both the BL Lacertæ objects and the quasars themselves remain highly enigmatical.

We cannot claim to know anything definite about the origin of the universe. The first modern-type theory – the 'big bang' – was proposed by the Belgian abbé Georges Lemaître in 1927; according to this theory all matter was created at one instant in a so-called primæval fireball, and expansion began, presumably to continue indefinitely. The Big Bang theory, elaborated by men such as George Gamow, was challenged in 1947 by H. Bondi and T. Gold, who proposed a 'steady-state' theory according to which the universe had no beginning and will have no end; as old galaxies die, new material in the form of hydrogen atoms is created spontaneously out of nothingness. There is also the oscillating universe theory, in which a phase of expansion is followed by a phase of contraction, and there is a new 'big bang' every 80 000 million years or so.

The steady-state theory has been abandoned by most authorities, since it does not fit the observed facts. It would mean that the distribution of galaxies would have been the same in the remote past as it is now (even though the galaxies themselves would be different from those which we know). Studying objects at distances of thousands of millions of light-years means, in effect, that we are looking back thousands of millions of years in time; and it seems that the distribution of remote radio sources is not the same as for objects closer to us, in which case the universe is not in a steady state.

We may even observe the remnants of the 'big bang'. In 1965 A. Penzias and R. Wilson, in America, were studying the sky with special radio equipment when they detected microwave radiation at 3·2 cm coming from all directions, and indicating a temperature of 3°K (3 degrees Kelvin, that is to say, three degrees above absolute zero). At Princeton University, R. Dicke and his team had actually been busy making a radio telescope to search for this very radiation. Dicke had calculated that the temperature of the universe would have been immensely high at the moment of creation – the 'big bang' – and would by now have dropped to an overall value of 3°K, which fitted in perfectly with the results by Penzias and Wilson.

Whether the overall expansion of the universe will continue indefinitely remains to be seen. Everything depends upon the amount of matter present. It has been calculated that the critical value is a mean of one hydrogen atom in 10^{23} cubic metres (about twice the volume of the Earth's globe). If there is less matter than this, the expansion cannot be stopped. Present results indicate that the density is lower than the critical value, so that expansion will continue, though the problem cannot be regarded as definitely solved.

Note, however, that no theory can explain how any matter came to be created in the first place, so that we are not genuinely discussing the **origin** of the universe; we are dealing with its **evolution**, which is a very different thing!

It seems that a few nebulous objects have been known since prehistoric times. This certainly applies to the Pleiades, which can hardly be overlooked. In his great work the *Almagest*, Ptolemy (circa AD 120–180) records the Pleiades, and also the Sword-Handle in Perseus (NGC 869 and 884), M.44 (Præsepe) and, with almost certain identification, the open cluster M.7 in Scorpio; probably Ptolemy himself discovered the last of these. We may also assume that ancient men of the southern hemisphere knew the Magellanic Clouds, though they did not come to the notice of European astronomers until about 1520.

From the year 1745 more and more objects were found, mainly by astronomers such as De Chéseaux, Legentil, Lacaille and, of course, the two great Frenchmen, Messier and Méchain. By 1781, when Messier published his Catalogue, 138 nebulous objects were known. The following list includes all the objects found before 1745:

		Discoverer
M.45	Pleiades	(Prehistoric)
NGC 869/884	Sword-Handle in Perseus	(Listed by Ptolemy)
M.44	Præsepe	(Listed by Ptolemy)
M.7	Open cluster in Scorpio	Ptolemy, c. AD 140
M.31	Andromeda Spiral	Al-Sûfi, c. 964
o Velorum	Open cluster	Al-Sûfi, c. 964
Large Cloud of Magellan		1519
Small Cloud of Magellan		1519
M.42	Great Nebula in Orion	N. Pieresc, 1610
M.22	Globular in Sagittarius	A. Ihle, 1665
	ω Centauri	Halley, 1677
M.8	Lagoon Nebula	Flamsteed, 1680
M.11	Wild Duck Cluster	G. Kirch, 1681
NGC 2244	12 Monocerotis	Flamsteed, 1690
M.41	Open cluster in Canis Major	Flamsteed, 1702
M.5	Globular in Serpens	G. Kirch, 1702
M.50	Open cluster in Monoceros	G. D. Cassini, c. 1711
M.13	Hercules Globular	Halley, 1714
M.43	Part of Orion Nebula	de Mairan, c. 1731
M.1	Crab Nebula	J. Bevis, 1731

This makes in all 21 objects. Many others had been previously listed, but subsequently found to be mere groups of stars rather than true clusters or nebulæ. One remarkable fact relates to the Great Spiral in Andromeda, which was recorded by the Persian astronomer Al-Sûfi. It was not noted again until 1612, when Simon Marius described it. Amazingly, it was completely overlooked by the greatest observer of pre-telescopic times, Tycho Brahe.

Messier's original catalogue ends with M.103. The later numbers are objects which, apart from the last, were discovered by Méchain (M.110 may have been discovered by Messier himself, but it is still more generally known as NGC 205, one of the companions to the Great Spiral in Andromeda). M.104 was added to the Messier catalogue by Camille Flammarion in 1921, on the basis of finding a handwritten note about it in Messier's own copy of his 1781 catalogue. M.105 to 107 were listed by H. S. Hogg in the 1947 edition of the catalogue; M.108 and 109 by Owen Gingerich in 1960. The naming of NCG 205 as M.110 was suggested by K. G. Jones in 1968, but does not seem to have received general acceptance.

M.	NGC	Constellation	Type	
1	1952	Taurus	Supernova remnant	Crab Nebula
2	7089	Aquarius	Globular	
3	5272	Canes Venatici	Globular	
4	6121	Scorpio	Globular	
5	5904	Serpens	Globular	
6	6405	Scorpio	Open cluster	
7	6475	Scorpio	Open cluster	
8	6523	Sagittarius	Nebula	Lagoon Nebula
9	6333	Ophiuchus	Globular	
10	6254	Ophiuchus	Globular	
11	6705	Scutum	Open cluster	Wild Duck cluster
12	6218	Ophiuchus	Globular	
13	6205	Hercules	Globular	Great Hercules cluster
14	6402	Ophiuchus	Globular	
15	7078	Pegasus	Globular	
16	6611	Serpens	Nebula and embedded cluster	
17	6618	Sagittarius	Nebula	Omega or Horseshoe Nebula
18	6613	Sagittarius	Open cluster	
19	6273	Ophiuchus	Globular	
20	6514	Sagittarius	Nebula	Trifid Nebula
21	6531	Sagittarius	Open cluster	
22	6656	Sagittarius	Globular	
23	6494	Sagittarius	Open cluster	
24	6603	Sagittarius	Open cluster	
25	I.4725	Sagittarius	Open cluster	
26	6694	Scutum	Open cluster	
27	6853	Vulpecula	Planetary	Dumbbell Nebula
28	6626	Sagittarius	Globular	
29	6913	Cygnus	Open cluster	
30	7099	Capricornus	Globular	
31	224	Andromeda	Spiral galaxy	Great Spiral
32	221	Andromeda	Elliptical galaxy	Companion to M.31
33	598	Triangulum	Spiral galaxy	Triangulum Spiral
34	1039	Perseus	Open cluster	
35	2168	Gemini	Open cluster	
36	1960	Auriga	Open cluster	
37	2099	Auriga	Open cluster	
38	1912	Auriga	Open cluster	
39	7092	Cygnus	Open cluster	
40	—	—	Missing. Possibly a comet?	
41	2287	Canis Major	Open cluster	
42	1976	Orion	Nebula	Great Nebula
43	1982	Orion	Nebula	Part of Orion Nebula
44	2632	Cancer	Open cluster	Præsepe
45	—	Taurus	Open cluster	Pleiades
46	2437	Puppis	Open cluster	
47*	2422	Puppis	Open cluster	
48*	2548	Hydra	Open cluster	
49	4472	Virgo	Elliptical galaxy	
50	2323	Monoceros	Open cluster	
51	5194	Canes Venatici	Spiral galaxy	Whirlpool Galaxy
52	7654	Cassiopeia	Open cluster	
53	5024	Coma Berenices	Globular	
54	6715	Sagittarius	Globular	
55	6809	Sagittarius	Globular	
56	6779	Lyra	Globular	
57	6720	Lyra	Planetary	Ring Nebula
58	4579	Virgo	Spiral galaxy	
59	4621	Virgo	Elliptical galaxy	
60	4649	Virgo	Elliptical galaxy	
61	4303	Virgo	Spiral galaxy	
62	6266	Ophiuchus	Globular	
63	5055	Canes Venatici	Spiral galaxy	
64	4826	Coma Berenices	Spiral galaxy	
65	3623	Leo	Spiral galaxy	
66	3627	Leo	Spiral galaxy	
67	2682	Cancer	Open cluster	Famous old cluster
68	4590	Hydra	Globular	
69	6637	Sagittarius	Globular	
70	6681	Sagittarius	Globular	
71	6838	Sagitta	Globular	
72	6981	Aquarius	Globular	
73	6994	Aquarius	Four faint stars (Not a cluster)	
74	628	Pisces	Spiral galaxy	
75	6864	Sagittarius	Globular	
76	650	Perseus	Planetary	

M.	NGC	Constellation	Type	
77	1068	Cetus	Spiral galaxy	
78	2068	Orion	Nebula	
79	1904	Lepus	Globular	
80	6093	Scorpio	Globular	
81	3031	Ursa Major	Spiral galaxy	
82	3034	Ursa Major	Irregular galaxy	Companion to M.31
83	5236	Hydra	Spiral galaxy	
84	4374	Virgo	Spiral galaxy	
85	4382	Coma Berenices	Spiral galaxy	
86	4406	Virgo	Elliptical galaxy	
87	4486	Virgo	Elliptical galaxy	Seyfert galaxy
88	4501	Coma Berenices	Spiral galaxy	
89	4552	Virgo	Elliptical galaxy	
90	4569	Virgo	Spiral galaxy	
91	—	—	Not identified. Possibly a comet?	
92	6341	Hercules	Globular	
93	2447	Puppis	Open cluster	
94	4736	Canes Venatici	Spiral galaxy	
95	3351	Leo	Barred spiral galaxy	
96	3368	Leo	Spiral galaxy	
97	3587	Ursa Major	Planetary	Owl Nebula
98	4192	Coma Berenices	Spiral galaxy	
99	4254	Coma Berenices	Spiral galaxy	
100	4321	Coma Berenices	Spiral galaxy	
101	5457	Ursa Major	Spiral galaxy	
102	—	—	Missing. Possibly a faint spiral in Draco, or else identical with M.101.	
103	581	Cassiopeia	Open cluster	
104	4594	Virgo	Spiral galaxy	Sombrero Hat Galaxy
105	3379	Leo	Elliptical galaxy	
106	4258	Ursa Major	Spiral galaxy	
107	6171	Ophiuchus	Globular	
108	3556	Ursa Major	Spiral galaxy	
109	3992	Ursa Major	Spiral galaxy	
110	205	Andromeda	Elliptical galaxy	Companion to M.31

THE CONSTELLATIONS

Cetus from John Flamsteed's Atlas Cœlestis published in London in 1729 (photograph by John R. Freeman & Co.)

THE CONSTELLATIONS

		1st mag	Stars to mag : 2.00	4.00	5.00	Area sq. deg.	Number of stars above mag. 5 per 100 square degrees (star density)
Andromeda	Andromeda	—	0	7	25	272	9·19
Antlia	The Airpump	—	0	0	4	239	1·67
Apus	The Bird of Paradise	—	0	2	6	206	2·91
Aquarius	The Water-bearer	—	0	7	31	980	3·16
Aquila	The Eagle	Altair	1	8	16	652	2·45
Ara	The Altar	—	0	7	10	237	4·21
Aries	The Ram	—	1	4	11	441	2·49
Auriga	The Charioteer	Capella	2	7	21	657	3·20
Boötes	The Herdsman	Arcturus	1	8	24	907	2·65
Cælum	The Graving Tool	—	0	0	2	125	1·60
Camelopardus	The Giraffe	—	0	0	11	757	1·45
Cancer	The Crab	—	0	1	6	506	1·19
Canes Venatici	The Hunting Dogs	—	0	1	7	465	1·51
Canis Major	The Great Dog	Sirius	4	10	26	380	6·84
Canis Minor	The Little Dog	Procyon	1	2	4	183	2·19
Capricornus	The Sea-Goat	—	0	5	16	414	3·86
Carina	The Keel	Canopus	3	14	40	494	8·10
Cassiopeia	Cassiopeia	—	0	7	23	598	3·85
Centaurus	The Centaur	Alpha Cen., Agena	2	14	49	1060	4·62
Cepheus	Cepheus	—	0	8	20	588	3·40
Cetus	The Whale	—	0	8	24	1232	1·95
Chamæleon	The Chameleon	—	0	0	6	132	4·55
Circinus	The Compasses	—	0	1	4	93	4·30
Columba	The Dove	—	0	4	9	270	3·33
Coma Berenices	Berenice's Hair	—	0	0	8	386	2·07
Corona Australis	The Southern Crown	—	0	0	7	128	5·47
Corona Borealis	The Northern Crown	—	0	3	10	179	5·59
Corvus	The Crow	—	0	5	6	184	3·26
Crater	The Cup	—	0	1	6	282	2·13
Crux Australis	The Southern Cross	Acrux, Beta Cru	3	5	13	68	19·12
Cygnus	The Swan	Deneb	1	11	43	804	5·35
Delphinus	The Dolphin	—	0	4	6	189	3·17
Dorado	The Swordfish	—	0	1	8	179	4·47
Draco	The Dragon	—	0	11	26	1083	2·40
Equuleus	The Foal	—	0	0	3	72	4·17
Eridanus	The River	Achernar	1	12	43	1138	3·78
Fornax	The Furnace	—	0	1	5	398	1·26
Gemini	The Twins	Pollux	3	13	23	514	4·47
Grus	The Crane	—	1	4	13	366	3·55
Hercules	Hercules	—	0	15	37	1225	3·02
Horologium	The Clock	—	0	1	2	249	0·80
Hydra	The Watersnake	—	1	9	32	1303	2·46
Hydrus	The Little Snake	—	0	3	9	243	3·70
Indus	The Indian	—	0	2	7	294	2·38
Lacerta	The Lizard	—	0	1	11	201	5·47
Leo	The Lion	Regulus	2	10	26	947	2·75
Leo Minor	The Little Lion	—	0	1	6	232	2·59
Lepus	The Hare	—	0	7	14	290	4·83
Libra	The Balance	—	0	5	13	538	2·42
Lupus	The Wolf	—	0	8	32	334	9·58

		1st mag	Stars to mag: 2.00	4.00	5.00	Area sq. deg.	Number of stars above mag. 5 per 100 square degrees (star density)
Lynx	The Lynx	—	0	2	12	545	2·20
Lyra	The Lyre	Vega	1	4	11	286	3·85
Mensa	The Table	—	0	0	0	153	0·00
Microscopium	The Microscope	—	0	0	4	210	1·90
Monoceros	The Unicorn	—	0	2	13	482	2·70
Musca Australis	The Southern Fly	—	0	4	11	138	7·97
Norma	The Rule	—	0	0	6	165	3·64
Octans	The Octant	—	0	1	4	291	1·37
Ophiuchus	The Serpent-bearer	—	0	12	36	948	3·80
Orion	Orion	Rigel, Betelgeux	5	15	42	594	7·07
Pavo	The Peacock	—	1	4	14	378	3·70
Pegasus	The Flying Horse	—	0	9	29	1121	2·59
Perseus	Perseus	—	1	10	34	615	5·52
Phœnix	The Phoenix	—	0	7	17	469	3·62
Pictor	The Painter	—	0	2	5	247	2·02
Pisces	The Fishes	—	0	3	24	889	2·70
Piscis Australis	The Southern Fish	Fomalhaut	1	1	7	245	2·86
Puppis	The Poop	—	0	11	42	673	6·24
Pyxis	The Compass	—	0	1	7	221	3·17
Reticulum	The Net	—	0	2	7	114	6·14
Sagitta	The Arrow	—	0	2	5	80	6·25
Sagittarius	The Archer	—	1	14	33	867	3·80
Scorpio	The Scorpion	Antares	3	17	38	497	7·64
Sculptor	The Sculptor	—	0	0	6	475	1·26
Scutum	The Shield	—	0	0	6	109	5·50
Serpens	The Serpent	—	0	9	17	637	2·67
Sextans	The Sextant	—	0	0	2	314	0·63
Taurus	The Bull	Aldebaran	2	14	44	797	5·52
Telescopium	The Telescope	—	0	1	4	252	1·59
Triangulum	The Triangle	—	0	2	3	132	2·27
Triangulum Australe	The Southern Triangle	—	1	3	6	110	5·45
Tucana	The Toucan	—	0	2	7	295	2·37
Ursa Major	The Great Bear	—	3	19	35	1280	2·73
Ursa Minor	The Little Bear	—	1	3	9	256	3·51
Vela	The Sails	—	2	10	30	500	6·00
Virgo	The Virgin	Spica	1	8	26	1294	2·01
Volans	The Flying Fish	—	0	3	7	141	4·96
Vulpecula	The Fox	—	0	0	10	268	3·73
Totals		21	50	455	1417		(Average: 3·8)

These counts do not include variable stars which can rise above the fifth magnitude, but whose average magnitude is below this limit. For instance, Mira Ceti is excluded even though its brightest maxima exceed magnitude 2.

FORMING THE CONSTELLATIONS

Ptolemy gave a list of 48 constellations: 21 northern, 12 Zodiacal and 15 southern, as follows:

Northern	Northern	Zodiacal	Southern
Ursa Minor	Aquila	Aries	Cetus
Ursa Major	Delphinus	Taurus	Orion
Draco	Equuleus	Gemini	Eridanus
Cepheus	Pegasus	Cancer	Lepus
Boötes	Andromeda	Leo	Canis Major
Corona Borealis	Triangulum	Virgo	Canis Minor
Hercules		Libra	Argo Navis
Lyra		Scorpio	Hydra
Cygnus		Sagittarius	Crater
Cassiopeia		Capricornus	Corvus
Perseus		Aquarius	Centaurus
Auriga		Pisces	Lupus
Ophiuchus			Ara
Serpens			Corona Australis
Sagitta			Piscis Australis

All these are still to be found in modern maps, though the huge, unwieldy Argo Navis has been divided up into Carina, Vela and Puppis.

Various constellations have been added since. With those that survive, the names have often been shortened; thus Piscis Volans, the Flying Fish, has become simply Volans, while Mons Mensæ, the Table Mountain, has become Mensa. There was also some confusion over two of Bayer's constellations, Apis (the Bee) and Avis Indica (the Bird of Paradise); modern maps give it as Apus. Surviving constellations were added as follows:

By Tycho Brahe, circa 1590:
Coma Berenices

By Bayer, 1603:
Pavo
Tucana
Grus
Phœnix
Dorado
Volans (originally Piscis Volans).
Hydrus
Chamæleon
Apus (originally Avis Indica)
Triangulum Australe
Indus

By Royer, 1679:
Columba (originally Columba Noachi, Noah's Dove)
Crux Australis

By Hevelius, 1690:
Camelopardus
Canes Venatici
Vulpecula (originally Vulpecula et Anser, the Fox and Goose)
Lacerta
Leo Minor
Lynx
Scutum Sobieskii
Monoceros
Sextans (originally Sextans Uraniæ, Urania's Sextant)

By La Caille, 1752:
Sculptor (originally Apparatus Sculptoris, the Sculptor's Apparatus)
Fornax (originally Fornax Chemica, the Chemical Furnace)
Horologium
Reticulum (originally Reticulus Rhomboidalis, the Rhomboidal Net)

Cælum (originally Cæla Sculptoris, the Sculptor's Tools)
Pictor (originally Equuleus Pictoris, the Painter's Easel)
Pyxis (originally Pyxis Nautica, the Mariner's Compass)
Antlia (originally Antlia Pneumatica, the Airpump)
Octans
Circinus
Norma (or Quadra Euclidis, Euclid's Square)
Telescopium
Microscopium
Mensa (originally Mons Mensæ, the Table Mountain)

The list of rejected constellations is very long. The most notable are as follows:

Proposed by Tycho Brahe, circa 1559:
Antinoüs

Proposed by Royer, 1679:
Nubes Major (the Great Cloud)
Nubes Minor (the Little Cloud)
Lilium (the Lily)

Proposed by Halley, circa 1680:
Robur Caroli (Charles' Oak)

Proposed by Flamsteed, circa 1700:
Mons Mænalus (the Mountain Mænalus)
Cor Caroli (Charles' Heart)

Proposed by Hevelius, 1690:
Triangulum Minor (the Little Triangle)
Cerberus

Proposed by Le Monnier, 1776:
Tarandus (the Reindeer)
Solitarus (the Solitaire)

Proposed by Lalande, 1776:
Messier

Proposed by Poczobut, 1777:
Taurus Poniatowski (Poniatowski's Bull)

Proposed by Hell, circa 1780:
Psalterium Georgianum (George's Lute).

Proposed by Bode, circa 1775:
Honores Frederici (the Honours of Frederick)
Sceptrum Brandenburgicum (the Sceptre of Brandenburg)
Telescopium Herschelii (Herschel's Telescope)

Globus Ærostaticus (the Balloon)
Quadrans Muralis (the Mural Quadrant)
Lochium Funis (the Log Line)
Machina Electrica (the Electrical Machine)
Officiana Typographica (the Printing Press)
Felis (the Cat).

There seem to have been two Muscas, one formed by La Caille to replace Bayer's Apis and the other (rejected) formed by Bode out of stars near Aries. One discarded group, Quadrans, has at least given its name to the Quadrantid meteor shower of early January.

LIST OF STARS OF MAGNITUDE 2·00 OR BRIGHTER

Stars above magnitude 1·4 are conventionally termed 'first magnitude' stars – that is to say, down the list as far as Regulus.

LIST OF STARS OF MAGNITUDE 2.00 OR BRIGHTER

Star			Apparent Magnitude	Absolute Magnitude	Luminosity Sun = 1	Spectrum	Distance Light years
1	Sirius	α Canis Majoris	−1·42	+1·4	26	A0	8·7
2	Canopus	α Carinæ	−0·72	−7·4?	80,000?	F0	650?
3		α Centauri	−0·27	+4·4	1·5	G4 + K5	4·3
4	Arcturus	α Boötis	−0·06	−0·3	115	K0	36
5	Vega	α Lyræ	0·04	+0·5	55	A0	27
6	Capella	α Aurigæ	0·05	−0·6	150	G0 + F	45
7	Rigel	β Orionis	0·08v	−7·1	60,000	B8p	900
8	Procyon	α Canis Minoris	0·37	+2·7	7	F5	11
9	Achernar	α Eridani	0·51	−2·3	720	B5	118
10	Betelgeux	α Orionis	var.	−5·6v	15,000	M2	520
11	Agena	β Centauri	0·63	−5·2	10,500	B1	490
12	Altair	α Aquilæ	0·77	+2·3	10	A5	16
13	Aldebaran	α Tauri	0·78v	−0·7	165	K5	68
14	Acrux	α Crucis	0·83	−3·9, −3·4	3200, 2000	B1 + B2	370
15	Antares	α Scorpionis	0·86v	−5·1	9600	M1 + A3	520
16	Spica	α Virginis	0·91v	−3·3	1800	B2	220
17	Pollux	β Geminorum	1·16	+1·0	34	K0	35
18	Fomalhaut	α Piscis Australis	1·19	+1·9	15	A3	23
19	Deneb	α Cygni	1·26	−7·1	60,000	A2p	1600
20		β Crucis	1·28	−4·6	6000	B1	490
21	Regulus	α Leonis	1·36	−0·7	170	B7	84
22	Adhara	ε Canis Majoris	1·48	−5·1	9600	B1	680
23	Castor	α Geminorum	1·58	+1·3, +2·3	26, 11	A0 + A5	45
24	Shaula	λ Scorpionis	1·60	−3·3	1800	B2	610
25	Bellatrix	γ Orionis	1·64	−4·2	4200	B7	470
26	Al Nath	β Tauri	1·65	−3·2	1700	A0	300
27	Miaplacidus	β Carinæ	1·67	−0·4	125	A0	86
28		γ Crucis	1·69	−2·5	870	M3	220
29	Alnilam	ε Orionis	1·70	−6·8	46,000	B0	1600
30	Alnair	α Gruis	1·76	+0·3	66	B5	64
31	Alnitak	ζ Orionis	1·79	−6·6	38,000	O9·5	1600
31	Alioth	ε Ursæ Majoris	1·79	+0·2	73	A0p	68
33	Mirphak	α Persei	1·80	−4·4	5000	F5	570
34	Dubhe	α Ursæ Majoris	1·81	−0·7	166	K0	105
34	Kaus Australis	ε Sagittarii	1·81	−1·1	240	A0	124
36	Wezea	δ Canis Majoris	1·85	−7·1	60,000	F8p	2100
37	Sargas	θ Scorpionis	1·86	−4·6	5500	F0	650
37	Menkarlina	β Aurigæ	1·86	−0·3	115	A0p	88
39	Alkaid	η Ursæ Majoris	1·87	−2·1	600	B3	210
40		γ Velorum	1·88	−4·1	3800	Oap	520
41	Atria	α Trianguli Australis	1·93	−0·1	96	K2	82
41	Alhena	γ Geminorum	1·93	−0·6	150	A0	105
43		α Pavonis	1·95	−2·9	1300	B3	310
43	Koo She	δ Velorum	1·95	+0·2	73	A0	76
45	Mirzam	β Canis Majoris	1·96v	−4·8	7200	B1	750
46	Avior	ε Carinæ	1·97	−3·1	1500	K0 + B	340
47	Alphard	α Hydræ	1·98	−0·3	115	K2	94
48	Algieba	γ Leonis	1·99	+0·1, +0·8	79, 42	K0 + G5	90
49	Polaris	α Ursæ Minoris	1·99v	−4·6	6000	F8	680
50	Hamal	α Arietis	2·00	+0·2	72	K2	76

Two other naked-eye stars can exceed magnitude 2. According to the Swedish astronomer Per Wargentin, Mira Ceti attained 1·2 at the maximum of 1772, though most maxima are below 2; the official maximum magnitude is usually given as 1·7. γ Cassiopeiæ is an irregular variable; it reached 1.6 in 1936, but at present (1978) is slightly below 2.

Excluding Canopus, for which estimates of distance and luminosity vary widely, the most luminous stars in this list are Rigel, Deneb and Wezea, each with about 60 000 times the luminosity of the Sun (though these values are naturally somewhat uncertain). The least luminous is α Centauri. The most remote star in the list is Wezea, at approximately 2100 light-years; the closest is α Centauri, at 4·3 light-years.

| — | Mira | o Ceti | var. | −0·5v. | 140v. | M5e | 103 |
| — | | γ Cassiopeiæ | var. | −0·3v. | 115v. | B0p | 96 |

CONSTELLATIONS AND STARS

The largest constellation is Hydra, with an area of 1303 square degrees. (Formerly it was exceeded by Argo Navis, but Argo has now been unceremoniously chopped up into a keel, sails and a poop.)

The smallest constellation is Crux Australis, with an area of 68 square degrees.

The constellation with the greatest number of stars above the second magnitude is Orion (5 stars).

The constellation with the greatest number of stars above the fourth magnitude is Ursa Major (19 stars).

The constellation with the greatest number of stars above the fifth magnitude is Centaurus (49 stars).

The only constellation with no star as bright as the fifth magnitude is Mensa. Next in order of 'dimness' come Cælum, Horologium and Sextans, each with only 2 stars above the fifth magnitude.

For stars above the fifth magnitude, the greatest 'star density' is Crux Australis, which averages out at 19·12 stars per 100 square degrees. Its nearest rival, Lupus, has a figure of only 9·58.

The least star density, excluding Mensa, is for Sextans (0·63 star per 100 square degrees).

The brightest star is Sirius (α Canis Majoris), magnitude —1·42.

The brightest star in the northern hemisphere of the sky is Arcturus (α Boötis), magnitude —0·06.

The only naked-eye star which is said to have a greenish tint is β Libræ – though most observers will certainly class it as white!

The reddest naked-eye star is probably μ Cephei, called by Herschel 'the Garnet Star', though optical aid is needed to bring out its colour well. The reddest first-magnitude stars are Antares, Betalgeux and Aldebaran. Among the reddest telescopic stars are R Leporis and U Cephei

The following data apply to the stars above magnitude 5·00, listed in the Catalogue which follows:

Most powerful stars, all about absolute magnitude —7:

ε Aurigæ, δ Canis Majoris, η Canis Majoris, o² Canis Majoris, ν Cephei, α Cygni (Deneb), β Orionis (Rigel), κ Orionis, ζ Puppis, ι¹ Scorpionis, b Velorum. α Carinæ (Canopus) is also very powerful, though published estimates of its absolute magnitude differ widely.

Least powerful stars are ε Indi (absolute magnitude +7·0), ε Eridani (+6·2) and o² Eridani (+6·1).

STAR CATALOGUE

The catalogue has been compiled from various sources; positions of the brightest stars (down to magnitude 4·75) are given for epoch 1977. Published values for absolute magnitudes and distances differ somewhat, and are subject to uncertainty; thus the distances given here have in general been 'rounded off', except for those stars whose distances are known with true precision. The principal double stars, variables, and nebular objects are also given, though it is by no means exhaustive and could be extended almost *ad infinitum*.

ANDROMEDA

(Abbreviation: And).

A large and important northern constellation; one of Ptolemy's 'originals'. It contains the Great Spiral, M.31. One of the leading stars in the constellation – Alpheratz or α Andromedæ – is actually a member of the Square of Pegasus, and was formerly known, more logically, as δ Pegasi.

In mythology, Andromeda was the beautiful daughter of King Cepheus and Queen Cassiopeia. Cassiopeia offended the sea-god Neptune by her boasting about Andromeda's beauty, which, she claimed, was greater than that of any sea-nymph. Neptune thereupon sent a sea-monster to ravage the kingdom, and the Oracle stated that the only solution was to chain Andromeda to a rock by the shore where she would be devoured by the monster. This was duly done, but the situation was redeemed by the hero Perseus, who was on his way home after killing the Gorgon, Medusa. Mounted upon his winged sandals, Perseus arrived in the nick of time, turned the monster to stone by the simple expedient of showing it Medusa's head, and then, in the best story-book tradition, married Andromeda. Cepheus, Cassiopeia and of course Perseus are to be found in the sky; the Gorgon's head is marked by the 'Demon Star' Algol, and the sea-monster has sometimes been identified with the constellation of Cetus.

Andromeda has no first-magnitude star, but there are seven above magnitude 4:

β	2·02	o	3·6v
α	2·06	51	3·77
γ	2·14	μ	3·94
δ	3·25		

BRIGHTEST STARS

Star	Name	Mag.	R.A. h m s			Dec. ° ' "			Abs. Mag.	Spec.	Dist. l/y
1 o		3·6v	23	00	54	+42	12	27	var	B6+A2	300
7		4·62	23	11	32	+49	17	09	+3·0	F0	65
16 λ		4·00	23	36	29	+46	20	20	+3·0	K0	85
17 ι		4·28	23	37	03	+43	08	46	−0·3	B8	270
19 κ		4·33	23	37	19	+44	12	13	+0·3	A0	200
21 α	Alpheratz	2·06	00	07	15	+28	48	08	−0·1	A0p	90
24 θ		4·44	00	15	56	+38	33	34	+1·2	A2	140
25 σ		4·51	00	17	10	+36	39	48	+0·5	A2	200
29 π		4·44	00	35	42	+33	35	54	−1·3	B3	450
30 ε		4·52	00	37	23	+29	11	23	+2·2	G6	90
31 δ		3·25	00	38	02	+30	43	17	−0·2	K2	160
35 ν		4·42	00	48	36	+40	57	34	−1·1	B3	400
37 μ		3·94	00	55	32	+38	22	49	+1·8	A2	85
38 η		4·62	00	56	02	+23	17	57	−0·6	G5	350
42 φ		4·28	01	08	13	+47	07	30	−1·5	B8	450
43 β	Mirach	2·02	01	08	30	+35	30	15	+0·2	M0	76
51		3·77	01	36	38	+48	31	02	+0·2	K0	170
57 γ	Almaak	2·14	02	02	33	+42	13	29	−2·4	K3	260

o is a β Lyræ type variable with a very small range (3·5—3·7) and a period of 1·6 days.

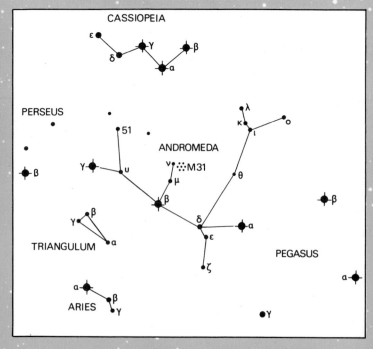

THE FOLLOWING STARS ARE ABOVE MAGNITUDE 5·00:

Star			Mag.	Abs. Mag.	Spectrum
3			4·91	+0·3	G8
8			4·99	+0·4	M0
46	ξ	Adhil	4·99	+1·5	G9
48	ω		4·96	+2·2	F2
53	τ		4·90	−0·6	B8
58			4·77	+1·6	A2
65			4·86	+0·1	K5

DOUBLES

Star	R.A. h m	Dec. ° ′	Mags.	P.A. Dist."	
π	00 35·7	+33 36	4·4, 8·7	174 36·1	Fixed.
γ	02 02·6	+42 13	2·3, 6·1	063 10·0	Orange, blue
γ²	02 02·6	+42 13	5·5, 6·3	Binary, 61 yrs; max sep 0"·6	

VARIABLES

Star	R.A. h m	Dec. ° ′	Range	Period, d.	Spec.	Type
SV	00 01·8	+39 49	7·7—14·3	316	M	Mira
T	00 19·8	+26 43	7·7—14·3	280	M	Mira
R	00 21·4	+38 18	5·9—14·9	409	S	Mira
W	02 14·3	+44 04	6·7—14·5	397	M	Mira
RW	00 44·6	+32 24	7·9—15·4	429	M	Mira

CLUSTERS AND NEBULÆ

Object		R.A. h m	Dec. ° ′	Type	Notes
NGC 7662 H.IV.18		23 23·4	+42 12	Planetary	Starlike; magnitude 14
I 7662		23 23·5	+42 14	Planetary	Mag. 8·9; central star mag. 12·5
NGC 7686		23 27·8	+48 51	Open cluster	Mag. 8·0
NGC 205		00 37·6	+41 25	Galaxy	Mag. 9. Type E6. Companion to M.31
NGC 224	M.31	00 40·0	+41 00	Galaxy	Great Spiral (type Sb). Naked-eye
NGC 221	M.32	00 40·0	+40 36	Galaxy	Mag. 8·7. Type E2 Companion to M.31
NGC 752		01 54·7	+37 25	Open cluster	Mag. 7·0. About 70 stars

ANTLIA

(Abbreviation: Ant).

A small southern constellation, covering 239 square degrees, added to the sky by Lacaille in 1752; its original name was Antlia Pneumatica. It contains no star brighter than magnitude 4·4, and no mythological legends are associated with it.

See chart for Carina

BRIGHTEST STARS

Star	Mag.	R.A. h m s	Dec. ° ′ ″	Abs. Mag.	Spec.	Dist.
ε	4·64	09 28 20	−35 51 17	−0·8	K2	400
α	4·42	10 26 09	−30 57 19	−0·6	K5	325
ι	4·70	10 55 41	−37 01 10	+0·4	K0	230

The only other star brighter than the 5th magnitude is θ ; mag. 4·98, absolute magnitude +2·8, spectrum F7p.

DOUBLE

Star	R.A. h m	Dec. ° ′	Mags.	P.A.	Dist.	
δ	10 27·3	−30 21	5·6, 9·7	226	11·0	Slow binary

VARIABLES

Star	R.A. h m	Dec. ° ′	Range	Period, d.	Spec.	Type
S	09 30·1	−28 24	6·4—6·8	0·64	A8	β Lyr type
U	10 35·1	−13 07	5·7—6·8	365	N	Semi-regular

NEBULÆ

Object	R.A. h m	Dec. ° ′	Type	Notes
NGC 2997	09 43·5	−30 58	Spiral galaxy	Mag. 11
NGC 3056	09 52·3	−28 04	Elliptical galaxy	Mag. 12·8
I 2522	09 53·1	−32 54	Spiral galaxy	Mag. 13
I 2537	10 01·7	−27 19	Spiral galaxy	Mag. 12·8
NGC 3132	10 04·9	−40 11	Planetary	Mag. 8·2; central star mag. 11
NGC 3175	10 12·4	−28 38	Spiral galaxy	Mag. 12
NGC 3223	10 19·4	−34 00	Spiral galaxy	Mag. 12
NGC 3281	10 29·7	−34 36	Spiral galaxy	Mag. 12·9
NGC 3347	10 40·5	−36 06	Spiral galaxy	Mag. 12.8

APUS

(Abbreviation: Aps).

Originally Avis Indica; it was introduced by Bayer in 1603. It lies in the far south, and has only two stars brighter than magnitude 4:

α 3·81 γ 3·90

See chart for Musca

BRIGHTEST STARS

Star	Mag.	R.A.			Dec.			Abs. Mag.	Spec.	Dist. l/y
		h	m	s	°	′	″			
α	3·81	14	45	04	−78	57	11	−0·5	K5	230
δ	4·2	16	12	48	−78	34	26	+0·7	M5	200
γ	3·90	16	30	01	−78	51	02	+1·4	K0	100
β	4·16	16	39	54	−77	28	27	+1·5	K0	100
ζ	4·74	17	19	41	−67	44	58	−0·3	K2	320

The magnitude of δ is given as the combination of δ¹ and δ² (respectively 4·8 and 5·2). Both the stars are of type M, and make up a very wide physically connected system.

The only other star above the fifth magnitude is η; magnitude 4·97, spectral type A2p.

VARIABLE

Star	R.A.		Dec.		Range	Period, d.	Spectrum	Type
	h	m	°	′				
θ	14	00·4	−76	33	6·4—8·6	119	M	Semi-regular

CLUSTER

Object	R.A.		Dec.		Type	Notes
	h	m	°	′		
I 4499	14	52·7	−82	02	Globular	Mag. 11·5

AQUARIUS

(Abbreviation: Aqr).

A Zodiacal constellation, and of course one of Ptolemy's 'originals'. Oddly enough there are no well-defined legends attached to it, though it has been associated with Ganymede, cup-bearer of the Olympian gods. There are 7 stars brighter than magnitude 4·00:

β 2·86	δ 3·28	ε 3·83
α 2·96	ζ 3·6	λ 3·84
	c² 3·80	γ 3·97

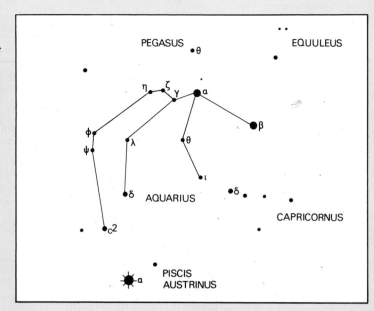

BRIGHTEST STARS

Star		Name	Mag.	R.A. h m s			Dec. ° ′ ″			Abs. Mag.	Spec.	Dist., l/y
2	ε	Albali	3·83	20 46 29			−09 34 38			+0·2	A0	170
3	k		4·60	20 46 35			−05 06 33			−0·6	M0	350
13	ν		4·52	21 08 24			−11 27 42			+0·7	K0	200
22	β	Sadalsuud	2·86	21 30 24			−05 40 07			−4·6	G0	1000
31	ο		4·66	22 02 11			−02 15 44			−0·3	B5p	320
34	α	Sadalmelik	2·96	22 04 39			−00 25 38			−4·6	G1	1080
33	ι		4·35	22 05 15			−13 58 37			−0·1	B8	250
43	θ	Ancha	4·32	22 15 40			−07 53 36			+0·5	K0	190
48	γ	Sadachiba	3·97	22 20 31			−01 29 55			+1·9	A0	85
52	π		4·64	22 24 09			+01 15 55			−3·5	B1p	1300
55	ζ		3·6	22 27 42			−00 07 59			+1·4, +1·2	F2+F2	140
62	η		4·13	22 34 14			−00 13 53			+0·7	B8	110
71	τ²		4·21	22 48 26			−13 42 33			−0·1	K5	230
73	λ		3·84	22 51 28			−07 41 49			−0·4	M0	230
76	δ	Scheat	3·28	22 53 29			−15 56 17			+1·2	A2	84
88	c²		3·80	23 08 17			−21 17 31			−0·5	K0	230
90	φ		4·40	23 13 11			−06 10 04			+0·1	M0	230
91	ψ¹		4·46	23 14 44			−09 12 28			+1·8	K0	120
93	ψ²		4·56	23 16 46			−09 18 10			−0·9	B5	400
98	b¹		4·20	23 21 49			−20 13 15			+1·2	K0	130
99	b²		4·52	23 24 53			−20 45 46			−0·3	K5	300
105	ω²		4·62	23 41 35			−14 40 00			+1·6	A0	130

THE FOLLOWING ADDITIONAL STARS ARE BRIGHTER THAN THE FIFTH MAGNITUDE:

Star		Mag.	Abs. Mag.	Spectrum
6	μ	4·80	+1·0	A8
23	ξ	4·78	+0·7	A5
57	σ	4·89	+1·0	A1
66	g¹	4·88	−0·3	K3
86	c¹	4·77	+1·3	G9
89	c³	4·94	+1·0	G2+A2
92	χ	5·0v	var	M5
101	b³	4·76	?	A0
104	A²	4·95	−2·7	G1

DOUBLES

Star	R.A. h m	Dec. ° ′	Mags.	P.A.	Distance ″
ζ	22 26·3	−00 17	4·4, 4·6	266	2·0 Slow binary
ψ¹	23 14·7	−09 12	4·5, 9·4	312	49 Fixed
94	23 16·5	−13 44	5·3, 7·5	348	13 Slow binary
τ¹	22 45·1	−14 19	5·7, 9·6	120	25·6 Optical

VARIABLES

Star	R.A. h m	Dec. ° ′	Range	Period d	Spec.	Type
W	20 43·9	−04 16	8·7—14·9	381	M	Mira
T	20 47·3	−05 20	7·2—14·2	202	M	Mira
X	22 15·9	−21 09	7·5—14·8	311	S	Mira
Z	22 49·7	−16 09	7·2—9·8	136	M	Semi-regular
S	22 54·6	−20 37	7·6—15·0	379	M	Mira
R	23 41·2	−15 34	5·8—11·5	387	M+pec	Mira

CLUSTERS AND NEBULÆ

Object		R.A. h m	Dec. ° ′	Type	
NGC 6981	M.72	20 50·7	−12 44	Globular	Mag. 8·6
NGC 7009	H.IV.1	21 01·4	−11 34	Planetary	Saturn Nebula, 25″ × 17″
NGC 7089	M.2	21 30·9	−01 03	Globular	Mag. 6·3; diameter 7″
NGC 7184		21 59·9	−21 04	Spiral galaxy	Mag. 12. Type Sb
NGC 7218		22 07·5	−16 54	Spiral galaxy	Mag. 12·7. Type Sc
NGC 7293		22 27·0	−21 06	Planetary	Mag. 6·5; brightest of all planetaries
NGC 7361		22 41·5	−30 19	Spiral galaxy	Mag. 12·8
NGC 7371		22 43·4	−11 16	Spiral galaxy	Mag. 12·9. Type Sb
NGC 7377		22 45·1	−22 35	Elliptical galaxy	Mag. 12·7. Type E1
NGC 7392		22 49·2	−20 53	Spiral galaxy	Mag. 12·6. Type Sb
NGC 7492		23 05·7	−15 54	Globular	Mag. 10·8
NGC 7585		23 15·4	−04 56	Spiral galaxy	Mag. 12·7
NGC 7606		23 16·5	−08 46	Spiral galaxy	Mag. 11·6. Type Sb
NGC 7721		23 36·2	−06 48	Spiral galaxy	Mag. 12·4. Type Sc
NGC 7723		23 36·4	−13 14	Spiral galaxy	Mag. 12·4. Type Sc

AQUILA

(Abbreviation: Aql).

One of the most distinctive of all the northern constellations, and, of course, an 'original'. Mythologically it represents an eagle which was sent by Jupiter to collect a Phrygian shepherd-boy, Ganymede, who was destined to become cup-bearer of the gods – following an unfortunate episode in which the former holder of the office, Hebe, tripped and fell during a particularly solemn ceremony.

The leading star is Altair. Altogether there are eight stars above magnitude 4:

α 0·77	δ 3·38
γ 2·67	λ 3·44
ζ 2·99	η 3·7 (max)
θ 3·31	β 3·90

BRIGHTEST STARS

Star			Mag.	R.A. h m s			Dec. ° ′ ″			Abs. Mag.	Spectrum	Distance l/y
13	ε		4·21	18 58 37			+15 02 15			+0·9	K0	145
12	i		4·15	19 00 30			−05 46 16			+0·4	K1	180
17	ζ	Dheneb	2·99	19 04 24			+13 49 47			+0·8	B9	90
16	λ	Althalimain	3·44	19 05 25			−04 55 00			−0·1	B9	160
30	δ		3·38	19 24 23			+03 04 11			+2·3	F0	53
38	μ		4·65	19 33 01			+07 19 52			+2·0	K0	105
41	ι		4·28	19 35 35			−01 20 11			−1·5	B5	450
50	γ	Tarazed	2·67	19 45 13			+10 33 31			−2·4	K2	340
53	α	Altair	0·77	19 40 43			+08 48 33			+2·3	A7	16·5
55	η		var	19 51 21			+00 56 53			var	G0v	900
60	β	Alshain	3·90	19 54 14			+06 21 03			+3·3	G8	42
65	θ		3·31	20 10 10			−00 53 16			−1·7	B9	330
71			4·51	20 37 12			−01 10 59			−0·3	K0	300

THERE ARE THREE FURTHER STARS ABOVE MAGNITUDE 5:

Star	Mag.	Abs. Mag.	Spectrum
32 ν	4·86	−4·5	F5
59 ξ	4·86	+1·1	K0
67 ρ	4·96	+1·2	A2

DOUBLE

Star	R.A. h m	Dec. ° ′	Mags.	P.A.	Dist.
π	19 46·3	+11 41	6·2, 6·8	111	1·5 Fixed

VARIABLES

Star	R.A. h m	Dec. ° ′	Range	Period	Spectrum	Type
V	19 01·7	−05 46	6·7—8·2	353	N	Semi-regular
R	19 04·0	+08 09	5·7—12·0	300	M	Mira
W	19 12·7	−07 08	7·8—14·2	490	S	Mira
U	19 26·7	−07 09	6·8—8·0	7·0	Gv	Cepheid
RT	19 35·6	+11 37	7·8—14·5	327	M	Mira
η	19 51·4	+00 54	3·7—4·7	7·2	F—G	Cepheid
RR	19 55·0	−02 01	7·8—14·5	394	M	Mira

CLUSTERS AND NEBULÆ

Object	R.A. h m	Dec. ° ′	Type	Notes
NGC 6709	18 49·1	+10 17	Open cluster	Mag. 8·1. Loose
NGC 6741	19 00·1	−00 31	Planetary	Mag. 11·7. Centre star, 17m
NGC 6751	19 03·2	−06 05	Planetary	Mag. 12·2. Centre star mag. 13·3
NGC 6755	19 05·3	+04 09	Open cluster	Mag. 8·3. Loose
NGC 6756	19 06·2	+04 35	Open cluster	Mag. 10·7. Fairly rich
NGC 6760	19 08·6	+00 57	Globular	Mag. 10·7
NGC 6772	19 12·0	−02 48	Planetary	Mag. 14·2. Centre star mag. 18
NGC 6781	19 16·0	+06 26	Planetary	Mag. 12·5. Centre star mag. 16
NGC 6790	19 20·4	+01 24	Planetary	Mag. 11·5. Centre star mag. 18
NGC 6803	19 28·9	+09 58	Planetary	Mag. 11·4. Centre star mag. 14
NGC 6804	19 29·2	+09 07	Planetary	Mag. 13·3. Centre star mag. 13·3
NGC 6807	19 32·1	+05 35	Planetary	Mag. 14. Centre star mag. 19
NGC 6891	20 12·8	+12 35	Planetary	Mag. 11·4. Centre star mag. 11·6

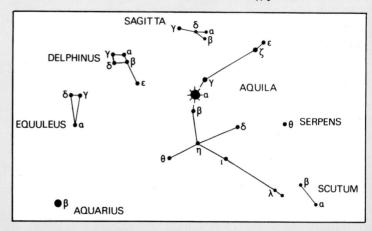

ARA

(Abbreviation: Ara).

An original constellation, though apparently without any definite legends attached to it. There are 7 stars above magnitude 4·00:

β	2·90	η	3·68
α	2·95	δ	3·79
ζ	3·16	θ	3·90
γ	3·32		

BRIGHTEST STARS

Star		Mag.	R.A.			Dec.			Abs. Mag.	Spec.	Dist. l/y
			h	m	s	°	′	″			
η		3·68	16	47	53	−59	00	14	+0·2	K5	160
ζ		3·16	16	54	48	−55	57	25	+0·9	K5	90
ε¹		4·15	16	57	45	−53	07	42	−1·1	K2	360
β		2·90	17	23	28	−55	30	39	−4·6	K2	1000
γ		3·32	17	23	32	−56	21	31	−3·3	B1	700
δ		3·79	17	29	07	−60	40	02	+0·5	B8	140
α	Choo	2·95	17	30	08	−49	51	38	−2·4	B3p	390
σ		4·63	17	31	56	−46	29	31		A0	
θ		3·90	18	04	55	−50	05	41	−5·2	B1p	1600

The only other star above the fifth magnitude is λ ; mag. 4·84, absolute magnitude +2·8, type F5.

VARIABLES

Star	R.A.		Dec.		Range	Period,	Spectrum	Type
	h	m	°	′		days		
R	16	35·6	−56	54	5·9—6·9	4·4	B9	Algol
RW	17	30·5	−57	07	8·7—12·1	4·4	A	Algol

CLUSTERS AND NEBULÆ

Object	R.A.		Dec.		Type	Notes
	h	m	°	′		
NGC 6167	12	30·6	−49	30	Open cluster	Mag. 6·4. Over 100 stars
NGC 6204	12	45·7	−46	56	Open cluster	Mag. 8·7. Over 25 stars
I 6188	16	35·9	−48	55	Nebula	Over 2000 l/y away
NGC 6193	16	37·6	−48	40	Open cluster	Mag. 5; 30 stars
NGC 6215	16	46·8	−58	55	Spiral galaxy	Mag. 11
NGC 6523	16	55·1	−52	38	Open cluster	Mag. 10. Over 70 stars
I 4642	17	07·6	−55	20	Planetary	Faint (below mag. 12)
I 4651	17	20·7	−49	54	Open cluster	Rich; about mag. 8.
NGC 6352	17	21·6	−67	01	Globular	Fairly easy. Mag. 8
NGC 6362	17	26·6	−67	01	Globular	Mag. 7; easy
NGC 6397	17	36·8	−53	39	Globular	Bright. Mag. 5

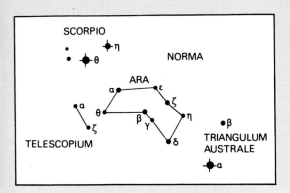

ARIES

(Abbreviation: Ari).

The first constellation of the Zodiac – though since the vernal equinox has shifted into Pisces, Aries should logically be classed as second! In mythology it represents a ram with a golden fleece, sent by the god Mercury (Hermes) to rescue the two children of the king of Thebes from an assassination plan by their step-mother. The ram, which had the remarkable ability to fly, carried out its mission; unfortunately the girl (Helle) lost her balance and fell to her death in that part of the sea now called the Hellespont, but the boy (Phryxus) arrived safely. After the ram's death, its golden fleece was hung in a sacred grove, from which it was later removed by the Argonauts commanded by Jason.

Aries is moderately conspicuous. There are four stars above the fourth magnitude:

α 2·00   c 3·68
β 2·68   γ 3·9

Of these, γ is an excellent double with two equal components – and it is surprising to find that it is an optical pair, not a binary.

See chart for Andromeda

BRIGHTEST STARS

Star			Mag.	R.A. h	m	s	Dec. °	'	"	Abs. Mag.	Spec.	Dist. l/y
5	γ	Mesartim	γ¹ 4·83 }3·9	01	52	19·1	+19	11	11	+1·5	A0p	150
			γ² 4·75							+1·1	A0p	170
6	β	Sheratan	2·68	01	53	25	+20	42	04	+1·7	A5	52
13	α	Hamal	2·00	02	05	56	+23	21	32	+0·2	K2	76
35			4·58	02	42	09	+27	36	52	−3·3	B3	1000
39			4·62	02	46	36	+29	00	24	+0·9	K1	180
41	c		3·68	02	48	41	+27	10	15	+0·4	B8	150
48	ε		4·64	02	57	57	+21	15	11	−0·8	A2	400
57	δ	Boteïn	4·53	03	10	22	+19	38	40	+0·9	K2	170

THERE ARE THREE MORE STARS ABOVE THE FIFTH MAGNITUDE:

Star		Mag.	Abs. Mag.	Spectrum
9	λ	4·83	+2·0	A6+G
58	ζ	4·95	+0·7	A0
62		4·70	+0·8	K5

DOUBLES

Star	R.A. h m	Dec. ° '	Mags.	P.A.	Dist. "	Notes
γ	01 52·3	+19 11	4·8, 4·7	359	8·2	Optical
ε	02 57·9	+21 15	5·2, 5·5	208	1·5	Slow binary

VARIABLES

Star	R.A. h m	Dec. ° '	Range	Period, days	Spec.	Type
V	02 12·3	+12 00	9·8—10·8	76	R	Semi-regular
R	02 13·3	+24 50	7·5—13·7	187	M	Mira
T	02 45·5	+17 18	7·5—11·3	320	M	Semi-regular
U	03 08·2	+14 36	6·4—15·2	371	M	Mira

NEBULÆ

Object	R.A. h m	Dec. ° '	Type	Notes
NGC 772	01 56·6	+18 46	Spiral galaxy	Mag. 10·9. Type Sb
NGC 821	02 05·6	+10 46	Elliptical galaxy	Mag. 12. Type E2
NGC 877	02 15·3	+14 19	Spiral galaxy	Mag. 12·4. Type Sc
NGC 976	02 31·2	+20 04	Spiral galaxy	Mag. 12·7. Type Sc
NGC 972	02 31·3	+29 06	Spiral galaxy	Mag. 12·1. Type Sc
NGC 1156	02 56·7	+25 03	Irregular galaxy	Mag. 11·8

AURIGA

(Abbreviation: Aur).

One of the most brilliant northern constellations, with Capella outstanding. Mythologically it honours Erechthonius, son of Vulcan, the blacksmith of the gods, who became King of Athens and also invented the four-horse chariot.

There are seven stars brighter than the fourth magnitude:

α 0·05   η 3·17
β 1·86   ε 3·3v
ι 2·64   δ 3·88
θ 2·65

In addition, Al Nath used to be called γ Aurigæ, but has now been given a free transfer, and is included in Taurus as β Tauri.

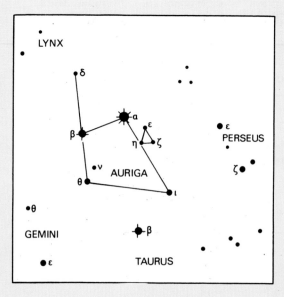

BRIGHTEST STARS

Star			Mag.	R.A. h	m	s	Dec. °	′	″	Abs. Mag.	Spec.	Dist. l/y
3	ι	Hassaleh	2·64	04	55	34	+33	07	57	−2·4	K2	330
7	ε		var.	05	00	23	+43	47	32	−6·1	F5p	1900
10	η		3·17	05	04	58	+41	12	22	−2·1	B3	370
13	α	Capella	0·05	05	15	04	+45	58	38	−0·6	G8+F	45
29	τ		4·64	05	47	39	+39	10	30	−0·1	K0	300
32	ν		4·18	05	49	58	+39	08	37	+0·2	K1	200
33	δ		3·88	05	57	43	+54	17	05	+0·6	K0	150
34	β	Menkarlina	1·86	05	57	55	+44	56	48	−0·3	A0p	88
37	θ		2·65	05	58	13	+37	12	44	+0·1	A0p	108
35	π		4·59	05	58	18	+45	56	11	−3·0	M3	850
44	κ		4·45	06	13	59	+29	30	27	+0·7	K0	180

THE FOLLOWING STARS ARE ABOVE THE FIFTH MAGNITUDE:

Star			Mag.	Abs. Mag.	Spectrum
4	ω		4·99	+1·0	A0
8	ζ	Sadatoni	var.	−5	K0+B1
9			4·99	+2·7	F3
11	μ		4·78	+1·8	A3
16			4·81	+0·4	K3
15	λ		4·85	+3·9	G0
25	χ		4·88	−6·2	B5
31	υ		4·99	−0·5	M1
30	ξ		4·92	+0·8	A2
46			4·9v	var	M0

DOUBLES

Star	R.A. h	m	Dec. °	′	Mags.	P.A.	Dist. ″	
ω	04	55·9	+37	49	5·0, 8·0	355	5·8	Slow binary
θ	05	58·2	+37	13	2·7, 7·5	320	3·0	Slow binary

VARIABLES

Star	R.A. h	m	Dec. °	′	Range	Period, d.	Spectrum	Type
ε	04	58·4	+43	45	3·3—4·2	9899	Fp	Eclipsing
ζ	04	59·0	+41	00	4·9—5·5	972	K+B7	Eclipsing
AE	05	13·0	+34	15	5·4—6·1	—	O9	Irregular
R	05	13·3	+53	32	6·7—13·7	459	M	Mira
U	05	38·9	+32	01	7·5—15·5	407	M	Mira
UU	06	33·1	+38	29	5·1—6·8	235	N	Semi-regular
RT	06	25·4	+30	32	5·4—6·5	3·73	F—G	Cepheid

CLUSTERS AND NEBULÆ

Object	R.A. h	m	Dec. °	′	Type	Notes
NGC 1664	04	47·4	+43	37	Open cluster	Mag. 7·5. Fairly rich
I 405	05	13·0	+34	16	Nebula	Associated with AE Aurigæ
NGC 1857	05	16·6	+39	18	Open cluster	Mag. 8·6. About 45 stars
NGC 1907	05	24·7	+35	17	Open cluster	Mag. 9·9
NGC 1912 M.38	05	25·3	+35	48	Open cluster	Mag. 7·4. Cruciform
NGC 1960 M.36	05	32·0	+34	07	Open cluster	Mag. 6·3
NGC 2099 M.37	05	49·1	+32	32	Open cluster	Mag. 6·2; superior to M.36
NGC 2281	06	45·8	+41	07	Open cluster	Mag. 6·7. About 30 stars
I 2149	05	52·6	+46	07	Planetary	Mag. 9·9; centre star mag. 14

BOÖTES

(Abbreviation: Boö).

An original constellation, dominated by Arcturus. There are various myths attached to it, but none is very definite. According to one version, Boötes was a herdsman who invented the plough drawn by two oxen, for which service he was transferred to the sky.

There are eight stars above magnitude 4·00:

α	—0·06	δ	3·47
ε	2·37	β	3·48
η	2·69	ρ	3·78
γ	3·05	ζ	3·86

BRIGHTEST STARS

Star		Mag.	R.A. h m s			Dec. ° ' "			Abs. Mag.	Spec.	Dist. l/y
4	τ	4·51	13	46	13	+17	33	57	+3·2	F5	52
5	υ	4·28	13	48	25	+15	54	23	—0·1	K5	250
8	η	2·69	13	53	38	+18	30	27	+2·7	G0	32
17	κ	4·60	14	12	42	+51	53	33	+0·7	A7	190
16	α Arcturus	—0·06	14	14	39	+19	17	47	—0·3	K0	36
19	λ	4·26	14	15	33	+46	11	19	+1·8	A0	95
23	θ	4·06	14	24	27	+51	57	07	+3·2	F8+M3	50
25	ρ	3·78	14	30	53	+30	28	02	+0·6	K0	140
27	γ Seginus	3·05	14	31	12	+38	24	14	+0·2	F0	118
28	σ	4·48	14	33	43	+29	50	24	+3·4	F0	50
30	ζ	3·86	14	40	06	+13	49	19	—0·4	A2	230
36	ε Izar	2·37	14	44	02	+27	09	58	0·0	K1	103
35	ο	4·69	14	44	13	+17	03	24	+0·9	K0	120
37	ξ	4·64	14	50	22	+19	11	29	+5·5	G5	23
42	β Nekkar	3·48	15	01	07	+40	28	35	+0·3	G5	140
43	ψ	4·67	15	03	30	+27	01	57	+0·4	K0	230
49	δ	3·47	15	14	37	+33	28	46	+0·3	K0	140
51	μ Alkalurops	4·47	15	23	40	+37	27	13	+2·1	F0	100

THE FOLLOWING STARS ARE ABOVE MAGNITUDE 5·00:

Star		Mag.	Abs. Mag.	Spectrum
21	ι	4·78	+2·7	A5
	A	4·83	+0·7	K1
20		4·97	+1·0	K3
29	π	4·94 + 5·81	—0·3, +0·6	A0+A0
41	ω	4·93	+1·5	K5
53	ν²	4·98	+1·4	A2

DOUBLES

Star	R.A. h m	Dec. ° '	Mags.	P.A.	Dist.	
κ	14 12·7	+51 54	4·6, 6·6	236	13·2	Fixed
ι	14 14·4	+51 36	4·8, 8·3	033	38·4	Fixed
π	14 38·4	+16 38	4·9, 5·8	108	5·6	P.A. increasing
ζ	14 40·1	+13 49	4·6, 4·6	313	1·2	Binary: 130 years
ε	14 44·0	+27 10	2·7, 5·1	338	2·9	Yellowish, bluish
ξ	14 50·4	+19 11	4·8, 6·9	350	6·7	Binary; 152 years
μ	15 23·7	+37 27	4·5, 6·7	171	108·8	Fixed, μ² is double (2″)

VARIABLES

Star	R.A. h m	Dec. ° '	Range	Period, d.	Spectrum	Type
S	14 21·2	+54 02	8·0—13·8	271	M	Mira
V	14 27·7	+39 05	7·0—11·3	258	M	Semi-regular
R	14 35·0	+26 57	6·7—12·8	223	M	Mira

CLUSTERS AND NEBULÆ

Object	R.A. h m	Dec. ° '	Type	Notes
NGC 5248	13 35·1	+09 08	Spiral galaxy	Mag. 11·3. Loose (Sc)
NGC 5466	14 03·2	+28 46	Globular cluster	Mag. 8·5
NGC 5557	14 16·4	+36 43	Elliptical galaxy	Mag. 11·6
NGC 5676	14 31·0	+49 41	Spiral galaxy	Mag. 11·2. Loose (Sc)
NGC 5689	14 33·7	+48 57	Spiral galaxy	Mag. 11·4

CÆLUM

(Abbreviation: Cae).

This entirely unremarkable constellation was introduced in 1752 by Lacaille, under the name of Cæla Sculptoris. It has no star brighter than the fourth magnitude, and only two which are brighter than the fifth.

See chart for Columba

BRIGHTEST STARS

Star	Mag.	R.A. h m s			Dec. ° ' "			Abs. Mag.	Spectrum	Dist. l/y
α	4·52	04	39	51	—41	54	19	+2·8	F2	70
γ	4·62	05	03	37	—35	30	45	+0·3	K0	230

DOUBLE

Star	R.A. h m	Dec. ° '	Mags.	P.A.	Dist.″	
γ	05 03·6	—35 33	4·6, 8·5	311	3·1	Slow binary

VARIABLE

Star	R.A. h m	Dec. ° '	Range	Period, d.	Spectrum	Type
R	04 38·7	—38 20	6·7—13·7	391	M	Mira

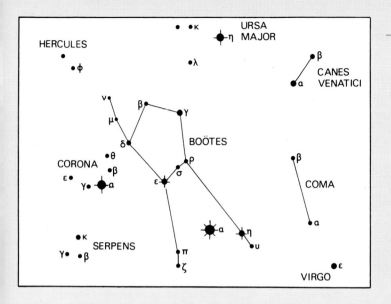

CAMELOPARDUS

(Abbreviation: Cam).

A very barren northern constellation. It was introduced to the sky by Hevelius in 1690, and some historians have maintained that it represents the camel which carried Rebecca to Isaac! It is interesting to note that several of the apparently faint stars are in fact highly luminous and remote; for instance α Cam, which is below the fourth magnitude, is well over 20 000 times more luminous than the Sun.

There are no stars in Camelopardus above the fourth magnitude.

See chart for Cassiopeia

BRIGHTEST STARS

Star		Mag.	R.A.			Dec.			Abs. Mag.	Spectrum	Dist. l/y
			h	m	s	°	′	″			
2	H (Boss 4113)	4·42	03	27	17	+59	51	54	−6·6	B9p	2700
	γ	4·67	03	48	01	+71	15	59	−0·3	A0	320
9	α	4·38	04	51	51	+66	18	26	−6·3	O9	3300
7		4·44	04	55	31	+53	43	07	−0·6	A2	320
10	β	4·22	05	01	27	+60	24	43	−5·2	G2p	1700
22	H (Boss 8020)	4·73	06	16	26	+69	19	48	+1·1	A0	170
24	H (Boss 9073)	4·75	06	56	53	+77	00	30	+0·8	K5	200

THE FOLLOWING STARS ARE ABOVE FIFTH MAGNITUDE:

Star	Mag.	Abs. Mag.	Spectrum
Boss 3947	4·76	−2·9	B3
Boss 4140	4·76	−6·7	A0p
Boss 4730	4·87	−0·1	B9
VZ	var.	var.	M4

VARIABLES

Star	R.A.		Dec.		Range	Period, d.	Spectrum	Type
	h	m	°	′				
U	03	37·5	+62	29	7·7—8·7	400	N	Semi-regular
T	04	35·2	+66	03	7·3—14·2	374	S	Mira
X	04	39·2	+75	00	7·4—13·7	143	M	Mira
VZ	07	20·6	+82	31	4·8—5·2	24?	M	Semi-regular
R	14	21·3	+84	04	7·9—14·4	270	S	Mira

CLUSTERS AND NEBULÆ

Object	R.A.		Dec.		Type	Notes
	h	m	°	′		
I 1501	04	02·6	+60	47	Planetary	Mag. 13; faint, oval
I 361	04	14·8	+58	11	Open cluster	Mag. 11·2
NGC 2146	06	10·7	+78	23	Spiral galaxy	Mag. 11·3
NGC 2336	07	16·2	+80	20	Spiral galaxy	Mag. 11. Type Sb
NGC 2403	07	32·0	+65	43	Spiral galaxy	Mag. 9. Loose (Sc)
NGC 2655	08	49·4	+78	25	Spiral galaxy	Mag. 11

CANCER

(Abbreviation: Cnc).

Cancer is an obscure constellation, redeemed only by the presence of two famous star-clusters, Præsepe and M.67. However, it lies in the Zodiac, and it has a legend attached to it. It represents a sea-crab which Juno, queen of Olympus, sent to the rescue of the multi-headed hydra which was doing battle with Hercules. Not surprisingly, Hercules trod on the crab, but as a reward for its efforts Juno placed it in the sky!

There is only one star above the fourth magnitude; this is β (3·76).

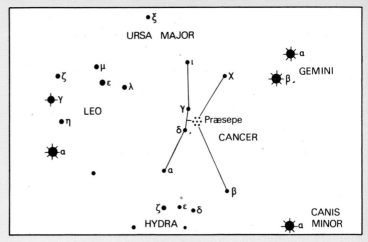

BRIGHTEST STARS

Star			Mag.	R.A.			Dec.			Abs. Mag.	Spectrum	Dist. l/y
				h	m	s	°	′	″			
16	ζ	Tegmine	4·7	08	09	21	+17	48	00	+3·2, +4·1	F7+G2	78
17	β		3·76	08	15	19	+09	15	15	−0·4	K2	220
43	γ	Asellus Borealis	4·73	08	42	01	+21	32	54	+0·5	A0	230
47	δ	Asellus Australis	4·17	08	43	26	+18	14	09	0·0	K0	220
48	ι¹		4·20	08	45	22	+28	50	29	+0·5	G5	170
65	α	Acubens	4·27	08	57	17	+11	56	37	+1·9	A3	100

There are no other stars above the fifth magnitude.

DOUBLE

Star	R.A.		Dec.		Mags.	P.A.	Dist.″
	h	m	°	′			
ζ	08	09·3	+17	48	5·1, 6·0	089	5·9

Each component is again double:

ζ¹ Mags. 5·7, 6·0; binary; 59·6 years; separation 1″
ζ² Mags. 6·3, 7·8; binary; 17·6 years; separation always below 0″·2

VARIABLES

Star	R.A.		Dec.		Range	Period, d.	Spectrum	Type
	h	m	°	′				
R	08	13·8	+11	53	6·2—11·8	362	M	Mira
V	08	18·9	+17	27	7·6—13·9	272	S	Mira
X	08	52·6	+17	25	5·9—7·3	170?	N	Semi-regular
T	08	53·8	+20	02	7·6—10·5	482	N	Semi-regular
W	09	06·9	+25	27	7·4—14·4	393	M	Mira

CLUSTERS AND NEBULÆ

Object		R.A.		Dec.		Type	Notes
		h	m	°	′		
NGC 2632	M.44	08	37·4	+20	00	Open cluster	Præsepe. Naked-eye
NGC 2672		08	46·6	+19	16	Elliptical galaxy	Mag. 12·2
NGC 2682	M.67	08	47·8	+12	00	Open cluster	Mag. 6·1
NGC 2775		09	07·7	+07	15	Spiral galaxy (Sa)	Mag. 10·7

CANES VENATICI
(Abbreviation: CVn)

One of Hevelius' constellations, dating only from his maps of 1690, and evidently representing two hunting dogs (Asterion and Chara) which are being held by the herdsman Boötes – possibly to stop them from chasing the two Bears round and round the celestial pole. The name of Cor Caroli was given to α^2 CVn by Edmond Halley in honour of King Charles I. Cor Caroli (mag. 2·90) is the only star above the fourth magnitude. It is also the prototype 'magnetic variable'.

See chart for Ursa Major

BRIGHTEST STARS

Star	Mag.	R.A. h	m	s	Dec. °	'	"	Abs. Mag.	Spectrum	Dist. l/y
8 β Chara	4·32	12	32	42	+41	28	36	+4·5	G0	33
12 α² Cor Caroli	2·90	12	55	00	+38	26	12	+0·1	A0p	120
20	4·66	13	16	33	+40	41	17	+0·4	F0	230
24	4·63	13	33	33	+49	07	40	+1·7	A3	120

ALSO ABOVE THE FIFTH MAGNITUDE:

Star	Mag.	Abs. Mag.	Spectrum
5	4·97	+1·2	G6
Boss 183569	4·96	+1·8	F2
25	4·92	+2·0	A3

DOUBLE

Star	R.A. h	m	Dec. °	'	Mags.	P.A.	Dist.	
α	12	55·0	+38	26	2·9, 5·4	228	19·7	Fixed

VARIABLES

Star	R.A. h	m	Dec. °	'	Range	Period, d	Spectrum	Type
Y	12	42·8	+45	43	5·2—6·6	158	N	Semi-regular
V	13	17·3	+45	47	6·8—8·8	192	M	Semi-regular
R	13	46·8	+38	47	7·3—12·9	328	M	Mira

CLUSTERS AND NEBULÆ

Object	R.A. h	m	Dec. °	'	Type	Notes
NGC 4143	12	07·1	+42	49	Elliptical galaxy	Mag. 11. Type E4
NGC 4151	12	08·0	+39	41	Galaxy	Mag. 11·3
NGC 4220	12	13·7	+48	10	Spiral galaxy	Mag. 12. Type Sa
NGC 4258 (M.106)	12	16·5	+47	35	Spiral galaxy	Mag. 8·6. Borders of CVn and UMa
NGC 4449	12	25·8	+44	22	Irregular galaxy	Mag. 9·2
NGC 4490	12	28·3	+41	55	Spiral galaxy	Mag. 9·7. Type Sc
NGC 5033	13	11·2	+36	51	Spiral galaxy	Mag. 10·3. Type Sb
NGC 5055 M.63	13	13·5	+42	17	Spiral galaxy	Mag. 9·5. Type Sb
NGC 5194/5 M.51	13	27·8	+47	27	Spiral galaxy	Mag. 8. Whirlpool galaxy
NGC 5272 M.3	13	39·9	+28	38	Globular cluster	Mag. 6·4
NGC 5395	13	56·5	+37	39	Spiral galaxy	Mag. 12·2. Type Sb

(NGC 4258 is one of the 'additional' Messier objects, included in Hogg's list of 1947.)

CANIS MAJOR

(Abbreviation: CMa).

An original constellation, representing one of Orion's hunting dogs. Though dominated by Sirius, it contains several other bright stars, of which one (Adhara) is only just below the first magnitude. Altogether these are 10 stars brighter than the fourth magnitude:

α	−1·42	σ²	3·02
ε	1·48	ζ	3·04
δ	1·85	δ	3·68
β	1·96v	ϰ	3·78
η	2·46	ω	3·83

It is interesting to note that three of these stars – Wezea, Aludra and o² – are extremely luminous, with at least 50 000 times as much output as the Sun. Compared with them, Sirius is extremely feeble, as it has only 26 times the Sun's luminosity.

BRIGHTEST STARS

Star			Mag.	R.A. h m s			Dec. ° ′ ″			Abs. Mag.	Spectrum	Dist. l/y
1	ζ	Phurad	3·04	06	19	28	−30	03	10	−2·4	B3	400
2	β	Mirzam	1·96v	06	21	44	−17	56	39	−4·8	B1	750
	λ		4·48	06	27	21	−32	33	56	−0·5	B5	320
4	ξ¹		4·35v	06	30	57	−23	24	06	−4·5	B1	1700
5	ξ²		4·54	06	34	08	−22	56	48	+0·6	A0	200
7	ν²		4·14	06	35	43	−19	14	11	+2·1	K1	85
8	ν³		4·65	06	36	55	−18	13	03	0·0	K0	270
9	α	Sirius	−1·42	06	44	11	−16	41	06	+1·45	A0	8·7
13	ϰ		3·78	06	49	01	−32	28	56	−2·3	B2p	550
15	EY		4·66	06	52	36	−20	11	46	−4·0	B1	1550
14	θ		4·25	06	53	10	−12	00	37	+0·9	K2	160
16	o¹		4·12	06	53	13	−24	09	20	−5·2	K2p	1600
19	π		4·62	06	54	40	−20	06	27	+1·9	F5	110
20	ι		4·39	06	55	09	−17	01	29	−4·3	B3	1400
21	ε	Adhara	1·48	06	57	46	−28	56	29	−5·1	B1	680
22	σ		3·68	07	00	51	−27	54	09	−1·1	K5	300
24	o²		3·02	07	02	06	−23	48	01	−7·1	B5p	3400
23	γ	Muliphen	4·07	07	02	46	−15	35	59	−0·9	B5	320
25	δ	Wezea	1·85	07	07	30	−26	21	27	−7·1	F8p	2100
27			4·66	07	13	21	−26	18	49	−1·8	B5p	650
28	ω		3·83	07	13	55	−26	44	02	−1·6	B3p	400
29	UW		4·6v	07	16	35	−24	27	59	var	O8	3300
30	τ		4·40	07	17	48	−24	54	48	−6·0	O9	3300
31	η	Aludra	2·46	07	23	14	−29	15	34	−7·1	B5p	2700

UW CMa is a Beta Lyræ-type eclipsing binary; period 4·4 days. Both components are of type O. The combined absolute magnitude is approximately −6. β CMa (Mirzam) is an intrinsic short-period variable, but the range is very slight, and the changes are not detectable with the naked eye. There are two other stars in Canis Major above the fifth magnitude:

Star	Mags.	Abs. Mag.	Spectrum
Boss 8694	4·97	−1·5	K5
Boss 9836	4·87	−0·6	B8

VARIABLES

Star	R.A. h m		Dec. ° ′		Range	Period, d.	Spectrum	Type
R	07	17·2	−16	18	6·2—6·8	1·14	A9	Algol
W	07	05·7	−11	51	6·9—7·5	—	N	Irregular

CLUSTERS AND NEBULÆ

Object		R.A. h m		Dec. ° ′		Type	Notes
NGC 2204		06	13·5	−18	35	Open cluster	Mag. 9. Fairly rich
NGC 2217		06	18·7	−27	14	Spiral galaxy	Mag. 12. Type SBa
I 2165		06	19·6	−12	57	Planetary	Mag. 12·5. Central star of mag. 17
NGC 2243		06	27·6	−31	15	Open cluster	Mag. 10. Small but quite rich
NGC 2287	M.41	06	44·9	−20	41	Open cluster	Mag. 4·6; naked eye
NGC 2354		07	12·2	−25	38	Open cluster	Mag. 9. Fairly rich
NGC 2360		07	15·4	−15	33	Open cluster	Mag. 10. Fairly rich
NGC 2362		07	16·6	−24	52	Open cluster	Mag. 10·5. Moderate

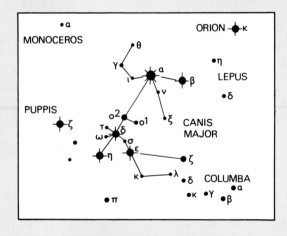

CANIS MINOR

(Abbreviation: CMi)

The second of Orion's two dogs.
There are two stars above the fourth
magnitude:

α 0·37   β 2·91

BRIGHTEST STARS

Star			Mag.	R.A.			Dec.			Abs. Mag.	Spectrum	Dist. l/y
				h	m	s	°	′	″			
3	β	Gomeisa	2·91	07	25	58	+08	20	05	−1·1	A0	200
4	γ		4·60	07	26	58	+08	58	16	+0·2	K0	250
10	α	Procyon	0·37	07	38	09	+05	16	56	+2·7	F5	11·3
Boss 10893												
(+2°1854)			4·52	08	01	07	+02	23	45	+0·7	K0	190

There are no other stars above magnitude 5. Procyon has a very faint and close
White Dwarf companion.

VARIABLES

Star	R.A.		Dec.		Range	Period, d.	Spectrum	Type
	h	m	°	′				
R	07	06·0	+10	06	7·4—11·6	338	S	Mira
S	07	30·0	+08	26	7·0—13·2	332	M	Mira

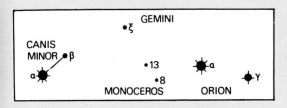

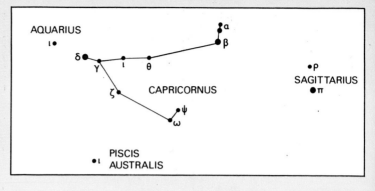

CAPRICORNUS

(Abbreviation: Cap).

A Zodiacal constellation, though by no means brilliant. It has been identified with the demigod Pan, but there seems to be no well-defined mythological associations. There are five stars above magnitude 4:

δ 2·92v	γ 3·80
β 3·06	ζ 3·86
α² 3·77	

α¹ and α² make up a naked-eye pair, separated by 376″ (position angle 291°), but the two components are not genuinely associated, and each is itself double; the fainter component of α² is also double!

BRIGHTEST STARS

Star			Mag.	R.A.			Dec.			Abs. Mag.	Spectrum	Dist. l/y
				h	m	s	°	′	″			
5	α¹	Al Giedi{	4·55	20	16	25·7	−12	34	38	−3·1	G0p	450
6	α²		3·77	20	16	50·0	−12	36	51	+1·2	G5	115
9	β	Dabih	3·06	20	19	47	−14	51	07	+0·1	G0+A0	130
16	ψ		4·26	20	44	48	−25	21	04	+3·7	F8	39
18	ω		4·24	20	50	31	−27	00	09	−1·0	M1	350
23	θ		4·19	21	04	43	−17	19	16	+0·6	A0	170
24	A		4·60	21	05	51	−25	05	41	+0·2	M0	250
32	ι		4·30	21	21	01	−16	55	45	+1·1	K0	140
34	ζ		3·86	21	25	25	−22	30	27	−2·2	G5	550
36	b		4·59	21	27	28	−21	54	14	+0·7	G5	190
39	ε		4·72	21	35	51	−19	33	55	−1·0	B5p	450
40	γ	Nashira	3·80	21	38	52	−16	45	45	+1·2	F0p	107
49	δ	Deneb al Giedi	2·92v	21	45	50	−16	13	40	+2·0	A5	50

δ has a range of between 2·88 and 2·95, so that its fluctuations are not detectable with the naked eye. There are four more stars in Capricornus above magnitude 5:

Star			Mag.	Abs. Mag.	Spectrum
8	ν	Alshat	4·84	+1·0	A0
11	ρ		4·96	+2·6	F1
22	η		4·93	+2·6	A4
43	χ		4·82	+0·8	G4

DOUBLES

Star	R.A.		Dec.		Mags.	P.A.	Dist.	
	h	m	°	′				
α¹	20	16·4	−12	35	4·5, 9·0	221	45·5	Optical
α²	20	16·8	−12	37	3·7, 10·6	158	7·1	B is again double; 1″·2

VARIABLE

Star	R.A.		Dec.		Range	Period, d.	Spectrum	Type
	h	m	°	′				
RT	20	14·1	−21	29	6·4—8·1	393	N	Semi-regular

CLUSTERS AND NEBULÆ

Object		R.A.		Dec.		Type	Notes
		h	m	°	′		
NGC 6907		20	22·1	−24	58	Spiral galaxy	Mag. 12. Type SBb
NGC 7099	M.30	21	37·5	−23	25	Globular cluster	Mag. 8·4

CARINA

(Abbreviation: Car).

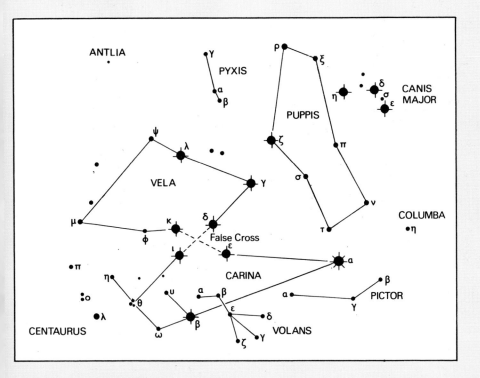

The keel of the ship Argo, in which Jason and his companions sailed upon their successful if somewhat unprincipled expedition to remove the Golden Fleece of the sacred ram (Aries) from its grove. Carina is the brightest part of Argo, and contains Canopus, which is second only to Sirius in brilliance. Canopus is a highly luminous star, but estimates of its power and distance vary considerably. One estimate makes it 80 000 times as luminous as the Sun; other estimates are much lower. Also in this constellation is η Carinæ, which during part of the last century rivalled Sirius, but which is now well below naked-eye visibility. It is unique, and at its peak may have been the most luminous star in the Galaxy. Excluding η, there are 14 stars in Carina above the fourth magnitude:

α	−0·72	υ	2·95	χ	3·48
β	1·67	ρ	3·22	l	3·6v
ε	1·97	q	3·30	u	3·88
ι	2·25	ω	3·33	c	3·98
θ	2·74	a	3·43		

BRIGHTEST STARS

Star		Mag.	R.A.			Dec.			Abs. Mag.	Spectrum	Dist. l/y
			h	m	s	°	′	″			
α	Canopus	−0·72	06	23	28	−52	41	00	−7?	F0	650?
N		4·44	06	34	29	−52	57	26	−1·0	A0	400
A		4·38	06	49	23	−53	35	46	+0·4	G5	200
χ		3·48	07	56	13	−52	55	22	−2·1	B3	430
ε	Avior	1·97	08	22	04	−59	26	18	−3·1	K0+B	340
d		4·42	08	40	08	−59	40	56	−4·3	B2	1550
f		4·63	08	46	09	−56	41	19	−1·6	B3	550
c		3·98	08	54	33	−60	33	36	−3·8	B8	1200
G		4·50	09	05	06	−72	30	51	+0·8	F5	180
a		3·43	09	10	23	−58	32	36	−2·9	B3	600
i		4·18	09	10	47	−62	13	36	−0·2	B3	250
β	Miaplacidus	1·67	09	12	58	−69	37	36	−0·4	A0	86
g		4·18	09	15	35	−57	26	56	−0·1	K5	230
ι	Tureis	2·25	09	16	30	−59	10	58	−4·6	F0	750
R		var.	09	31	42	−62	44	28	var.	M5	—
h		4·20	09	33	48	−59	07	52	−2·3	B5	650
m		4·67	09	38	44	−61	13	41	+0·4	B9	230
l		var.	09	44	39	−62	24	22	−5·5v	G0	2700
υ		2·95	09	46	33	−64	58	11	−2·1	F0	340
ω		3·33	10	13	13	−69	55	42	−1·5	B8	300
q		3·3v	10	16	21	−61	13	20	−4·6	K5	1300
I		4·08	10	23	58	−73	55	10	+3·6	F5	42
s		4·08	10	27	04	−58	37	37	—	F0	—
p		3·2v	10	31	14	−61	34	19	−2·3	B5p	430
r		4·54	10	34	44	−57	26	37	−0·5	K5	330
t²		4·73	10	37	55	−59	04	06	−0·1	K5	300
θ		2·74	10	42	10	−64	16	44	−4·0	B0	700
w		4·49	10	42	42	−60	27	04	+0·2	K5	230
u		3·88	10	52	36	−58	44	11	+1·7	K0	90
x		4·02	11	07	39	−58	51	21	—	F8p	—
y		4·73	11	11	39	−60	11	52	−5·7	F5p	3200

CARINA

THE FOLLOWING STARS ARE ABOVE MAGNITUDE 5:

Star	Mag.	Abs. Mag.	Spectrum
Q	4·92	−0·8	M0
D¹	4·96	−2·4	B3
B	4·80	+3·5	F5
e²	4·80	+0·2	G6
p	4·94	—	F1
k	4·86	+1·5	G4
K	4·94	−0·5	A2
Boss 14480	4·90	—	M1
z¹	4·76	+1·1	G5

DOUBLE

Star	R.A. h m	Dec. ° ,	Mags.	P.A.	Dist.	
υ	09 46·6	−64 58	3·1, 6·0	126	4·6	Fixed

VARIABLES

Star	R.A. h m	Dec. ° ,	Range	Period, d.	Spectrum	Type
η	10 43·1	−59 25	−0·8—7·9	—	Peculiar	Irregular
U	10 55·8	−59 28	6·4—8·4	38·8	F-G	Cepheid
R	09 31·0	−62 41	3·9—10·0	381	M	Mira
l	09 44·7	−62 24	3·4—4·8	35·2	F-G	Cepheid
S	10 07·8	−61 18	4·5—9·9	150	M	Mira

CLUSTERS AND NEBULÆ

Object	R.A. h m	Dec. ° ,	Type	Notes
NGC 2516	07 59·7	−60 44	Open cluster	Naked-eye. Rich
NGC 2808 Δ263	09 11·0	−64 39	Globular	Mag. 6. Rich
NGC 2867	09 20·0	−58 06	Planetary	Mag. 10. Oval
NGC 3114	10 01·1	−59 53	Open cluster	Mag. 4·5. Fairly rich
NGC 3136	10 04·5	−67 08	Galaxy	Mag. 12·4. Elliptical galaxy
I 2581	10 25·4	−57 23	Open cluster	Mag. 5. Fairly rich
NGC 3372 Δ309	10 43·0	−59 25	Nebula	Keyhole Nebula, round η
NGC 3532 Δ323	11 04·3	−58 24	Open cluster	Fairly rich

CASSIOPEIA

(Abbreviation: Cas).

An original constellation – one of the most distinctive in the sky. Mythologically, Cassiopeia was Andromeda's mother and wife of Cepheus; it was her boasting which led to the unfortunate contretemps with Neptune's sea-monster.

There are seven stars above magnitude 4. Of these, γ (Cih) is variable, and it is also suspected that α (Shedir) is variable over a small range – possibly 2·1 to 2·4, though opinions differ. x, which looks obscure, is extremely luminous.

There are 7 stars brighter than magnitude 4:

α	2·16v?	ε	3·33
γ	2·2v	η	3·47
β	2·26	ζ	3·72
δ	2·67		

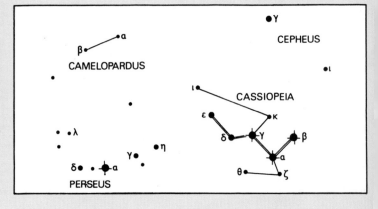

BRIGHTEST STARS

Star			Mag.	R.A.			Dec.				Abs. Mag.	Spectrum	Dist. l/y
				h	m	s	°	′	″				
7	ρ		var.	23	53	17	+57	22	37	var.	F8p	800
11	β	Chaph	2·26	00	08	00	+59	01	43	+1·6	F5	45
15	ϰ		4·24	00	31	44	+62	48	38	−6·9	B0	3200
17	ζ		3·72	00	35	44	+53	46	34	−3·2	B3	800
18	α	Shedir	2·16v?	00	39	15	+56	25	01	−1·1	K0	150
22	o		4·70	00	43	30	+48	09	51	−2·7	B2	800
24	η	Achird	3·47	00	47	45	+57	41	58	+4·8	F8	18
27	γ	Cih	var.	00	55	22	+60	35	53	var.	B0p	96
33	θ	Marfak	4·52	01	09	45	+55	02	00	+0·1	A5	250
37	δ	Ruchbah	2·67	01	24	22	+60	07	17	+2·1	A5	43
45	ε	Segin	3·33	01	52	48	+63	33	45	−2·7	B3	500
48	A		4·61	02	00	07	+70	48	05	+1·7	A3	120
50	ι		4·06	02	01	32	+72	18	57	+1·0	A2	130
			4·57	02	27	14	+67	18	17	+0·9	A5p	180

THE FOLLOWING STARS ARE BRIGHTER THAN MAGNITUDE 5:

Star		Mag.	Abs. Mag.	Spectrum
1		4·93	−4·4	B1
	AR	4·7v	var.	B3
8	σ	4·93	−3·5	B1
14	λ	4·88	−0·9	B8
19	ξ	4·85	−3·1	B3
26	υ¹	4·95	0·0	K2
36	ψ	4·96	+0·5	G8
39	χ	4·88	+0·9	G6
Boss 3759		4·89	+1·4	A2

DOUBLES

Star	R.A.		Dec.		Mags.	P.A.	Dist.	
	h	m	°	′				
σ	23	56·5	+55	29	5·1, 7·2	332	3·1	V. slow binary
η	00	47·8	+57	42	3·6, 7·5	293	10·1	Binary; over 500 yrs.
γ	00	55·4	+60	36	2v, 11·0	248	2·1	Fixed
ψ	01	22·4	+67	52	5·0, 9·8	112	26·2	B is double (3″)
ι	02	27·2	+67	18	4·7, 7·0, 7·1	240, 116	2·3, 8·2	Little change

VARIABLES

Star	R.A.		Dec.		Range	Period, d.	Spectrum	Type
	h	m	°	′				
V	23	09·5	+59	25	7·3—12·8	228	M	Mira
ρ	23	53·3	+57	23	4·1—6·2	?	F	?
R	23	55·9	+51	07	5·5—13·0	431	M	Mira
α	00	39·3	+56	25	2·1—2·4?	?	K	Suspected
U	00	43·6	+47	58	7·9—15·4	429	M	Mira
γ	00	55·4	+60	36	1·6—3·2	—	Bp	Irregular
S	01	16·0	+72	21	7·9—15·2	611	S	Mira
RZ	02	44·3	+69	26	6·4—7·8	1·2	A	Cepheid
SU	02	47·5	+68	41	5·9—6·3	1·9	F-G	Cepheid

CLUSTERS AND NEBULÆ

Object		R.A.		Dec.		Type	Notes
		h	m	°	′		
NGC 7654	M.52	23	22·0	+61	19	Open cluster	Mag. 7·3. Not distinctive
NGC 7789		23	54·5	+56	26	Open cluster	Mag. 9·6. About 200 stars
NGC 40		00	10·2	+72	15	Planetary	Mag. 10·2. Oval; 11·4m central star
NGC 103		00	22·6	+61	03	Open cluster	Mag. 10·8. Rich
NGC 129		00	27·0	+59	57	Open cluster	Mag. 10. Fairly rich
NGC 133		00	28·4	+63	04	Open cluster	Mag. 9. Fairly rich
NGC 146		00	30·3	+63	01	Open cluster	Mag. 10. Fairly rich
NGC 225	(H.VIII.78)	00	40·6	+61	31	Open cluster	Mag. 9. Fine rich cluster
NGC 281		00	50·4	+56	19	Nebula	Distance over 5000 light-years
NGC 381		01	05·2	+61	18	Open cluster	Mag. 9. Fairly rich
NGC 475	(H.VII.42)	01	06·0	+58	03	Open cluster	Rich
NGC 457		01	15·9	+58	04	Open cluster	Mag. 7·5. Over 100 stars
NGC 559		01	26·1	+63	02	Open cluster	Mag. 7·3. Fairly rich
NGC 581	(M.103)	01	29·9	+60	27	Open cluster	Mag. 7·4. Loose and poor
NGC 663	(H.VI.31)	01	42·5	+61	00	Open cluster	Mag. 7·1. Visible in finder
I.289		03	06·2	+61	02	Planetary	Mag. 12·3. Central star mag. 15

CENTAURUS

(Abbreviation: Cen).

A brilliant southern constellation (one of Ptolemy's originals), containing many important objects, including the nearest of all the bright stars (α) and the superb globular cluster ω. There are 14 stars brighter than the fourth magnitude.

α	−0·27	ε	2·33	κ	3·15
β	0·63	ζ	2·56	λ	3·25
θ	2·06	δ	2·56v	ν	3·42
γ	2·17	ι	2·76	d	3·96
η	2·33v	μ	3·08v		

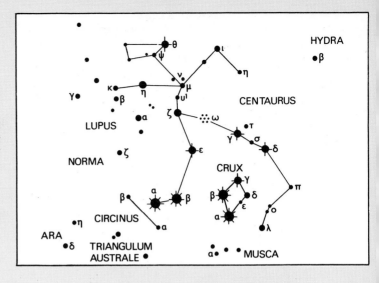

BRIGHTEST STARS

Star		Mag.	R.A. h	m	s	Dec. °	′	″	Abs. Mag.	Spectrum	Dist. l/y
π		4·26	11	20	00	−54	22	13	−0·7	B5	320
λ		3·15	11	34	46	−62	53	33	−2·1	B9	370
65G (Boss 16147)		4·22	11	45	27	−61	03	22	−1·9	G0	550
j		4·52	11	48	36	−63	50	59	−1·2	B5	460
B		4·71	11	50	02	−45	03	04	+1·0	K4	180
δ		2·56v	12	07	13	−50	36	00	−2·7	B3p	370
ρ		4·20	12	10	30	−52	14	46	−1·3	B3	400
σ		4·16	12	56	21	−50	06	33	−1·6	B3	460
τ		4·02	12	36	30	−48	25	13	+0·6	A2	150
1		4·79	12	37	10	−39	41	45	+0·5	B8p	230
γ	Menkent	2·17	12	40	18	−48	50	21	−0·5	A0	160
w		4·65	12	41	21	−48	41	33	+0·4	K0	230
e		4·35	12	51	52	−48	49	26	+0·2	K2	210
n		4·34	12	52	13	−40	03	35	+2·8	A5	65
ξ²		4·40	13	05	37	−49	47	20	−3·2	B3	1000
ι		2·76	13	19	21	−36	35	48	+1·1	A2	71
J		4·62	13	21	12	−60	52	24	−0·6	B5	360
m		4·50	13	22	31	−64	25	17	—	G4	—
d		3·96	13	29	46	−39	17	39	−2·1	K0	550
ε		2·33	13	38	29	−53	21	18	−3·9	B1	570
l		4·36	13	44	26	−32	55	59	+2·6	F5	72
M		4·68	13	45	15	−51	19	23	+0·2	K0	250
2 g		4·40	13	48	10	−34	20	29	+3·0	M6	62
ν		3·42	13	48	11	−41	34	43	−3·4	B2	750
μ		3·08v	13	48	17	−42	21	53	−2·7	B2p	470
3 k		4·72	13	50	33	−32	53	09	−1·7	B5	600
ζ		2·56	13	54	10	−47	10	50	−3·4	B2p	520
294G (Boss 18845)		4·68	13	56	02	−63	34	47	+0·8	K0	190
φ		4·05	13	56	56	−41	59	38	−2·4	B3	500
υ¹		4·17	13	57	19	−44	41	49	−2·4	B3	700
υ²		4·39	14	00	21	−45	29	51	−3·8	F5	1200
β	Agena	0·63	14	02	16	−60	16	04	−5·2	B1	490
χ		4·54	14	04	42	−41	04	30	−2·4	B3	700
θ		2·06	14	05	23	−36	15	45	+1·0	K0	55
v		4·41	14	18	47	−56	17	10	−3·9	B5	1000
ψ		4·17	14	19	13	−37	47	05	—	A0	—
a		4·55	14	21	41	−39	24	44	−1·6	B5	550
η		2·33v	14	34	07	−42	03	43	−3·0	B3p + A2	390
α		−0·27	14	38	06	−60	44	44	+4·39	G4 + K5	4·3
b		4·09	14	40	35	−37	47	30	−1·6	B3	450
c¹		4·13	14	42	18	−35	04	48	0·0	K0	220
κ		3·15	14	57	43	−42	01	01	−2·7	B3	470

Strangely, α has no official proper name. It has been called Al Rijil, and also Toliman, but astronomers in general prefer to call it simply α Centauri.

Star	Mag.	Abs. Mag.	Spectrum
o¹	4·96	+1·1	G4
A	4·82	+1·0	B8
Boss 16037	4·88	+1·5	G6
Boss 16576	4·81	−0·7	B5
f	4·96	−2·1	B3
Boss 17866	4·76	+0·6	B8
Boss 17869	4·89	+3·2	G3
4 h	4·76	−1·3	B7
Boss 19099	4·78	+0·7	G5

DOUBLES

Star	R.A. h m	Dec. ° ′	Mags.	P.A.	Dist. ″
γ	12 40·3	−48 50	3·1, 3·2	—	Binary, period 85 years Close
η	14 34·1	−42 04	2·6, 13·5	270	5·6 Optical
J	13 21·2	−60 52	4·6, 6·5	343	60·5 Easy
β	14 02·3	−60 16	0·6, 9·0	255	1·4 Fixed
α	14 38·1	−60 45	0·0, 1·7	—	Binary period 80 years Wide and very easy

VARIABLES

Star	R.A. h m	Dec. ° ′	Range	Period, d.	Spectrum	Type
RS	11 18·3	−61 36	7·8—13·9	164	M	Mira
X	11 46·7	−41 28	7·0—13·9	315	M	Mira
U	12 30·7	−54 23	7·2—14·0	220	M	Mira
RV	13 34·3	−56 13	7·0—10·8	446	N	Mira
T	13 38·9	−33 21	5·5—9·0	91	M	Semi-regular
R	14 12·9	−59 41	5·4—11·8	547	M	Mira

CLUSTERS AND NEBULÆ

Object	R.A. h m	Dec. ° ′	Type	Notes
NGC 3766 △ 289	11 33·9	−61 20	Open cluster	Mag. 5; easy in binoculars
NGC 3918	11 47·8	−56 24	Planetary	Mag. 8·4
NGC 3882	11 43·6	−56 05	Nebula	
NGC 5139 ω	13 23·7	−47 03	Globular cluster	Finest of all globulars
NGC 5286	13 43·0	−51 07	Globular cluster	Mag. 8·5
NGC 5460	14 04·5	−48 05	Open cluster	Mag. 6·5. Fairly rich

CEPHEUS

(Abbreviation: Cep).

A rather undistinguished constellation, though it contains δ Cephei – the prototype Cepheid. Mythologically, Cepheus was Andromeda's father and Cassiopeia's wife. There are 8 stars above the 4th magnitude – though one of these, the 'Garnet Star' μ, is an irregular variable which is generally rather below this limit:

α	2·44	η	3·41
β	3·14v	δ	3·5v
γ	3·20	μ	3·6v
ζ	3·31	ι	3·68

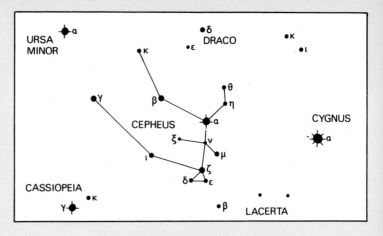

BRIGHTEST STARS

Star			Mag.	R.A.			Dec.			Abs. Mag.	Spectrum	Dist. l/y
				h	m	s	°	′	″			
1	κ		4·40	20	09	39	+77	38	45	−0·6	B9	320
2	θ		4·28	20	29	13	+62	55	11	+1·8	A5	100
6H	(Boss 28956)		4·63	20	41	48	+57	30	01	+2·8	G0	75
3	η		3·41	20	44	51	+61	45	11	+2·7	K0	46
5	α	Alderamin	2·44	21	18	03	+62	29	31	+1·4	A5	52
8	β	Alphirk	3·14v	21	28	23	+70	27	50	−4·1	B1	980
	μ		var.	21	42	50	+58	40	53	var.	M2	1100
10	ν		4·46	21	44	49	+61	01	08	−7·0	A2p	3300
17	ξ	Kurdah	4·57	22	03	09	+64	31	13	+2·3	F7+G	90
21	ζ		3·31	22	10	05	+58	05	33	−4·6	K0	1200
23	ε		4·22	22	14	13	+56	56	00	+2·3	F0	75
27	δ		var.	22	28	21	+58	18	08	−4v	G	1300
32	ι		3·68	22	48	54	+66	05	04	+1·3	K1	95
33	π		4·56	23	07	12	+75	16	07	−0·2	G5	300
35	γ	Alrai	3·20	23	38	26	+77	30	35	+2·2	K1	51
43H			4·50	01	05	28	+86	08	24	—	K0	—

THERE ARE FOUR MORE STARS ABOVE MAGNITUDE 5:

Star	Mag.	Abs. Mag.	Spectrum
9	4·87	−5·9	B2p
11	4·85	+0·1	K1
Boss 31999	4·96	+0·2	K5
34	4·90	+1·2	G7

DOUBLES

Star	R.A.		Dec.		Mags.	P.A.	Dist. ″	
	h	m	°	′				
κ	20	09·7	+77	39	4·4, 8·2	122	7·4	Slow binary
β	21	28·4	+70	28	3·2, 8·2	250	13·7	Fixed
ξ	22	03·2	+64	31	4·6, 6·6	280	7·2	Slow binary
o	23	16·6	+67	50	5·0, 7·3	208	2·7	Binary
δ	22	28·4	+58	18	var, 7·5	192	41·0	Fixed

VARIABLES

Star	R.A.		Dec.		Range	Period, d.	Spectrum	Type
	h	m	°					
T	21	08·9	+68	17	5·4—11·0	389	M	Mira
S	21	35·9	+78	24	7·4—12·9	488	N	Mira
VV	21	55·2	+63	23	6·7—7·5	7430	M+B	Eclipsing
δ	22	28·4	+58	18	3·51—4·42	5·37	F-G	Cepheid
W	22	34·5	+58	10	6·9—8·6	1100?	K	Semi-regular
μ	21	42·8	+58	41	3·6—5·1	—	M	Irregular
U	00	57·7	+81	36	6·6—9·8	2·5	B+G	Algol

CLUSTERS AND NEBULÆ

Object	R.A.		Dec.		Type	Notes
	h	m	°	′		
NGC 7380	22	44·9	+57	49	Open cluster	Mag. 9. Fairly rich
I 1470	23	03·2	+59	59	Planetary	Mag. 8; central star mag. 12
NGC 7510	22	44·9	+57	49	Open cluster	Mag. 9. Fairly rich
NGC 7748	23	42·7	+69	28	Nebula	

CETUS

(Abbreviation: Cet).

One of the largest of all constellations; sometimes associated with the sea-monster of the Perseus legend, at others relegated to the status of a harmless whale. It contains the prototype long-period variable Mira, which can occasionally rise to magnitude 1·7, but which spends most of its period below naked-eye visibility. Cetus abounds in faint galaxies.

Excluding Mira, there are 8 stars above the fourth magnitude:

β 2·02	τ 3·50
α 2·54	ι 3·75
η 3·47	θ 3·83
γ 3·48	ζ 3·92

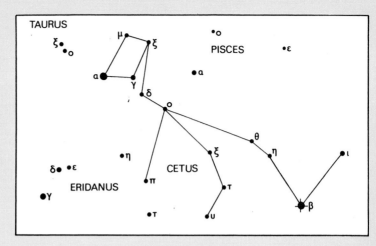

BRIGHTEST STARS

Star			Mag.	R.A. h	m	s	Dec. °	′	″	Abs. Mag.	Spectrum	Dist. l/y
2			4·62	00	02	37	−17	27	30	0·0	A0	270
7			4·68	00	13	31	−19	03	17	+0·4	M1	230
8	ι	Baten Kaitos Shemali	3·75	00	18	18	−08	56	45	−0·1	K0	540
16	β	Diphda	2·02	00	42	29	−18	06	26	+0·8	K0	57
31	η		3·47	01	07	29	−10	17	55	+1·0	K3	102
45	θ		3·83	01	22	55	−08	17	48	+1·1	K0	110
52	τ		3·50	01	43	03	−16	03	10	+5·7	K0	11·9
55	ζ	Baten Kaitos	3·92	01	50	22	−10	26	36	+0·6	K0	140
59	υ		4·18	01	58	58	−21	11	02	−0·1	M0	230
65	ξ¹		4·54	02	11	50	+08	44	39	+0·1	G5	250
68	o	Mira	var.	02	18	14	−03	04	37	−0·5v	M5	103
73	ξ²		4·34	02	26	59	+08	21	44	+0·6	A0	180
82	δ		4·04	02	38	21	+01	14	04	−2·9	B2	800
86	γ	Alkaffal-jidhina	3·48	02	42	10	+03	08	38	+2·0	A2+F7	68
89	π		4·39	02	43	05	−13	57	04	−0·6	B5	330
87	μ		4·36	02	43	45	+10	01	19	+1·9	F0	100
91	λ		4·69	02	58	32	+08	49	13	−0·5	B5	360
92	α	Menkar	2·54	03	01	08	+04	00	15	−0·2	M0	130

THERE ARE SIX MORE STARS ABOVE MAGNITUDE 5:

Star		Mag.	Abs. Mag.	Spectrum
17	φ¹	4·93	+0·8	G6
20		4·92	−0·8	M0
53	χ	4·77	+2·9	F1
72	ρ	4·90	+1·3	B9
76	σ	4·82	+2·2	F3
96	κ	4·96	+5·1	G5

DOUBLES

Star	R.A. h	m	Dec. °	′	Mags.	P.A.	Distance ″
37	01	11·9	−08	11	5·2, 7·8	331	49·6 Fixed
66	02	10·2	−02	38	5·7, 7·7	232	16·3 Near Mira
o	02	18·2	−03	05	var, 10·0	131	0·8 Binary
γ	02	42·2	+03	02	3·7, 6·4	293	3·0 Slow binary

VARIABLES

Star	R.A. h	m	Dec. °	′	Range	Period, d.	Spectrum	Type
W	23	59·6	−14	58	7·1—14·6	351	S	Mira
S	00	21·5	−09	36	7·6—14·7	320	M	Mira
UV	01	36·4	−18	13	6·8—12·9	—	dM	Flare star
o	02	18·2	−03	05	1·7—10·1	332	M	Mira
R	02	23·6	−00	24	7·2—14·0	166	M	Mira
U	02	31·3	−13	22	6·8—13·4	235	M	Mira

CETUS

CLUSTERS AND NEBULÆ

Object	R.A. h m	Dec. ° '	Type	Notes
NGC 45	00 11·4	−23 27	Spiral galaxy	Mag. 12
NGC 210	00 38·0	−14 09	Spiral galaxy	Mag. 12. Type Sb
NGC 246	00 44·6	−12 09	Planetary	Mag. 8·5. Oval. Central star mag. 11
NGC 247	00 44·6	−21 01	Spiral galaxy	Mag. 9·5
NGC 428	01 10·4	+00 43	Spiral galaxy	Mag. 11·7. Type Sc
NGC 578	01 28·0	−22 56	Spiral galaxy	Mag. 11·4. Type Sc
NGC 584	01 28·8	−07 07	Elliptical galaxy	Mag. 11·4. Type E4
NGC 779	01 57·2	−06 12	Barred spiral galaxy	Mag. 11·3
NGC 908	02 20·8	−21 27	Spiral galaxy	Mag. 11·5. Type Sc
NGC 936	02 25·1	−01 22	Barred spiral galaxy	Mag. 11·2. Type SBa
NGC 1052	02 38·6	−08 28	Elliptical galaxy	Mag. 11·2. Type E2
NGC 1068 M.77	02 40·1	−00 14	Seyfert galaxy	Mag. 8·9. Radio source
NGC 1073	02 41·2	+01 10	Barred spiral galaxy	Mag. 11·4. Type SBc
NGC 1087	02 43·9	−00 42	Spiral galaxy	Mag. 11·4. Type Sc

CHAMÆLEON

(Abbreviation: Cha).

A small southern constellation of no particular note; there are no legends attached to it, and there are no stars above the fourth magnitude.

See chart for Musca

BRIGHTEST STARS

Star	Mag.	R.A. h m s	Dec. ° ' "	Abs. Mag.	Spectrum	Dist. l/y
α	4·08	08 19 07	−76 51 02	+2·5	F5	65
θ	4·26	08 21 20	−77 24 51	+0·6	K0	170
γ	4·10	10 35 13	−78 29 37	−2·9	M0	800
δ 2	4·62	10 45 36	−80 25 27	−0·9	B3	400
β	4·38	12 17 02	−79 11 25	0·0	B5	250

The only other star down to the fifth magnitude is ϰ ; mag. 5.00, spectrum K6.

DOUBLE

Star	R.A. h m	Dec. ° '	Mags.	P.A.	Dist."	
ε	11 57·1	−77 57	5·5, 6·3	183	1·1	Binary

NEBULA

Object	R.A. h m	Dec. ° '	Type	Notes
NGC 3195	10 10·1	−80 37	Planetary	Faint object

CIRCINUS

(Abbreviation: Cir).

A very small southern constellation, in the area of α and β Centauri. It is associated with no legends. The only star above magnitude 4 is α (3·18).

See chart for Centaurus

BRIGHTEST STARS

Star	Mag.	R.A. h m s	Dec. ° ' "	Abs. Mag.	Spectrum	Dist. l/y
α	3·18	14 40 43	−64 52 50	+1·6	F0	66
β	4·16	15 15 47	−58 43 13	—	A3	—
γ	4·54	15 21 37	−59 14 34	−0·1	B5+F8	270

There is one more star above magnitude 5; this is ε (mag. 4·84, absolute mag. −5·2, spectrum K4). It is extremely luminous, and more than 3000 light-years away.

DOUBLES

Star	R.A. h m	Dec. ° '	Mags.	P.A.	Dist.
α	14 40·7	−64 53	3·4, 8·8	235	15·8. P.A. decreasing
γ	15 21·6	−59 15	5·2, 5·3	051	1·3. Slow binary

CLUSTERS AND NEBULÆ

Object	R.A. h m	Dec. ° '	Type	Notes
NGC 5315	13 50·2	−66 16	Planetary	Very faint (mag. 13)
NGC 5715	14 39·8	−57 20	Open cluster	Mag. 10. At least 30 stars
NGC 5823	15 01·9	−55 24	Open cluster	Mag. 9. 80 stars; quite rich
NGC 5925	15 23·9	−54 21	Open cluster	Mag. 8·3. Moderate cluster

COLUMBA

(Abbreviation: Col).

A 'modern' constellation dating from 1679, when it was introduced to the sky by Royer. Apparently it represents the dove which Noah released from the Ark; it was originally called Columba Noachi. There are four stars above magnitude 4:

α 2·64	ε 3·92
β 3·12	δ 3·98

δ Columbæ was formerly called 3 Canis Majoris.

BRIGHTEST STARS

Star		Mag.	R.A. h m s	Dec. ° ′ ″	Abs. Mag.	Spectrum	Dist. l/y
ε		3·92	05 30 26	−35 29 10	−0·5	K0	250
α	Phakt	2·64	05 38 51	−34 05 06	−0·6	B5p	140
β	Wezen	3·12	05 50 11	−35 46 33	0·0	K1	140
γ		4·36	05 56 45	−35 17 06	−2·4	B3	650
η		4·03	05 58 28	−42 48 57	−0·4	K0	250
ϰ		4·51	06 15 46	−35 07 57	+0·5	K0	200
δ		3·98	06 21 19	−33 25 28	+0·3	G5	180

THERE ARE TWO FURTHER STARS ABOVE MAGNITUDE 5:

Star	Mag.	Abs. Mag.	Spectrum
ο	4·91	+1·7	K0
λ	4·89	+0·1	B5

VARIABLE

Star	R.A. h m	Dec. ° ′	Range	Period, d.	Spectrum	Type
T	05 17·5	−33 45	6·6—12·7	225	M	Mira

CLUSTERS AND NEBULÆ

Object	R.A. h m	Dec. ° ′	Type	Notes
NGC 1851	05 12·4	−40 05	Globular	Mag. 8·1
NGC 1792	05 30·5	−38 04	Spiral galaxy	Mag. 10·7
NGC 2188	06 08·3	−34 05	Irregular glaxy	Mag. 13

From the British Isles, Columba is only partly visible; of the brighter stars in it, η never rises.

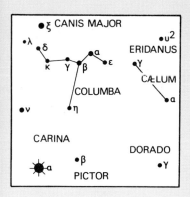

COMA BERENICES

(Abbreviation: Com).

At first glance this constellation gives the impression of being a vast, dim cluster. This is not the case; Coma has no star above magnitude 4·2, but it abounds in faint ones, and there are many telescopic galaxies.

Though the constellation is not 'original', there is a legend attached to it. When Ptolemy Euergetes, King of Egypt, set out in an expedition against the Assyrians, his wife Berenice vowed that if he returned safely she would cut off her lovely hair and place it in the temple of Venus. The King returned; the Queen kept her vow, and Jupiter placed the shining tresses in the sky.

See chart for Boötes

BRIGHTEST STARS

Star		Mag.	R.A.			Dec.			Abs. Mag.	Spectrum	Dist. l/y
			h	m	s	°	′	″			
15 γ		4·56	12	25	51	+28	23	26	−0·2	K0	300
42 α	Diadem	4·22	13	08	55	+17	38	43	+4·0	F4	60
43 β		4·32	13	10	51	+27	59	21	+4·8	G0	26

THERE ARE FIVE FURTHER STARS ABOVE MAGNITUDE 5:

Star	Mag.	Abs. Mag.	Spectrum
11	4·91	+1·4	G6
12	4·73	+0·7	F2
23	4·78	−0·4	A0
36	4·96		M0
41	4·90	−0·3	K5

VARIABLE

Star	R.A.		Dec.		Range	Period, d.	Spectrum	Type
	h	m	°	′				
R	12	01·7	+19	04	7·3—14·6	362	M	Mira

CLUSTERS AND NEBULÆ

Object		R.A.		Dec.		Type	Notes
		h	m	°	′		
NGC 4192	M.98	12	11·3	+15	11	Galaxy	Mag. 10·7. Type Sb
NGC 4212		12	13·1	+14	11	Galaxy	Mag. 12. Type Sc
NGC 4251		12	15·7	+28	27	Galaxy	Mag. 10·2. Type Sa
NGC 4254	M.99	12	16·3	+14	42	Galaxy	Mag. 10. Type Sc
NGC 4274		12	17·4	+29	53	Galaxy	Mag. 11. Type Sb
NGC 4278		12	17·7	+29	34	Galaxy	Mag. 10. Type E1
NGC 4321	M.100	12	20·4	+16	06	Galaxy	Mag. 10·6. Type Sc
NGC 4350		12	21·4	+16	58	Galaxy	Mag. 12. Type E7
NGC 4382	M.85	12	22·8	+18	28	Galaxy	Mag. 9. Type Ep
NGC 4394		12	23·4	+18	29	Galaxy	Mag. 11. Type SBb
NGC 4414		12	24·0	+31	30	Galaxy	Mag. 10. Type Sc
NGC 4448		12	25·8	+28	54	Galaxy	Mag. 11·4. Type Sb
NGC 4459		12	26·5	+14	15	Galaxy	Mag. 11. Type E2
NGC 4494		12	28·9	+26	03	Galaxy	Mag. 10. Type E1
NGC 4501	M.88	12	29·5	+14	42	Galaxy	Mag. 10. Type Sb
NGC 4548		12	32·9	+14	46	Galaxy	Mag. 11. Type SBb
NGC 4559		12	33·5	+28	14	Galaxy	Mag. 11. Type Sc
NGC 4565	H.V.24	12	33·9	+26	16	Galaxy	Mag. 10. Type Sb
NGC 4571	M.91?	12	34·3	+14	28	Galaxy	Mag. 12. Identification uncertain
NGC 4689		12	45·2	+14	01	Galaxy	Mag. 11·5. Type Sb
NGC 4793		12	52·3	+29	13	Galaxy	Mag. 12. Type Sc
NGC 4826	M.64	12	54·3	+21	57	Galaxy	Mag. 9. Type Sb. 'Black-eye' Galaxy
NGC 5012		13	09·3	+23	11	Galaxy	Mag. 11. Type Sb
NGC 5024	M.53	13	10·5	+18	26	Globular	Mag. 7·6
NGC 5053		13	13·9	+11	57	Globular	Mag. 10·5

M.91 is one of the 'missing' Messier objects. It may have been a passing comet, but there is at least a chance that it is identical with NGC 4571.

CORONA AUSTRALIS

(Abbreviation: CrA).

An original constellation. It has no bright stars, but is easy to recognize because of its distinctive shape.

See chart for Sagittarius

BRIGHTEST STARS

Star	Mag.	R.A. h	m	s	Dec. °	'	"	Abs. Mag.	Spectrum	Dist. l/y
θ	4·69	18	31	56	−42	19	47	−1·1	G5	450
γ	4·26	19	04	56	−37	05	47	+3·0	F8	60
δ	4·66	19	06	49	−40	31	56	+0·5	K0	220
α	4·12	19	07	59	−37	56	25	+1·6	A2	100
β	4·16	19	08	31	−39	22	38	−1·1	G5	360

THERE ARE TWO FURTHER STARS ABOVE MAGNITUDE 5:

One of which (ε) is a W Ursæ Majoris type variable with a very small magnitude range and a period of 0·6 day.

Star	Mag.	Abs. Mag.	Spectrum
ε	4·9v	+1·9	F0
ζ	4·85	+1·7	A0

DOUBLES

Star	R.A. h	m	Dec. °	'	Mags.	P.A.	Dist.	
ϰ	18	29·9	−38	46	5·9, 6·5	359	21·6	Fixed
γ	19	04·9	−37	06	5·0, 5·1	054	2·7	Binary

CLUSTERS AND NEBULÆ

Object	R.A. h	m	Dec. °	'	Type	Notes
NGC 6541	18	04·4	−43	44	Globular	Mag. 6
NGC 6726-7	18	58·4	−36	58	Nebula	
NGC 6729	18	58·4	−37	02	Nebula	Variable; associated with R CrA
I 4812	19	01·8	−37	08	Nebula	
I 1297	19	14·0	−39	42	Planetary	Mag. 13; appears almost stellar

CORONA BOREALIS

(Abbreviation: CrB)

A small but very distinctive constellation representing a crown given by Bacchus to Ariadne, the daughter of King Minos of Crete. There are three stars above magnitude 4; α is an eclipsing binary with a very small range:

α 2.23v   γ 3·93

β 3·72

See chart for Boötes

BRIGHTEST STARS

Star			Mag.	R.A. h	m	s	Dec. °	'	"	Abs. Mag.	Spectrum	Dist. l/y
3	β	Nusakan	3·72	15	26	55	+29	10	51	+1·2	F0p	100
4	θ		4·17	15	32	02	+31	25	57	0·0	B5	220
5	α	Alphekka	2·23v	15	33	45	+26	47	16	+0·4	A0	76
8	γ		3·93	15	41	49	+26	21	52	+0·7	A0	140
10	δ		4·73	15	48	40	+26	08	05	+1·3	G5	155
13	ε		4·22	15	56	41	+26	56	26	+0·1	K0	170
19	ξ		4·72	16	21	14	+30	56	32	+0·6	K0	220

THERE ARE THREE FURTHER STARS ABOVE MAGNITUDE 5:

Star		Mag.	Abs. Mag.	Spectrum
11	ϰ	4·77	+1·9	K1
14	ι	4·91	+1·3	A1
16	τ	4·94	+2·2	K1

DOUBLES

Star	R.A. h	m	Dec. °	'	Mags.	P.A.	Dist."	
η	15	21·1	+30	28	5·7, 6·0		Binary, period 42 years; dist. <1"·1
ζ	15	37·5	+36	48	5·1, 6·0	305	6·3	Fixed
σ	16	12·8	+33	59	5·7, 6·7	229	6·2	Opening

VARIABLES

Star	R.A. h	m	Dec. °	'	Range	Period, d.	Spectrum	Type
U	15	16·1	+31	50	7·0—8·3	3·5	A	Algol
S	15	19·4	+31	33	6·6—14·0	361	M	Mira
R	15	46·5	+28	19	5·8—15	—	Gp	Prototype R CrB
V	15	47·7	+39	43	6·9—12·2	358	N	Mira
T	15	57·4	+26	04	2·0—10·8	—	Q+M	Recurrent nova
W	16	13·6	+37	54	7·8—14·3	238	M	Mira

T Coronæ (the 'Blaze Star') showed outbursts in 1866 and in 1946.

CORVUS

(Abbreviation: Crv).

An original group. When the god Apollo became enamoured of Coronis, mother of the great doctor Æsculapius, he sent a crow to watch her and report on her behaviour. To be candid, the crow's report was decidedly adverse; but Apollo rewarded the bird with a place in the sky!

Corvus is distinctive, since its leading stars form a quadrilateral. There are four stars above magnitude 4:

γ 2·59   δ 2·97
β 2·66   ε 3·04

Curiously, the star lettered α is more than a magnitude fainter than any of these.

See chart for Hydra

BRIGHTEST STARS

Star			Mag.	R.A.			Dec.			Abs. Mag.	Spectrum	Dist. l/y
				h	m	s	°	′	″			
1	α	Alkhiba	4·18	12	07	17	−24	36	23	+2·8	F2	60
2	ε		3·04	12	08	59	−22	29	51	−0·2	K0	140
4	γ	Minkar	2·59	12	14	40	−17	25	12	−3·1	B8	450
7	δ	Algorel	2·97	12	28	43	−16	23	36	+0·1	A0	125
8	η		4·42	12	30	56	−16	04	28	+3·1	F0	60
9	β	Kraz	2·66	12	33	14	−23	16	31	+0·1	G5	110

There are no other stars above the fifth magnitude.

DOUBLE

Star	R.A.		Dec.		Mags.	P.A.	Dist.″	
	h	m	°	′				
δ	12	28·8	−16	24	3·0, 8·4	212	24·2	Fixed

VARIABLE

Star	R.A.		Dec.		Range	Period, d.	Spectrum	Type
	h	m	°	′				
R	12	17·0	−18	59	6·7–14·4	317	M	Mira

NEBULÆ

Object	R.A.		Dec.		Type	Notes
	h	m	°	′		
NGC 4027	11	57·0	−18	59	Galaxy	Mag. 12. Type Sc
NGC 4361	12	21·9	−18	29	Planetary	Mag. 11; central star mag. 12·8. Oval

CRATER

(Abbreviation: Crt).

Like Corvus, a small constellation adjoining Hydra; it has been identified with the wine-goblet of Bacchus. The only star above the fourth magnitude is δ (3·82).

See chart for Hydra

BRIGHTEST STARS

Star			Mag.	R.A.			Dec.			Abs. Mag.	Spectrum	Dist. l/y
				h	m	s	°	′	″			
7	α	Alkes	4·20	10	58	42	−18	10	53	+0·7	K0	160
11	β		4·52	11	10	34	−22	42	20	+2·9	A2	70
12	δ		3·82	11	18	14	−14	39	34	+0·8	K0	130
15	γ		4·14	11	23	47	−17	33	47	+0·8	A5	150

THERE ARE TWO FURTHER STARS ABOVE THE FIFTH MAGNITUDE:

Star		Mag.	Abs. Mag.	Spectrum
21	θ	4·81	+0·2	B2
27	ζ	4·90	+1·0	G8

DOUBLE

Star	R.A.		Dec.		Mags.	P.A.	Dist.″	
	h	m	°	′				
γ	11	23·8	−17	34	4·1, 9·5	097	5·2	Fixed

NEBULÆ

Object	R.A.		Dec.		Type	Notes
	h	m	°	′		
NGC 3511	11	00·8	−22	50	Galaxy	Mag. 12. Type Sc
NGC 3513	11	01·1	−22	58	Galaxy	Mag. 12. Type Sc
NGC 3672	11	22·5	−09	32	Galaxy	Mag. 12·4. Type Sb
NGC 3887	11	44·6	−16	35	Galaxy	Mag. 12. Type Sc

CRUX AUSTRALIS

(Abbreviation: Cru).

Though Crux is the smallest constellation in the entire sky, it is also one of the most famous. Before Royer introduced it, in 1679, it had been included in Centaurus. Strictly speaking it is more like a kite than a cross. As well as its brilliant stars it contains the glorious 'Jewel Box' cluster, and also the dark nebula known as the Coal Sack.

There are five stars above the fourth magnitude:

α 0·83   δ 2·78v
β 1·28   ε 3·57
γ 1·69

Of the four main stars, any casual glance will show the difference between γ, a red giant, and the other three, which are hot and bluish-white.

See chart for Centaurus

BRIGHTEST STARS

Star		Mag.	R.A.			Dec.			Abs. Mag.	Spectrum	Dist. l/y
			h	m	s	°	′	″			
θ		4·5	12	01	54	−63	11	26	+0·5	A5	200
η		4·30	12	05	43	−64	29	28	+2·7	F0	65
δ		2·78v	12	13	58	−58	37	36	−3·4	B3	570
ζ		4·26	12	17	14	−63	52	52	−2·4	B3	670
ε		3·57	12	20	10	−60	16	48	0·0	K2	170
α	Acrux	0·83	12	25	22	−62	58	39	−3·9, −3·4	B1+B1	370
γ		1·69	12	29	56	−56	59	25	−2·5	M3	220
ι		4·68	12	44	20	−60	51	39	+1·8	K0	120
β	Mimosa	1·28	12	46	26	−59	34	08	−4·6	B1	490
μ		4·3	12	53	17	−57	03	31	−2·4	B3+B3	670

There are two further stars above magnitude 5, of these θ², mag 4·98, is a close optical companion of θ¹. The other is Boss 17352; mag 4·86, absolute mag −2·4, spectrum B5. μ² (mag 5·45) is physically associated with μ¹.

DOUBLES

Star	R.A.		Dec.		Mags.	P.A.	Dist."	
	h	m	°	′				
γ	12	29·9	−56	59	1·6, 6·7	031	110·6	Optical
α	12	25·4	−60	59	1·4, 1·9	114	4·7	Third star in field
ι	12	44·3	−60	52	4·7, 7·8	027	26·4	Optical

VARIABLES

Star	R.A.		Dec.		Range	Period, d.	Spectrum	Type
	h	m	°	′				
T	12	18·6	−62	00	7·0—7·7	6·7	G	Cepheid
R	12	20·9	−61	21	6·9—8·0	5·8	G	Cepheid
S	12	51·4	−58	10	6·6—7·7	4·7	G	Cepheid

CLUSTERS

Object		R.A.		Dec.		Type	Notes
		h	m	°	′		
NGC 4052		12	00·6	−62	54	Open cluster	Mag. 9. About 50 stars
NGC 4337		12	21·2	−57	50	Open cluster	Mag. 10. Fairly rich
NGC 4349		12	21·4	−61	37	Open cluster	Mag. 8. Rich; over 100 stars
NGC 4463		12	27·1	−64	30	Open cluster	Mag. 8·5. Over 20 stars
NGC 4609		12	39·4	−62	42	Open cluster	Mag. 10. Over 20 stars
NGC 4755 △ 301		12	50·7	−60	05	Open cluster	κ Crucis—the 'Jewel Box'

CYGNUS

(Abbreviation: Cyg).

Cygnus is one of the richest constellations in the sky; it is often nicknamed the Northern Cross – certainly it is much more nearly cruciform than is Crux. Various legends are associated with it. According to one, the group was placed in the sky to honour a swan into which Jupiter once transformed himself when on a visit to the wife of the King of Sparta!

There are eleven stars above the fourth magnitude. To these must be added the red variable χ, which can rise above magnitude 4 at times.

α 1·26   β 3·07   ι 3·94
γ 2·22   ·ζ 3·25   o² 3·95
ε 2·46   τ 3·82   χ 3·98
δ 2·87   ξ 3·92

BRIGHTEST STARS

Star			Mag.	R.A. h	m	s	Dec. °	′	″	Abs. Mag.	Spectrum	Dist. l/y
1	κ		3·98	19	16	36	+53	19	38	+0·7	K0	150
10	ι		3·94	19	29	09	+51	40	56	0·0	A2	200
6	β	Albireo	3·07	19	29	50	+27	54	45	−2·4, −0·1	K0+A0	410
13	θ		4·64	19	35	51	+50	10	10	−1·5	F5	550
12	φ		4·79	19	37	24	+30	02	13	−0·2	K0	320
18	δ		2·87	19	44	17	+45	04	35	−1·7	A0	270
	χ		var.	19	49	43	+32	51	28	var.	M7	230
21	η		4·03	19	55	29	+35	01	27	+0·2	K0	190
32	o ²		3·95	20	12	56	+46	40	26	−2·5	K0+B8	650
33			4·32	20	12	53	+56	30	00	+0·8	A3	160
32			4·16	20	14	47	+47	38	46	−1·9	B8	550
34	P		var.	20	15	57	+37	52	35	var.	Pec.	4600
37	γ	Sadr	2·22	20	21	26	+40	11	08	−4·6	F8p	750
39			4·60	20	22	59	+32	07	06	+0·3	K5	230
41			4·09	20	28	30	+30	17	40	−1·4	F5p	400
50	α	Deneb	1·26	20	40	41	+45	12	04	−7·1	A2p	1600
52			4·34	20	44	45	+30	38	19	+0·5	K0	190
53	ε	Gienah	2·46	20	45	19	+33	53	14	+0·7	K0	74
54	λ		4·47	20	46	33	+36	24	33	−1·0	B5	400
57			4·68	20	52	28	+44	18	12		B3	
58	ν		4·04	20	56	21	+41	04	55	−0·4	A0	250
62	ξ		3·92	21	04	08	+43	50	22	−2·2	K5	550
64	ζ		3·25	21	12	00	+30	08	10	−2·2	K0	390
65	τ		3·82	21	13	55	+37	57	04	+2·2	F0	70
67	σ		4·28	21	16	33	+39	18	07	−6·2	A0p	3300
66	υ		4·42	21	17	01	+34	48	14	−0·4	B3p	300
73	ρ		4·22	21	33	09	+45	29	39	+0·1	K0	430
78	μ		4·4	21	43	09	+28	38	34	+4·5, +3·3	F5+F5	65
81	π ²		4·26	21	45	59	+49	12	26	−2·7	B3	800

THERE ARE 14 FURTHER STARS ABOVE THE FIFTH MAGNITUDE:

Star			Mag.	Abs. Mag.	Spectrum
2			4·86	−2·1	B5
8			4·85	−0·9	B5
22			4·87	−1·2	B6
24	ψ		4·90	−0·1	A0
28	b²		4·82	−1·3	B3
30			4·96	+0·5	A0
29	b³		4·98	+1·4	A3
45	ω¹		4·89	−1·6	B3
47			4·85	−3·6	K4+A3
55			4·89	−7·0	B3
59	f¹		4·86	−4·0	B1
63	f²		4·88	−1·2	K6
72			4·98	+0·2	G7
80	π¹	Azelfafage	4·78	−1·3	B3

DOUBLES

Star	R.A. h	m	Dec. °	′	Mags.	P.A.	Dist.″	
β	19	29·8	+27	55	3·2, 5·4	055	34·6	Fixed Yellow, blue
δ	19	44·3	+45	05	3·0, 6·5	246	2·1	Binary: 321 years
γ	20	21·4	+40	11	2·3, 9·6	196	141·7	Optical
61	21	04·7	+38	30	5·5, 6·3	140	27·0	Widening

VARIABLES

Star	R.A. h	m	Dec. °	′	Range	Period, d.	Spectrum	Type
CH	19	23·2	+50	09	6·6—7·8	97	M	Semi-regular
R	19	35·5	+50	05	6·5—14·2	426	S	Mira
RT	19	42·2	+48	39	6·4—12·7	190	M	Mira
χ	19	49·7	+32	51	3·3—14·2	407	S	Mira
Z	20	00·0	+49	54	7·6—14·7	264	M	Mira
P	20	15·9	+37	53	3—6	—	Pec	Prototype P Cyg
CN	20	16·9	+59	38	7·3—14·0	199	M	Mira
U	20	18·1	+47	44	6·7—11·4	465	N	Mira
V	20	39·7	+47	58	7·7—13·9	421	N	Mira
X	20	41·4	+35	24	6·5—8·2	16·4	F-G	Cepheid
W	21	34·1	+45	09	5·0—7·6	±130	M	Semi-regular?
T	20	45·2	+34	11	5·0—5·5	—	K	Irregular?
WY	21	46·9	+44	01	7·6—14·9	304	M	Mira

CLUSTERS AND NEBULÆ

Object		R.A. h m	Dec. ° ′	Type	Notes
NGC 6826		19 43·4	+50 24	Planetary	Mag. 9; centre star mag. 10·8
NGC 6871		20 04·0	+35 38	Open cluster	Mag. 5·6 (27 Cygni). 60 stars
NGC 6884		20 08·8	+46 19	Planetary	Mag. 13. Very faint centre star
NGC 6913	M.29	20 22·2	+38 21	Open cluster	Mag. 7. Sparse; about 20 stars
NGC 7027		21 05·1	+42 02	Planetary	Mag. 10. Centre star mag. 17
NGC 7086		21 29·8	+51 22	Open cluster	Mag. 9·4. About 50 stars
NGC 7092		21 30·4	+48 13	Open cluster	Mag. 5·2. About 25 stars
NGC 6946		21 33·9	+59 58	Galaxy	Mag. 10. Type Sc
NGC 6914		20 23·4	+42 10	Nebula	Near γ
NGC 6960		20 43·6	+30 32	Nebula	Cirrus Nebula (52 Cygni)
I 5067		20 46·9	+44 11	Nebula	Pelican Nebula, near Deneb
NGC 6992-5		20 54·3	+31 30	Nebula	Cirrus Nebula
I 5146		21 51·3	+47 02	Nebula	Cocoon Nebula

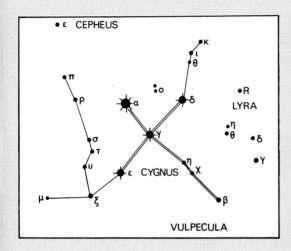

DELPHINUS
(Abbreviation: Del.)

A small but compact constellation; one of Ptolemy's originals. It honours the dolphin which carried the great singer Arion to safety, after he had been thrown overboard by the crew of the ship carrying him home after winning all the prizes in a competition. The curious names of α and β were allotted by one Nicolaus Venator, for reasons which are obvious!

There are four stars above the fourth magnitude:

β 3·72   ε 3·98
α 3·86   γ 3·9 (4·5 + 5·5)

See chart for Aquila

BRIGHTEST STARS

Star			Mag.	R.A.			Dec.			Abs. Mag.	Spectrum	Dist. l/y
				h	m	s	°	′	″			
2	ε		3·98	20	32	10	+11	13	39	−0·6	B5	270
4	ζ		4·69	20	34	17	+14	35	51	+0·8	A2	190
6	β	Rotanev	3·72	20	36	31	+14	31	04	+1·4	F3	95
9	α	Svalocin	3·86	20	38	37	+15	50	01	−0·7	B8	270
11	δ		4·53	20	42	26	+14	59	41	+0·1	A5	250
12	γ		3·9	20	45	39	+16	02	39	+2·8, +1·8	F6+G5	110

There are no other stars above the fifth magnitude.

DOUBLE

Star	R.A.		Dec.		Mags.	P.A.	Dist.″
	h	m	°	′			
γ	20	45·7	+16	03	4·5, 5·5	269	10·4 Fixed

VARIABLES

Star	R.A.		Dec.		Range	Period, d.	Spectrum	Type
	h	m	°	′				
R	20	12·5	+08	56	7·6—13·7	285	M	Mira
EU	20	35·6	+18	06	6·0—6·9	±60	M	Semi-regular
HR	20	40·0	+18	59	3·7—12?	—	Q	Nova
U	20	43·2	+17	54	5·6—7·5	—	M	Irregular
V	20	45·5	+19	09	8·1—15·5	534	M	Mira

The exceptional slow nova HR Delphini (1967) is included here because its pre-nova magnitude was 12, and it is unlikely to decline much below that value.

CLUSTERS AND NEBULÆ

Object	R.A.		Dec.		Type	Notes
	h	m	°	′		
NGC 6905	20	20·2	+19	57	Planetary	Mag. 12; centre star mag. 14
NGC 7006	20	59·1	+16	00	Globular	Mag. 10

DORADO

(Abbreviation: Dor).

A 'modern' southern constellation. The only star above the fourth magnitude is α (3·28), but Dorado contains part of the Large Magellanic Cloud (LMC), in which is the magnificent Tarantula looped nebula round 30 Doradûs.

See chart for Reticulum

BRIGHTEST STARS

Star	Mag.	R.A. h	m	s	Dec. °	′	″	Abs. Mag.	Spectrum	Dist. l/y
γ	4·36	04	15	27	−51	32	30	+3·6	F5	46
α	3·28	04	33	31	−55	05	24	−1·2	A0p	260
β	var.	05	33	26	−62	30	15	var.	F6—G2	900
δ	4·52	05	44	44	−65	44	38	+1·2	A5	300
Boss 7477 (−63°498)	4·53	05	53	56	−63	05	47	+1·4	K0	140

THERE ARE THREE FURTHER STARS ABOVE THE FIFTH MAGNITUDE:

Star	Mag.	Abs. Mag.	Spectrum
ζ	4·76	+4·4	F4
θ	4·78	+0·7	K6
η²	4·88	−2·1	M4

VARIABLES

Star	R.A. h	m	Dec. °	′	Range	Period, d.	Spectrum	Type
β	05	33·4	−62	30	4·5—5·7	9·84	F-G	Cepheid
R	04	36·2	−62	11	7·1—8·1	338	M	Semi-regular

CLUSTERS AND NEBULÆ

Object	R.A. h	m	Dec. °	′	Type	Notes
NGC 1566	04	18·9	−55	04	Galaxy	Mag. 10·5. Spiral
NGC 1617	04	30·6	−54	42	Galaxy	Mag. 12. Spiral
NGC 1672	04	44·9	−59	20	Galaxy	Mag. 11·4. Spiral
NGC 2070 △ 142	06	39·1	−69	09	Nebula	In LMC. Tarantula Nebula, round 30 Dor. Naked-eye

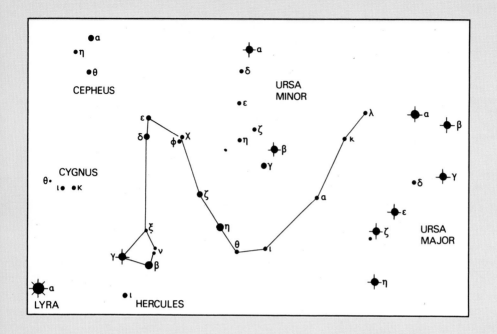

DRACO

(Abbreviation: Dra).

A long, sprawling northern group. In mythology it has been indentified either with the dragon killed by Cadmus before the founding of the city of Bœotia, or with the dragon which guarded the golden apples in the Garden of the Hesperides. Thuban (α) was the pole star in ancient times.

There are eleven stars above the fourth magnitude:

γ	2·22	ξ	3·20	ϰ	3·88
η	2·71	ι	3·28	ξ	3·90
β	2·77	α	3·64	ε	3·99
δ	3·06	χ	3·69		

BRIGHTEST STARS

Star			Mag.	R.A.			Dec.			Abs. Mag.	Spectrum	Dist. l/y
				h	m	s	°	′	″			
Boss 13174	1H		4·58	09	34	04	+81	25	31	−0·2	K2	300
1	λ	Giansar	4·06	11	30	07	+69	27	09	+0·2	M0	190
5	ϰ		3·88	12	32	33	+69	54	33	−0·9	B5p	300
11	α	Thuban	3·64	14	03	48	+64	28	50	−0·5	A0p	220
12	ι	Edasich	3·28	15	24	26	+59	02	33	+0·8	K0	100
13	θ		4·11	16	01	28	+58	37	25	+2·5	F8	68
14	η	Aldhibain	2·71	16	23	41	+61	33	49	+0·9	G8	76
22	ζ	Aldhibah	3·20	17	08	43	+65	44	30	−3·2	B6	620
23	β	Alwaid	2·77	17	29	56	+52	19	01	−2·1	G0	310
32	ξ	Juza	3·90	17	53	09	+56	52	32	+1·4	K0	105
33	γ	Etamin	2·22	17	56	06	+51	29	27	−0·6	K5	117
43	φ		4·24	18	21	04	+71	19	35	−0·6	A0p	330
44	χ		3·69	18	21	27	+72	43	25	+4·1	F8	26
57	δ	Taïs	3·06	19	12	33	+67	37	22	+0·2	G9	125
60	τ		4·63	19	15	59	+73	18	54	+0·4	K0	230
58	π		4·63	19	20	36	+65	40	19	+1·0	A2	170
63	ε	Tyl	3·99	19	48	15	+70	12	43	−0·4	K0	250
67	ρ		4·66	20	02	43	+67	48	39	−1·1	K0	460

THERE ARE 15 FURTHER STARS ABOVE THE FIFTH MAGNITUDE:

Star			Mag.	Abs. Mag.	Spectrum	Star			Mag.	Abs. Mag.	Spectrum
10	i		4·77	−0·2	M3	31	ψ	Dziban	4·3	+3·3, +4·2	F5, F6
Boss 21467			4·96	+1·8	A5	42			4·99	+1·0	K1
18	g		5·00	+0·2	K2	45	d		4·95	−4·5	F8
Boss 22584			4·88	+3·0	F1	47			4·78	−0·7	G8
19	h		4·82	+4·1	F7	Boss 25935			4·97	+1·0	G4
24	ν¹	} Kuma	4·98	+2·1	A8	52			4·91	−0·1	K0
25	ν²		4·95	+2·1	A4	61	σ		4·78	+6·1	G8
28	ω		4·87	+2·8	F4						

σ is one of the nearer stars, at 17·9 light-years. μ (mag. 5·01) is named Alrakis.

DOUBLES

Star	R.A.		Dec.		Mags.	P.A.	Dist."	
	h	m	°	′				
η	16	23·7	+61	34	2·7, 8·8	142	6·1	Slow binary
μ	17	04·3	+54	32	5·8, 5·8	081	2·2	Widening
ν	17	31·3	+55	13	4·9, 5·0	312	62·0	Wide physical pair
ψ	17	42·8	+72	11	4·9, 6·1	016	30·3	Fixed
ε	19	48·3	+70	08	4·0, 7·1	012	3·3	Slow binary

VARIABLES

Star	R.A.		Dec.		Range	Period, d.	Spectrum	Type
	h	m	°	,				
TW	15	33·1	+64	04	7·7—10·0	2·8	A+K	Algol
T	17	55·7	+58	14	7·2—13·5	422	N	Mira
R	16	32·5	+66	51	6·9—13·0	246	M	Mira
S	16	41·9	+55	00	8·2—9·4	136	M	Semi-regular

CLUSTERS AND NEBULÆ

Object	R.A.		Dec.		Type	Notes
	h	m	° .	,		
NGC 4125	12	05·7	+65	27	Galaxy	Mag. 10. Type E5
NGC 5907	15	14·6	+56	31	Galaxy	Mag. 11. Type Sb
NGC 5985	15	38·6	+59	30	Galaxy	Mag. 11·4. Type Sb
NGC 6503	17	49·9	+70	10	Galaxy	Mag. 9·6. Type Sb
NGC 6543 H.IV.37	17	58·6	+66	38	Planetary	Mag. 9. Centre star mag. 9·6
NGC 6643	18	21·2	+74	33	Galaxy	Mag. 11·3. Type Sc

EQUULEUS

(Abbreviation: Eql).

A very obscure and small constellation, but one of the 'originals'. It represents a foal given by Mercury to Castor, one of the Heavenly Twins. There is no star as bright as the fourth magnitude. The name often applied to α has a decidedly modern flavour!

See chart for Aquila

BRIGHTEST STARS

Star	Mag.	R.A.			Dec.			Abs. Mag.	Spectrum	Dist. l/y
		h	m	s	°	,	"			
7 δ	4·61	21	13	24	+09	55	02	+3·6	F5	52
8 α Kitalpha	4·14	21	14	43	+05	09	22	+0·8	F8+A3	300

The only other star above the fifth magnitude is γ (4·76); abs. mag. +1·0 spectrum F1p.

TRIPLE STAR

Star	R.A.		Dec.		Mags.	P.A.s	Dists."	
	h	m	°	,				
ε	20	56·6	+04	06	5·7, 7·0, 7·1	322, 072	0·9, 10·9	A and B form a binary (101 years)

ERIDANUS

(Abbreviation: Eri).

An immensely long constellation, extending from Achernar in the far south as far as Kursa, near Orion. Achernar is the only brilliant star, though θ (Acamar) was ranked as the first magnitude in ancient times, and there are suggestions that it has faded. Mythologically, Eridanus is the river Po – and this was the river into which the youth Phæthon was plunged when he had obtained permission to drive the Sun-chariot for a day, and had lost control of it, so that Jupiter was forced to strike him down with a thunderbolt.

There are twelve stars above magnitude 4:

α 0·51	υ⁴ 3·59	ε 3·81
β 2·79	δ 3·72	υ² 3·88
θ 2·92	χ 3·73	σ⁴ 3·95
γ 3·01	φ 3·78	1 3·89

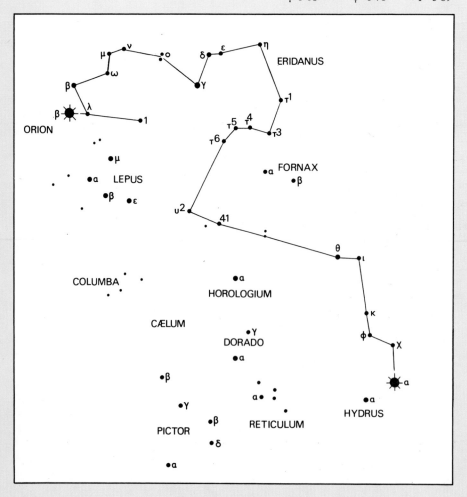

Star		Mag.	R.A.			Dec.			Abs. Mag.	Spectrum	Dist. l/y
			h	m	s	°	′	″			
α	Achernar	0·51	01	36	54	−57	20	54	−2·3	B5	118
χ		3·73	01	55	06	−51	43	05	+2·4	G5	57
φ		3·78	02	15	43	−51	36	49	+0·1	B8	180
κ		4·44	02	26	11	−47	48	08	−1·7	B5	540
s		4·53	02	38	58	−42	59	08		A2	
ι		4·06	02	39	48	−39	56	57	+1·4	K0	105
1	τ¹	4·61	02	44	05	−18	39	54	+3·7	F5	49
3	η Azha	4·05	02	55	21	−08	59	06	+0·9	K0	135
θ	Acamar	2·92	02	57	26	−40	23	32	+0·6, +1·7	A2+A2	65
11	τ³	4·16	03	01	25	−23	42	36	+3·0	A3	55
16	τ⁴ Angetenar	3·95	03	18	32	−21	50	14	−1·3	M3	360
	e 82G	4·30	03	19	03	−43	09	12	+5·2	G5	20
18	ε	3·81	03	31	54	−09	31	55	+6·2	K0	10·7
19	τ⁵	4·32	03	32	49	−21	42	21	−0·1	B8	250
	y	4·58	03	36	18	−40	20	47	+0·1	K0	240
	h	4·64	03	42	01	−37	22	56	+0·7	K2	200
23	δ Rana	3·72	03	42	12	−09	50	14	+4·0	K0	29
26	π	4·64	03	45	06	−12	10	11	−1·1	M2	460
27	τ⁶	4·33	03	45	54	−23	18	50	+3·1	F8	59
	g	4·24	03	48	38	−36	15	58	+0·3	K0	200
34	γ Zaurak	3·01	03	57	00	−13	34	13	−0·5	K5	160
36	τ⁹	4·69	03	58	59	−24	04	41	+0·1	A0p	270
38	o¹ Beid	4·14	04	10	47	−06	53	38	+1·6	F2	105
40	o² Keid	4·48	04	14	16	−07	41	10	+6·1	G5	16
41	υ⁴	3·59	04	17	04	−33	51	05	−0·1	B9	180
43	υ³	4·06	04	23	13	−34	04	02	−1·4	K5	400
50	υ¹	4·59	04	32	39	−29	48	37	+0·9	K0	180
52	υ² Theemini	3·88	04	34	42	−30	36	24	−0·9	K0	300
48	ν	4·12	04	35	13	−03	23	47	−3·5	B2	1000
53	l Sceptrum	3·98	04	37	10	−13	20	46	+1·5	K0	100
54		4·54	04	39	29	−19	42	46	−0·5	M4	330
57	μ	4·18	04	44	21	−03	17	38	−1·7	B5	470
61	ω	4·45	04	51	49	−05	29	19	+1·3	F0	140
67	β Kursa	2·79	05	06	46	−05	06	56	+0·9	A3	78
69	λ	4·34	05	08	06	−08	46	53	−3·2	B3	930

THERE ARE NINE FURTHER STARS ABOVE MAGNITUDE 5:

Star		Mag.	Abs. Mag.	Spectrum
2	τ²	4·81	+0·4	K0
13	ζ Zibal	4·90	+1·8	A3
17	v	4·80	0·0	B9
	f	4·86	+1·8	B8
33	τ⁸	4·76	−1·7	B5
32	w	4·95	+0·2	G4
45		4·97	−0·8	K4
64	S	4·9	+1·6	F0
65	ψ	4·81	−1·7	B8

DOUBLES

Star		R.A.		Dec.		Mags.	P.A.	Dist."	
		h	m	°	′				
θ		02	57·4	−40	24	3·4, 4·4	088	8·5	P.A. increasing
40	o²	04	14·3	−07	41	4·5, 9·4	105	82·8	B is again double

VARIABLES

Star	R.A.		Dec.		Range	Period, d	Spectrum	Type
	h	m	°	′				
Y	02	03·0	−57	23	10·7—13·5	308	M	Mira
Z	02	45·5	−12	40	6·4—7·8	80	M	Semi-regular
T	03	53·1	−24	11	7·4—13·2	252	M	Mira

NEBULÆ

Object	R.A.		Dec.		Type	Notes
	h	m	°	′		
NGC 1084	02	43·5	−07	47	Galaxy	Mag. 11. Type Sc
NGC 1187	03	00·4	−23	04	Galaxy	Mag. 11. Type SBc
NGC 1232	03	07·5	−20	46	Galaxy	Mag. 10·5. Type Sc
NGC 1291	03	15·5	−41	17	Galaxy	Mag. 10·2. Elliptical
NGC 1300	03	17·5	−19	35	Galaxy	Mag. 11·5. Type SBb
NGC 1309	03	19·8	−15	35	Galaxy	Mag. 11·4. Type Sc
NGC 1332	03	24·1	−21	31	Galaxy	Mag. 11. Type E7
NGC 1395	03	36·3	−23	11	Galaxy	Mag. 11·4. Type E3
NGC 1407	03	37·9	−18	44	Galaxy	Mag. 10·6. Type E0
I 2118	05	04·5	−07	17	Nebula	Adjoining Rigel

FORNAX

(Abbreviation: For).

A southern group, originally Fornax Chemica (the Chemical Furnace). It has no bright stars, but is notable for containing a large number of faint galaxies.

The only star above the fourth magnitude is α (3·95).

See chart for Eridanus

BRIGHTEST STARS

Star			Mag.	R.A. h m s			Dec. ° ′ ″			Abs. Mag.	Spectrum	Dist. l/y
ν			4·74	02	03	30	—29	24	07	—1·0	A0p	460
β			4·50	02	48	10	—32	29	52	+1·2	K0	150
α			3·95	03	11	08	—29	04	23	+3·1	F8	45

THERE ARE TWO FURTHER STARS ABOVE THE FIFTH MAGNITUDE:

Star		Mag.	Abs. Mag.	Spectrum
ω		4·95	+0·2	B9
δ		4·93		B5

DOUBLE

Star	R.A. h m	Dec. ° ′	Mags.	P.A.	Dist.″	
ω	02 31·6	—28 27	4·9, 8·7	245	10·8	Slow binary

NEBULÆ

Object	R.A. h m	Dec. ° ′	Type	Notes
NGC 986	02 31·6	—39 15	Galaxy	Mag. 12. Spiral
NGC 1097	02 44·3	—30 29	Galaxy	Mag. 11. Type SB
NGC 1201	03 62·0	—26 15	Galaxy	Mag. 11·7. Type Sa
NGC 1302	03 17·7	—26 14	Galaxy	Mag. 11. Type SBa
NGC 1326	03 22·0	—36 39	Galaxy	Mag. 12. Type SB
NGC 1344	03 26·7	—31 14	Galaxy	Mag. 11·6. Elliptical
NGC 1350	03 29·1	—33 38	Galaxy	Mag. 11·8. Spiral
NGC 1365	03 31·8	—36 18	Galaxy	Mag. 11. Type SB
NGC 1380	03 34·6	—35 09	Galaxy	Mag. 11·4. Elliptical
NGC 1385	03 35·2	—26 40	Galaxy	Mag. 11·5. Type Sc
NGC 1399	03 36·6	—35 37	Galaxy	Mag. 11. Elliptical
NGC 1398	03 36·8	—26 30	Galaxy	Mag. 10·4. Type SBb
NGC 1404	03 37·0	—35 45	Galaxy	Mag. 11·5. Elliptical

GEMINI

(Abbreviation: Gem).

A brilliant Zodiacal constellation. In mythology, Castor and Pollux were the twin sons of the King and Queen of Sparta. Pollux was immortal, but Castor was not. When Castor was killed, Pollux pleaded with the gods to be allowed to share his immortality; so Castor was brought back to life, and both youths placed in the sky.

Today Pollux is the brighter of the two stars, but in ancient times Castor was recorded as being the more brilliant. If any change has occurred (and this is by no means certain) it is more likely to have been in the late-type Pollux than in Castor, which is a multiple system. There are 13 stars in Gemini above the fourth magnitude:

β 1·16	η 3·1 (max)	λ 3·65
α 1·58	ξ 3·38	κ 3·68
γ 1·93	δ 3·51	ζ 3·7
μ 2·92v	θ 3·64	. (max)
ε 3·00		ι 3·89

BRIGHTEST STARS

Star			Mag.	R.A. h m s			Dec. ° ′ ″			Abs. Mag.	Spectrum	Dist. l/y
1			4·30	06	02	47	+23	15	57	+1·5	G5	120
7	η	Propus	var.	06	13	33	+22	30	52	—0·6v	M3	200
13	μ	Tejat	2·92v	06	21	38	+22	31	34	—0·6	M0	160
18	ν		4·06	06	27	39	+20	13	38	—0·2	B5	230
24	γ	Alhena	1·93	06	36	26	+16	25	09	—0·6	A0	105
27	ε	Mebsuta	3·00	06	42	35	+25	09	15	—4·6	G5	1080
30			4·65	06	42	45	+13	15	04	—0·6	K0	360
31	ξ	Alzirr	3·38	06	44	03	+12	55	14	+1·9	F5	64
34	θ		3·64	06	51	20	+33	59	21	+0·5	A2	140
38	e		4·70	06	53	24	+13	12	25	+2·8	F0	170
43	ζ	Mekbuda	var.	07	02	48	+20	36	13	var.	G0v	820
46	τ		4·48	07	09	44	+30	16	57	—0·3	K0	600
54	λ		3·65	07	16	50	+16	34	52	+1·8	A2	75
55	δ	Wasat	3·51	07	18	49	+22	01	26	+2·2	A8	60
60	ι		3·89	07	24	22	+27	50	35	+1·0	K0	250
62	ρ		4·18	07	27	42	+31	49	46	+2·8	F0	125
66	α	Castor	1·58	07	33	12	+31	56	15	+1·3, +2·3	A0+A5	45
69	υ		4·22	07	34	34	+26	56	45	+0·1	K5	430
75	σ		4·26	07	41	56	+28	56	16	—1·6	K0	460
77	κ		3·68	07	43	07	+28	04	50	+1·0	G5	140
78	β	Pollux	1·16	07	43	58	+28	04	50	+1·0	K0	35

THERE ARE TWO FURTHER STARS ABOVE THE FIFTH MAGNITUDE:

Star		Mag.	Abs. Mag.	Spectrum
71	o	4·92	+1·3	F1
83	φ	4·99	+1·3	A4

DOUBLES

Star	R.A. h m	Dec. ° '	Mags.	P.A.	Dist."	
η	06 13·6	+22 31	var, 8·8	278	1·3	Slow binary
μ	06 21·6	+22 32	2·9, 9·8	141	122·5	Optical
λ	07 16·8	+16 35	3·7, 10·0	033	10·0	Fixed
δ	07 18·8	+22 02	3·5, 8·1	211	6·8	Optical
α	07 33·2	+31 56	1·97, 2·95			Binary, 350 years; closing
						Castor C (YY Gem) lies at 73″
ϰ	07 43·1	+24 27	3·7, 9·5	236	6·8	Slow binary

VARIABLES

Star	R.A. h m	Dec. ° '	Range	Period, d	Spectrum	Type
η	06 13·6	+22 31	3·1—3·9	±233	M	Semi-regular
W	06 32·1	+15 22	6·9—7·9	7·9	G	Cepheid
X	06 43·9	+30 20	7·6—13·6	264	M	Mira
ζ	07 02·8	+20 36	3·7—4·3	10·2	G	Cepheid
R	07 04·3	+22 47	6·0—14·0	370	S	Mira
V	07 20·3	+13 12	7·8—14·4	275	M	Mira
T	07 46·3	+23 52	8·0—15·0	288	S	Mira
U	07 52·1	+22 08	8·8—14·4	(c.103)	Gp	Prototype U Gem

CLUSTERS AND NEBULÆ

Object	R.A. h m	Dec. ° '	Type	Notes
NGC 2129	06 58·1	+23 18	Open cluster	Mag. 7·2
NGC 2168 M.35	06 05·8	+24 21	Open cluster	Mag. 5·3; naked-eye
NGC 2392	07 26·2	+21 01	Planetary	Mag. 8·3. Centre star mag. 10·5

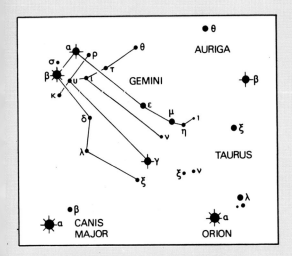

GRUS

(Abbreviation: Gru).

The most distinctive of the 'Southern Birds'. The contrast in colour between α and β is very marked.

There are four stars above the fourth magnitude:

α 1·76   γ 3·03
β 2·11v   ε 3·69

β has a very small range (2·11 to 2·23), so that its fluctuations are not detectable with the naked eye.

BRIGHTEST STARS

Star		Mag.	R.A. h m s	Dec. ° ′ ″	Abs. Mag.	Spectrum	Dist. l/y
γ		3·03	21 52 36	−37 28 08	−3·1	B8	540
λ		4·60	22 04 47	−39 39 00	+0·3	K2	250
α	Alnair	1·76	22 06 51	−47 04 06	+0·3	B5	64
δ¹		4·02	22 27 57	−43 36 31	−0·2	G5	230
δ²		4·31	22 28 27	−43 51 45		M4	
β	Al Dhanab	2·11v	22 41 22	−47 00 01	−2·5	M3	280
ε		3·69	22 47 14	−51 25 58	+1·7	A2	75
ζ		4·18	22 59 35	−52 52 21	+0·9	G5	150
θ		4·35	23 05 39	−43 38 23	+0·8	F5	165
ι		4·10	23 09 07	−45 21 59	+0·9	K0	140

THERE ARE THREE FURTHER STARS ABOVE THE FIFTH MAGNITUDE:

Star	Mag.	Abs. Mag.	Spectrum
μ¹	4·86	−1·6	G2
ρ	4·89	+0·9	G5
η	4·86	+0·6	G8

DOUBLE

Star	R.A. h m	Dec. ° ′	Mags.	P.A.	Dist.″	
θ	23 05·7	−43 38	4·4, 7·0	052	1·4	Slow binary

VARIABLES

Star	R.A. h m	Dec. ° ′	Range	Period, d	Spectrum	Type
R	21 45·3	−47 09	7·4—14·9	333	M	Mira
S	22 23·0	−48 41	6·0—15·0	410	M	Mira

CLUSTERS AND NEBULÆ

Object	R.A. h m	Dec. ° ′	Type	Notes
NGC 7213	22 06·2	−47 25	Galaxy	Mag. 12. Elliptical
NGC 7410	22 52·1	−39 56	Galaxy	Mag. 12. Spiral
NGC 7418	22 53·8	−37 17	Galaxy	Mag. 11·8. Spiral
NGC 7552	23 13·5	−42 53	Galaxy	Mag. 12. Spiral
NGC 7582	23 15·8	−42 38	Galaxy	Mag. 11·8. Spiral
NGC 7590	23 16·3	−42 31	Galaxy	Mag. 12. Spiral

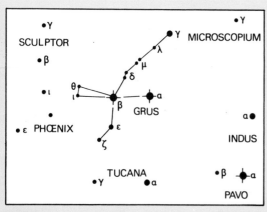

GRUS

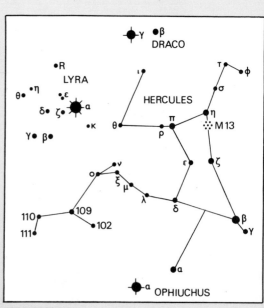

HERCULES

BRIGHTEST STARS

Star		Mag.	R.A. h	m	s	Dec. °	'	"	Abs. Mag.	Spectrum	Dist. l/y
1 χ		4·61	15	51	55	+42	30	45	+3·5	G0	55
6 υ		4·64	16	02	07	+46	05	49	−0·1	B9	300
11 φ		4·26	16	08	04	+44	59	30	0·0	A0	230
22 τ		3·91	16	19	05	+46	21	54	−0·7	B5	270
20 γ		3·79	16	20	57	+19	12	14	+0·6	F0	140
24 ω	Cujam	4·53	16	24	24	+13	04	58	+1·8	A0p	115
27 β	Kornephoros	2·78	16	29	16	+21	32	11	+0·3	K0	103
35 σ		4·25	16	33	24	+42	28	54	+1·1	A0	140
40 ζ	Rutilicus	2·81	16	40	27	+31	38	30	+3·1	G0	30
44 η		3·46	16	42	08	+38	57	48	+2·1	K0	62
58 ε		3·92	16	59	27	+30	57	28	+0·7	A0	140
64 α	Rasalgethi	var.	17	13	39	+14	24	52	−2·3, −0·7	M3+F8	410
65 δ	Sarin	3·14	17	14	08	+24	51	51	+0·8	A2	96
67 π		3·13	17	14	17	+36	49	59	−2·4	K5	400
68 u		4·6v	17	16	31	+33	07	22	var.	B3+B5	460
75 ρ		4·52	17	22	55	+37	09	55	−0·1	A0	270
76 λ	Masym	4·48	17	29	51	+26	07	35	+0·2	K0	230
85 ι		3·79	17	38	51	+46	01	02	−2·7	B3	650
86 μ		3·42	17	45	36	+27	43	58	+3·6	G5	30
91 θ		3·99	17	55	30	+37	15	09	−1·5	K1	410
92 ξ		3·82	17	56	55	+29	14	58	+0·3	K0	160
94 ν		4·48	17	57	40	+30	11	25	−0·7	F0	360
93		4·71	17	59	05	+16	45	04	−0·5	K0	360
103 o		3·83	18	06	41	+28	45	31	−0·6	A0	250
102		4·32	18	07	49	+20	48	36	−3·3	B3	1080
109		3·92	18	22	46	+21	45	31	+0·6	K0	150
110		4·26	18	44	43	+20	31	27	+2·5	F5	60
111		4·37	18	46	03	+18	09	21	+2·2	A3	88
113		4·56	18	53	49	+22	36	59	+0·4	G0+A3	220

HERCULES

(Abbreviation: Her).

A large but by no means brilliant constellation, commemorating the great hero of mythology. The most interesting objects are the red super-giant α and the globular clusters M.13 and M.92; M.13 is the brightest globular in the northern hemisphere of the sky, and is surpassed only by the southern ω Centauri and 47 Tucanæ.

Hercules contains fifteen stars above the fourth magnitude:

β 2·78		μ 3·42		o 3·83	
ζ 2·81		η 3·46		τ 3·91	
α 3 v		ι 3·79		109 3·92	
π 3·13		γ 3·79		ε 3·92	
δ 3·14		ξ 3·82		θ 3·99	

THERE ARE SEVEN FURTHER STARS ABOVE THE FIFTH MAGNITUDE:

Star	Mag.	Abs. Mag.	Spectrum
30 g	var.	var.	M6
29 h	4·92	−0·6	M0
52	4·86	+0·9	A4p
60	4·91	+1·3	A3
69 e	4·80	+0·8	A1
95	4·7	−0·6, −0·6	G3+A1
106	4·98	+0·4	M0

χ (magnitude 5·3) has a proper name: Marsik.

DOUBLES

Star	R.A. h	m	Dec. °	'	Mags.	P.A.	Dist."	
ζ	16	40·5	+31	39	3·1, 5·6	var.	Below 2. Binary: 34 years	
α	17	13·6	+14	25	var, 5·4	109	4·6	Fixed. Red, green
δ	17	14·1	+24	52	3·1 8·8	216	10·0	Optical
ρ	17	22·9	+37	10	4·5, 5·5	317	4·0	Slow binary
μ	17	45·6	+27	44	3·4, 9·9	247	33·5	Slow binary

VARIABLES

Star	R.A. h	m	Dec. °	'	Range	Period, d	Spectrum	Type
R	16	04·0	+18	30	8·2—15·0	319	M	Mira
RU	16	08·2	+25	12	6·9—14·3	485	M	Mira
U	16	23·6	+19	00	7·0—13·4	406	M	Mira
g	16	27·0	+41	59	4·6—6·0	±70	M	Semi-regular
SS	16	30·4	+06	58	8·5—13·2	107	M	Mira
α	17	13·6	+14	25	3·0—4·0	±100	M	Semi-regular
u	17	16·5	+33	07	4·6—5·3	2·1	B	β Lyræ
RS	17	19·6	+22	58	7·4—12·9	220	M	Mira
T	18	07·2	+31	01	7·1—13·6	165	M	Mira

CLUSTERS

Object		R.A. h	m	Dec. °	'	Type	Notes
NGC 6205	M.13	16	39·9	+36	33	Globular	Mag. 5·7; naked-eye
NGC 6341	M.92	17	15·6	+43	12	Globular	Mag. 6·1. Little inferior to M.13

HOROLOGIUM

(Abbreviation: Hor).

A very obscure southern constellation. The only star above the fourth magnitude is α: magnitude 3·83, R.A. 04h 13m 16s, dec. —42° 20′ 53″, absolute magnitude +0·5, spectrum K0, distance 150 light-years.

The only other star above the fifth magnitude is δ; magnitude 4·85, absolute magnitude +2·8, spectrum F0.

See chart for Eridanus

VARIABLES

Star	R.A. h m	Dec. ° ′	Range	Period, d	Spectrum	Type
R	02 52·3	—50 06	4·7—14·3	403	M	Mira
T	02 59·5	—50 50	7·2 —13·7	217	M	Mira

CLUSTERS AND NEBULÆ

Object	R.A. h m	Dec. ° ′	Type	Notes
NGC 1261	03 10·9	—55 25	Globular	Mag. 8·5
NGC 1493	03 55·9	—46 21	Galaxy	Mag. 11·5. Spiral
NGC 1532	04 10·2	—33 00	Galaxy	Mag. 11·8. Spiral

HYDRA

(Abbreviation: Hya).

The largest constellation in the sky, representing the hundred-headed monster which lived in the Lernæan marshes until it was killed by Hercules. It extends for over 6 hours of R.A. α (Alphard) is often known as 'the Solitary One', because there are no other bright stars anywhere in the neighbourhood.

Hydra contains 9 stars above the fourth magnitude:

α 1·98   ν 3·12   ξ 3·72
γ 2·98   π 3·25   λ 3·83
ζ 3·11   ε 3·39   θ 3·84

BRIGHTEST STARS

Star	Mag.	R.A. h m s	Dec. ° ′ ″	Abs. Mag.	Spectrum	Dist. l/y
4 δ	4·18	08 36 30	+05 46 52	+1·2	A0	130
5 σ	4·54	08 37 37	+03 25 10	—0·1	K0	270
7 η	4·32	08 42 05	+03 20 42	—1·2	B3	400
12 D	4·44	08 45 20	—13 28 00	—0·3	G5	300
11 ε	3·39	08 45 37	—06 30 01	+0·6	F8	140
13 ρ	4·42	08 47 16	+05 55 11	—0·4	A0	300
16 ζ	3·11	08 54 14	+06 01 48	—1·1	K0	220
22 θ	3·84	09 13 13	+02 24 28	+0·5	A0	150
30 α Alphard	1·98	09 26 30	—08 33 45	—0·3	K2	94
32 τ ²	4·50	09 30 52	—01 05 14	+1·7	A3	115
35 ι	4·10	09 38 44	—01 02 33	+0·6	K0	160
I	4·74	09 40 17	—23 29 28	—1·4	B2p	550
39 υ ¹	4·29	09 50 25	—14 44 34	+0·3	K0	200
40 υ ²	4·72	10 04 03	—12 57 27	+0·3	B8	250
41 λ	3·83	10 09 31	—12 14 42	—0·1	K0	200
42 μ	4·06	10 25 02	—16 43 25	—0·1	K5	220
ν	3·12	10 48 32	—16 04 41	—0·2	K2	150
ξ	3·72	11 31 55	—31 44 09	+0·2	G5	160
β	4·40	11 51 48	—33 47 08	—0·2	B9	270
46 γ	2·98	13 17 43	—23 03 21	+0·3	G8	113
R	var.	13 28 30	—23 10 05	var.	M7	
49 π	3·25	14 05 07	—26 34 38	+1·2	K0	84
58 E	4·63	14 49 00	—27 52 11	—0·4	K2	330

THERE ARE NINE FURTHER STARS ABOVE THE FIFTH MAGNITUDE:

Star	Mag.	Abs. mag.	Spectrum
9	4·98	+2·0	G8
F	4·70	—4·1	G4
26	4·94	+0·1	G5
27 P	4·97	+0·7	G9
G	4·94	+0·7	K3
31 τ ¹	4·78	+4·0	F4
38 ϰ	4·96	—0·5	B4
o	4·88	+0·3	B8
51 k	4·93	+1·1	K3

F Hydræ is also known as 31 Monocarotis.

DOUBLES

Star	R.A. h m	Dec. ° ′	Mags.	P.A.	Dist.″	
ε	08 45·6	+06 30	3·5, 6·9	269	2·9	A is a close binary (15 years)
β	11 51·8	—33 47	4·8, 5·6	001	1·2	P.A. increasing

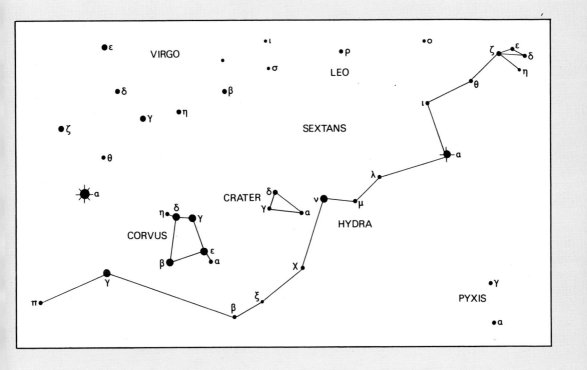

VARIABLES

Star	R.A.		Dec.		Range	Period, d	Spectrum	Type
	h	m	°	'				
RT	08	27·2	—06	09	7·1—10·2	253	M	Semi-regular
S	08	51·0	+03	16	7·4—13·3	257	M	Mira
T	08	53·2	—08	57	7·2—13·2	289	M	Mira
X	09	33·4	—14	28	8·0—13·6	302	M	Mira
U	10	35·1	—13	07	4·8—5·8	—	N	Irregular
V	10	49·2	—20	59	6·0—12·5	533	N	Mira
R	13	28·5	—23	10	4·0—10·0	386	M	Mira
W	13	46·2	—28	07	7·5—11·4	382	M	Semi-regular
RU	14	08·5	—28	39	7·2—14·3	334	M	Mira

CLUSTERS AND NEBULÆ

Object	R.A.		Dec.		Type	Notes
	h	m	°	'		
NGC 2548 M.48	08	11·2	—05	38	Open cluster	Mag. 5·5
NGC 2610	08	31·2	—15	38	Planetary	Mag. 13·6. Centre star mag. 15·7
NGC 2784	09	10·1	—23	58	Galaxy	Mag. 11·8. Elliptical (type E8)
NGC 3109	10	00·8	—25	55	Galaxy	Mag. 11·2. Irregular
NGC 3242 H.IV.27	10	22·3	—18	23	Planetary	Mag. 9. Centre star mag. 11·4
NGC 3585	11	10·9	—26	29	Galaxy	Mag. 11. Elliptical (type E5)
NGC 3904	11	46·7	—29	02	Galaxy	Mag. 12. Elliptical
NGC 3923	11	48·5	—28	33	Galaxy	Mag. 11. Elliptical
NGC 4590 M.68	12	36·8	—26	29	Globular	Mag. 8·2
NGC 5061	13	15·3	—26	36	Galaxy	Mag. 12. Elliptical (type E2)
NGC 5236 M.83	13	34·3	—29	37	Galaxy	Mag. 10. Type Sc

HYDRUS

(Abbreviation: Hyi).

A constellation in the far south – remarkably lacking in interesting objects. There are three stars above the fourth magnitude:

β 2·78   α 2·84   γ 3·30

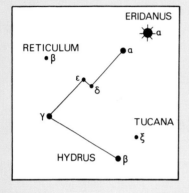

BRIGHTEST STARS

Star	Mag.	R.A. h	m	s	Dec. °	′	″	Abs. Mag.	Spectrum	Dist. l/y
β	2·78	00	24	37	−77	22	41	+3·7	G1	21
η²	4·72	01	54	23	−67	45	19	+1·8	K0	125
α	2·84	01	58	05	−61	40	35	+2·9	F0	31
δ	4·26	02	21	21	−68	45	33	+2·5	A2	72
ε	4·26	02	39	15	−68	21	39	+0·4	B9	190
ν	4·70	02	50	36	−75	09	24	−0·1	K2	300
γ	3·30	03	47	34	−74	18	24	−1·5	M2	300

THERE ARE TWO FURTHER STARS ABOVE THE FIFTH MAGNITUDE:

Star	Mag.	Abs. Mag.	Spectrum
λ	4·96	+1·5	M1
ζ	4·90	−1·6	A2

INDUS

(Abbreviation: Ind).

An undistinguished little constellation, but there are two stars above the fourth magnitude:

α 3·11   β 3·72

See chart for Grus

BRIGHTEST STARS

Star	Mag.	R.A. h	m	s	Dec. °	′	″	Abs. Mag.	Spectrum	Dist. l/y
α	3·11	20	36	01	−47	22	10	+1·1	K0	84
η	4·70	20	42	26	−52	00	03	+2·2	A7	100
β	3·72	20	53	06	−58	32	18	−0·9	K0	270
θ	4·60	21	18	18	−53	32	34	+2·7	A5	75
δ	4·56	21	56	26	−55	05	53	+0·7	F0	190
ε	4·74	22	01	41	−56	52	39	+7·0	K5	11·4

The only other star above the fifth magnitude is ζ ; mag. 4·90, absolute mag. 0·6, spectrum type M1.

VARIABLES

Star	R.A. h	m	Dec. °	′	Range	Period, d	Spectrum	Type
S	20	52·7	−54	31	7·4—14·5	400	M	Mira
T	21	16·9	−45	14	7·7—9·4	320	N	Semi-regular

NEBULÆ

Object	R.A. h	m	Dec. °	′	Type	Notes
NGC 7049	21	15·6	−48	47	Galaxy	Mag. 12. Elliptical
NGC 7090	21	32·9	−54	47	Galaxy	Mag. 12. Spiral
NGC 7205	22	05·1	−57	40	Galaxy	Mag. 12. Spiral

LACERTA

(Abbreviation: Lac).

An obscure constellation with only
one star above the fourth magnitude
(α, 3·85).

BRIGHTEST STARS

Star		Mag.	R.A.			Dec.			Abs. Mag.	Spectrum	Dist, l/y
			h	m	s	°	′	″			
Boss 31104	1 H	4·64	22	12	56	+39	36	19	0·0	K2	270
1		4·22	22	15	01	+37	38	20	−0·8	K0	330
2		4·66	22	20	07	+46	25	32	−0·6	B5	650
3 β		4·58	22	22	42	+52	07	06	+1·0	K0	170
4		4·64	22	23	37	+49	21	52	−3·0	B8p	1000
5		4·61	22	28	37	+47	35	38	−3·0	K0+A0	1000
6		4·54	22	29	32	+43	00	37	−2·0	B3	650
7 α		3·85	22	30	23	+50	10	08	+1·6	A0	90
11		4·64	22	39	33	+44	09	40	0·0	K0	270

THERE ARE TWO FURTHER STARS ABOVE THE FIFTH MAGNITUDE:

Star	Mag.	Abs. Mag.	Spectrum
9	4·83	+1·5	A5
10	4·91	−4·8	O9

10 Lacertæ is a highly luminous star, lying at a distance of well over 2000 light-years.

VARIABLE

Star	R.A.		Dec.		Range	Period, d	Spectrum	Type
	h	m	°	′				
S	22	26·8	+40	04	7·6—13·9	240	M	Mira

CLUSTERS AND NEBULÆ

Object	R.A.		Dec.		Type	Notes
	h	m	°	′		
NGC 7209	22	01·8	+46	16	Open cluster	Mag. 7·6. About 50 stars
NGC 7243 H.VIII.75	22	13·1	+49	38	Open cluster	Mag. 7·4. About 30 stars
I 5217	22	21·9	+50	43	Planetary	Mag. 12·6 (centre star mag. 14·6). Oval

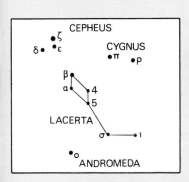

LEO

(Abbreviation: Leo).

An important Zodiacal constellation – mythologically, the Nemæan lion which was killed by Hercules. There are ten stars above the fourth magnitude:

α 1·36	δ 2·57	η 3·58
γ 1·99	ε 2·99	o 3·76
β 2·14	θ 3·34	ρ 3·85
	ζ 3·46	

β (Denebola) was ranked as of the first magnitude by Ptolemy and others, and there is therefore a suspicion that it has faded, though the evidence is not conclusive.

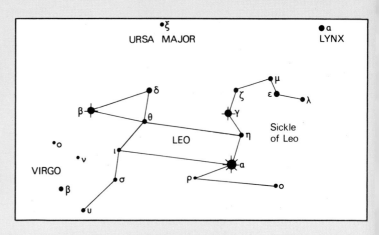

BRIGHTEST STARS

Star			Mag.	R.A.			Dec.			Abs. Mag.	Spectrum	Dist. l/y
				h	m	s	°	′	″			
1	κ		4·61	09	23	23	+26	16	40	−0·4	K0	330
4	λ	Alterf	4·48	09	30	28	+23	03	56	+0·8	K5	180
14	o	Subra	3·76	09	39	59	+09	59	35	+0·1	F5+A3	175
17	ε	Asad Australis	2·99	09	44	37	+24	00	19	−2·1	G0p	340
24	μ	Rassalas	4·10	09	51	31	+26	06	40	+0·6	K0	160
30	η		3·58	10	06	08	+16	52	13	−5·3	A0p	1900
31		A	4·58	10	06	44	+10	06	21	+0·1	K2	250
32	α	Regulus	1·36	10	07	12	+12	04	31	−0·7	B7	84
36	ζ	Adhafera	3·46	10	15	28	+23	31	29	−0·5	F0	130
41	γ	Algieba	1·99	10	18	46	+19	57	12	+0·1	K0p+G5	90
47	ρ		3·85	10	31	39	+09	25	13	−6·0	B0p	2700
54			4·51	10	54	26	+24	52	02	+0·5	A0	200
60		b	4·42	11	01	09	+20	17	53	+0·4	A0	200
63	χ		4·66	11	03	53	+07	27	19	+1·0	F0	175
68	δ	Zosma	2·57	11	12	56	+20	38	40	+0·6	A3	82
70	θ	Chort	3·34	11	13	05	+15	33	00	+1·1	A2	90
74	φ		4·58	11	15	32	−03	31	52	+1·3	A5	150
77	σ		4·13	11	20	01	+06	09	00	+0·1	B9	400
78	ι		4·03	11	22	47	+10	39	02	+2·4	F4	68
92	υ		4·47	11	35	49	−00	42	08	+0·6	G8	190
93			4·54	11	46	51	+20	20	28	+2·2	F8	95
94	β	Denebola	2·14	11	47	56	+14	41	42	+1·5	A3	43

THERE ARE FOUR FURTHER STARS ABOVE THE FIFTH MAGNITUDE:

Star	Mag.	Abs. Mag.	Spectrum
29 π	4·89	+0·3	M2
40	4·97	+3·5	F5
61 p²	4·97	−0·3	M1
72	4·87	−0·6	M2

DOUBLES

Star	R.A.		Dec.		Mags.	P.A.	Dist."	
	h	m	°	′				
κ	09	23·4	+26	17	4·6, 9·7	208	2·6	Slow binary
α	10	07·2	+12	05	1·4, 7·6	307	176·5	Fixed
γ	10	18·8	+19	57	2·6, 3·8	122	4·3	Binary (407 years). Widening

VARIABLE

Star	R.A.		Dec.		Range	Period, d	Spectrum	Type
	h	m	°	′				
R	09	44·9	+11	40	5·4—10·5	313	M	Mira

NEBULÆ

Object		R.A. h m	Dec. ° ′	Type	Notes
NGC 2903		09 29·3	+21 44	Galaxy	Mag. 9. Type Sb
NGC 2964		09 40·0	+32 05	Galaxy	Mag. 11. Type Sc
NGC 3193		10 15·7	+22 09	Galaxy	Mag. 11·5. Type E0
NGC 3226		10 20·7	+20 09	Galaxy	Mag. 11·4. Type E2
NGC 3227		10 20·7	+20 07	Galaxy	Mag. 11·4. Type Sb
NGC 3338		10 39·5	+14 00	Galaxy	Mag. 11·3. Type Sb
NGC 3351	M.95	10 41·3	+11 58	Galaxy	Mag. 10·4. Type SBb
NGC 3367		10 44·0	+14 01	Galaxy	Mag. 12. Type Sc
NGC 3368	M.96	10 44·1	+12 05	Galaxy	Mag. 9. Type Sa. Near M.95
NGC 3412		10 48·3	+13 41	Galaxy	Mag. 10·4. Type E5
NGC 3489		10 57·7	+14 10	Galaxy	Mag. 10·5. Type E6
NGC 3521		11 03·2	+00 14	Galaxy	Mag. 9·5. Type Sc
NGC 3593		11 12·0	+13 06	Galaxy	Mag. 11·3. Type Sb
NGC 3607		11 14·3	+18 20	Galaxy	Mag. 9·6. Type E1
NGC 3623	M.65	11 16·3	+13 22	Galaxy	Mag. 9·5. Type Sa
NGC 3627	M.66	11 17·6	+13 16	Galaxy	Mag. 8·8. Type Sb
NGC 3640		11 18·5	+03 31	Galaxy	Mag. 10·7. Type E1
NGC 3686		11 25·1	+17 30	Galaxy	Mag. 11·4. Type Sc
NGC 3810		11 38·4	+11 45	Galaxy	Mag. 10·8. Type Sc
NGC 3900		11 46·6	+27 17	Galaxy	Mag. 11·5. Type Sb

LEO MINOR

(Abbreviation: LMi).

A small and obscure constellation. The only star above the fourth magnitude is 46 (mag. 3·92).

See chart for Ursa Major

BRIGHTEST STARS

Star		Mag.	R.A. h m s	Dec. ° ′ ″	Abs. Mag.	Spectrum	Dist. l/y
10		4·62	09 32 53	+36 29 45	+1·0	G0	170
21		4·47	10 06 08	+35 21 09	+1·6	A5	120
31	β	4·41	10 26 37	+36 49 13	+0·6	K0	190
46	Præcipua	3·92	10 52 05	+34 20 01	+1·4	K0	100

THERE ARE TWO FURTHER STARS ABOVE THE FIFTH MAGNITUDE:

Star	Mag.	Abs. Mag.	Spectrum
30	4·83	+2·5	F3
37	4·77	−3·8	G2

VARIABLES

Star	R.A. h m	Dec. ° ′	Range	Period, d	Spectrum	Type
R	09 42·6	+34 45	6·3—13·2	372	M	Mira
S	09 50·8	+35 10	7·9—14·3	234	M	Mira

GALAXY

Object	R.A. h m	Dec. ° ′	Type	Notes
NGC 2859	09 21·3	+33 44	Galaxy	Mag. 10·7. Type SBa

LEPUS

(Abbreviation: Lep).
An original constellation. In mythology, it was said that Orion was particularly fond of hunting hares – and so a hare was placed beside him in the sky.

Lepus is fairly distinctive. There are seven stars above the fourth magnitude:

α 2·58   ε 3·21   η 3·77
β 2·81   μ 3·29   γ 3·80
          ζ 3·67

BRIGHTEST STARS

Star			Mag.	R.A.			Dec.			Abs. Mag.	Spectrum	Dist. l/y
				h	m	s	°	′	″			
2	ε		3·21	05	04	32	−22	23	59	−0·4	K5	170
3	ι		4·54	05	11	16	−11	53	40	−0·2	B8	300
5	μ		3·29	05	11	57	−16	13	50	−2·1	B9	390
4	κ		4·46	05	12	13	−12	58	00	+0·2	B8	230
6	λ		4·29	05	18	34	−13	11	55	−4·5	B1	1700
	Boss 6559											
	(−21° 1135)		4·73	05	19	31	−21	15	39	−0·3	A0	330
9	β	Nihal	2·81	05	27	18	−20	46	34	+0·1	G0	113
11	α	Arneb	2·58	05	31	46	−17	50	13	−4·6	F0	900
13	γ		3·80	05	43	33	−22	27	17	+4·1	F8	27
14	ζ		3·67	05	45	58	−14	49	45	+1·6	A2	85
15	δ		3·90	05	50	23	−20	52	48	+1·1	K0	120
16	η		3·77	05	55	24	−14	10	15	+2·8	F0	50
18	θ		4·67	06	05	10	−14	55	56	+0·8	A0	190

The only other star above the fifth magnitude is Boss 6142 (mag. 4·99, absolute mag. +1·4, spectrum B9).

DOUBLES

Star	R.A.		Dec.		Mags.	P.A.	Dist.″	
	h	m	°	′				
κ	05	12·2	−12	58	4·5, 7·5	258	2·6	Fixed
β	05	27·3	−20	47	2·8, 11·0	308	2·6	Slow binary
γ	05	43·6	−22	27	3·8, 6·4	351	94·9	Fixed

VARIABLES

Star	R.A.		Dec.		Range	Period, d	Spectrum	Type
	h	m	°	′				
R	04	57·3	−14	53	5·9–10·5	432	N	Mira
T	05	02·7	−21	58	7·4–13·5	368	M	Mira

CLUSTERS AND NEBULÆ

Object		R.A.		Dec.		Type	Notes
		h	m	°	′		
NGC 1904	M.79	05	22·2	−24	34	Globular	Mag. 7·9
I 418		05	25·4	−12	44	Planetary	Mag. 12. Centre star mag. 11
NGC 1964		05	31·2	−21	59	Galaxy	Mag. 12. Type Sb

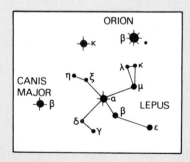

LIBRA

(Abbreviation: Lib).

One of the Zodiacal constellations. It is, however, decidedly obscure. It was originally known as Chelæ Scorpionis (the Scorpion's Claws). Some Greek legends associate it, though rather vaguely, with Mochis, the inventor of weights and measures.

There are five stars above the fourth magnitude:

β 2·61   σ 3·41
α² 2·76   υ 3.78
               τ 3·80

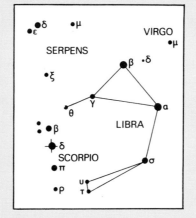

BRIGHTEST STARS

Star			Mag.	R.A.			Dec.			Abs. Mag.	Spectrum	Dist. l/y
				h	m	s	°	′	″			
9 α²	Zubenelgenubi		2·76	14	49	40	−15	57	05	+1·2	A3	66
16			4·59	14	56	02	−04	15	28	+2·5	A7	85
20 σ	Zubenalgubi		3·41	15	02	47	−25	11	47	+0·9	M3	105
24 ι			4·66	15	10	58	−19	42	34	+0·2	B9	250
27 β	Zubanelchemale		2·61	15	15	49	−09	18	10	−0·6	B8	140
38 γ	Zubenelhakrabi		4·02	15	34	18	−14	43	02	+1·4	K0	110
39 υ			3·78	15	35	41	−28	03	48	+0·7	K2	135
40 τ			3·80	15	37	18	−29	42	24	−1·6	B3	400
46 θ			4·34	15	52	34	−16	39	57	+1·2	K0	140
48			4·68	15	56	57	−14	13	02	−0·8	B3p	400

σ Libræ was formerly included in Scorpio, as γ Scorpionis. β Libræ has often been described as being the only single naked-eye star to show a definite green colour, though most people will certainly describe it as white!

THERE ARE THREE MORE STARS ABOVE THE FIFTH MAGNITUDE:

Star	Mag.	Abs. Mag.	Spectrum
19 δ	var.	var.	A1
37	4·83	+1·8	K1
43 κ	4·96	+1·0	M0

DOUBLES

Star	R.A.		Dec.		Mags.	P.A.	Dist."	
	h	m	°	′				
μ	14	46·6	−13	57	5·8, 6·7	355	1·7	Slow binary
ι	15	11·0	−19	43	4·7, 9·7	111	58·6	Fixed

VARIABLES

Star	R.A.		Dec.		Range	Period, d	Spectrum	Type
	h	m	°	′				
δ	14	58·3	−08	19	4·8 —6·1	2·3	A1	Algol
Y	15	09·1	−05	49	7·6—14·7	275	M	Mira
S	15	18·5	−20	13	8·0—13·0	193	M	Mira
RU	15	30·5	−15	09	7·4—14·2	317	M	Mira

CLUSTERS AND NEBULÆ

Object	R.A.		Dec.		Type	Notes
	h	m	°	′		
NGC 5728	14	39·6	−17	03	Galaxy	Type SBb. Mag. 12
NGC 5897	15	14·5	−20	50	Globular	Mag. 11

LUPUS

(Abbreviation: Lup).

An original constellation, though no definite legends seem to be attached to it. There are eight stars above the fourth magnitude:

α 2·32   ζ 3·42
β 2·69   η 3·45
γ 2·80   ø¹ 3·59
δ 3·24   ε 3·74

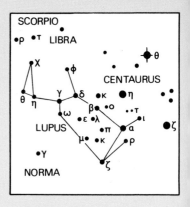

BRIGHTEST STARS

Star		Mag.	R.A.			Dec.			Abs. Mag.	Spectrum	Dist. l/y
			h	m	s	°	′	″			
ι		4·10	14	17	59	−45	57	26	−1·7	B3	460
τ ¹		4·65	14	24	43	−45	07	22	−1·1	B3	460
τ ²		4·49	14	24	45	−45	16	51	—	F1	?
σ		4·60	14	31	07	−50	21	37	−2·4	B2	780
ρ		4·14	14	36	24	−49	19	51	−0·1	B5	230
α	Men	2·32	14	40	28	−47	17	41	−3·3	B2	430
o		4·49	14	50	12	−43	29	07	+0·2	B5	230
β	Ke Kouan	2·69	14	57	05	−43	02	47	−3·4	B2	540
π		4·02	15	03	37	−46	57	58	−1·0	B5	330
λ		4·39	15	07	21	−45	11	46	−2·4	B3	740
x		3·7	15	10	24	−48	39	19	+0·7, +2·6	B9+A0	160
ζ		3·42	15	10	42	−52	01	00	+1·2	K0	90
2 f		4·43	15	16	29	−30	04	08	−0·2	K0	270
μ		4·36	15	17	00	−47	47	43	+0·5	B8	190
δ		3·24	15	19	55	−40	34	08	−3·4	B2	680
φ¹		3·59	15	20	24	−36	10	58	−1·0	K5	270
ε		3·74	15	21	11	−44	36	41	−2·4	B3	540
φ²		4·69	15	21	45	−36	46	51	−2·4	B3	800
k		4·68	15	23	54	−38	39	24	−1·1	A0	470
γ		2·80	15	33	40	−41	05	38	−2·7	B3	570
ω		4·27	15	36	34	−42	29	47	0·0	K5	230
3 ψ¹		4·63	15	38	22	−34	20	28	+0·9	K0	180
g		4·69	15	39	40	−44	35	22	+3·5	F5	58
5 χ		4·11	15	49	33	−33	33	41	−0·5	B9	270
η		3·45	15	58	40	−38	20	07	−2·7	B3	570
θ		4·33	16	05	09	−36	44	37	−2·4	B3	800

THERE ARE FIVE OTHER STARS ABOVE THE FIFTH MAGNITUDE:

Star		Abs. Mag.		Spectrum
	e	4·92	−2·8	B5
1	i	4·95	−0·5	F0
	d	4·84	−2·4	B3
4	ψ²	4·82	−0·9	B6
	ξ	4·6	+1·9, +2·2	A0+A0
Boss 21548		4·97	−1·6	B5

DOUBLES

Star	R.A.		Dec.		Mags.	P.A.	Dist. ″	
	h	m	°	′				
x	15	10·4	−48	39	4·1, 6·0	144	27·0	Fixed
η	15	58·7	−38	20	3·5, 7·7	021	14·2	Slow binary

CLUSTERS AND NEBULÆ

Object	R.A.		Dec.		Type	Notes
	h	m	°	′		
I 4406	14	19·3	−43	55	Planetary	Mag. 11. Oval
NGC 5593	14	22·4	−54	35	Open cluster	Rather sparse
NGC 5643	14	29·4	−43	59	Galaxy	Spiral. Mag. 12
NGC 5749	14	45·3	−54	19	Open cluster	Mag. 9
NGC 5824	15	00·9	−32	53	Globular	Mag. 10
NGC 5822	15	01·6	−54	09	Open cluster	Mag. 6·4
NGC 5882	15	13·3	−45	27	Planetary	Mag. 10·5
NGC 5927	15	24·4	−50	29	Globular	Mag. 9
NGC 5986	15	42·8	−37	37	Globular	Mag. 9

LYNX

(Abbreviation: Lyn).

A very ill-defined and obscure northern constellation. It was added to the sky by Hevelius, and has no mythological associations.

There are two stars above the fourth magnitude:

α 3·17   38 3·82

See chart for Ursa Major

BRIGHTEST STARS

Star	Mag.	R.A. h m s	Dec. ° ' "	Abs. Mag.	Spectrum	Dist. l/y
2	4·42	06 17 41	+59 01 14	+1·9	A0	105
15	4·54	06 55 23	+58 17 12	+0·3	G0	230
21	4·45	07 25 04	+49 15 24	0·0	A0	250
31	4·43	08 21 20	+43 15 35	+0·3	K5	220
Boss 12565	4·71	09 05 08	+38 32 37	-0·3	G5	330
38	3·82	09 17 29	+36 53 47	+1·2	A2	107
40 α	3·17	09 19 43	+34 29 11	-0·5	K5	180

THERE ARE FIVE MORE STARS ABOVE THE FIFTH MAGNITUDE:

Star	Mag.	Abs. Mag.	Spectrum
12	4·89	+1·2	A2
16	4·80	+0·8	A2
24	4·96	+1·2	A2
27	4·87	-0·1	A2
Boss 13221	4·99	+0·6	G8

VARIABLES

Star	R.A. h m	Dec. ° '	Range	Period, d	Spectrum	Type
R	06 57·2	+55 24	7·2—14·0	370	S	Mira
Y	07 24·6	+46 06	6·9—7·4	110	M	Semi-regular

CLUSTERS AND NEBULÆ

Object	R.A. h m	Dec. ° '	Type	Notes
NGC 2419	07 34·8	+39 00	Globular	Mag. 11·5
NGC 2683	08 49·6	+33 38	Galaxy	Type Sb. Mag. 9·6
NGC 2776	09 08·9	+45 11	Galaxy	Type Sc. Mag. 11·7

LYRA

(Abbreviation: Lyr).

A small constellation, but a very interesting one; it is graced by the presence of the brilliant blue Vega, as well as the prototype eclipsing binary β Lyræ, the quadruple ε Lyræ, and the 'Ring Nebula' M.57. Mythologically it represents the harp which Apollo gave to the great musician Orpheus.

There are several stars above the fourth magnitude. Of these, one (β) is the famous variable; the combined magnitude of the quadruple ε is about 3·9, though keen-sighted people can see the two main components as separated. The stars are:

α 0·04   β 3·4 (max)
γ 3·25   ε 3·9 (combined)

CYGNUS

LYRA

BRIGHTEST STARS

Star	Mag.	R.A. h m s	Dec. ° ' "	Abs. Mag.	Spectrum	Dist. l/y
1 κ	4·34	18 19 05	+36 03 14	+0·1	K0	230
3 α Vega	0·04	18 36 12	+38 45 44	+0·5	A0	26·5
4 + 5 ε	3·9	18 43 39	+39 35 21	+2·2, +0·5	A4+A2	200
6 + 7 ζ	4·1	18 44 01	+37 34 53	+1·2, +2·8	A3+A3	140
10 β Sheliak	var.	18 49 16	+33 20 10	-4·6v	B2+B8p	1300
12 δ²	4·3v	18 53 44	+36 52 12	var.	M4	360
13 R	var.	18 54 40	+43 54 59	var.	M3	540
14 γ Sulaphat	3·25	18 58 07	+32 39 31	-2·1	A0p	370
20 η Aladfar	4·46	19 13 01	+39 06 27	-2·5	B3	800
21 θ	4·46	19 15 36	+38 05 37	-1·0	K0	400

There is one other star above the fifth magnitude: Boss 25721, mag. 4·92 absolute mag. +1·2, spectrum K1.

μ (magnitude 5·1) has the proper name of Al Athfar.

DOUBLES

Star	R.A. h m	Dec. ° '	Mags.	P.A.	Dist."	
η	19 13·2	+39 06	4·5, 8·7	082	28·2	Fixed
ε	18 43·7	+39 35	4·7, 4·5	172	207·8	Naked-eye pair
ε¹	—	—	5·1, 6·0	002	2·8	
ε²	—	—	5·1, 5·4	101	2·3	
ζ	18 44·0	+37 35	4·3, 5·9	150	43·7	Fixed
β	18 49·3	+33 20	var.,°7·8	149	46·6	Optical

VARIABLES

Star	R.A. h m	Dec. ° '	Range	Period, d	Spectrum	Type
W	18 13·2	+36 39	7·5—13·0	196	M	Mira
T	18 30·7	+36 58	7·8—9·6	—	R	Irregular
β	18 49·3	+33 20	3·4—4·0	12·9	Bp	Prototype β Lyr.
δ²	18 53·7	+36 52	4·3—4·7	?	M4	?
R	18 54·7	+43 55	4·0—5·0	±47	M	Semi-regular
RR	19 23·9	+42 41	7·0—8·0	0·6	A-F	Prototype RR Lyr.

CLUSTERS AND NEBULÆ

Object	R.A. h m	Dec. ° '	Type	Notes
NGC 6720 M.57	18 51·7	+32 58	Planetary	Ring Nebula. Mag. 9. Central star, mag. 14
NGC 6779 M.56	19 14·6	+30 05	Globular	Mag. 8·2. Rich area

MENSA

(Abbreviation: Men).

A very dim constellation, introduced by Lacaille in 1752 under the name of Mons Mensæ (the Table Mountain). A small part of the Large Magellanic Cloud extends into it. There are no stars brighter than the fifth magnitude, and no objects to be listed. For the record, the brightest star is γ: R.A. 05h 34m, dec —76° 23′, magnitude 5·06. Next comes α (5·14).

See chart for Musca

MICROSCOPIUM

(Abbreviation: Mic).

A small southern constellation. γ (4·71) is the brightest star.

See chart for Grus

BRIGHTEST STAR

Star	Mag.	R.A.			Dec.			Abs. Mag.	Spectrum	Dist. l/y
		h	m	s	°	′	″			
γ	4·71	20	59	57	—32	20	41	+0·4	G5	230

THERE ARE THREE OTHER STARS DOWN TO THE FIFTH MAGNITUDE:

Star	Mag.	Abs. Mag.	Spectrum
α	5·00	—0·2	G6
ε	4·79	+2·1	A0
θ[1]	4·92	—0·6	A2

γ was formerly known as 1 PsA. and ε as 4 PsA.

VARIABLE

Star	R.A.		Dec.		Range	Period, d	Spectrum	Type
	h	m	°	′				
T	20	24·9	—28	26	7·7—9·6	347	M	Semi-regular

MONOCEROS

(Abbreviation: Mon).

Not an ancient constellation, and though it represents the fabled unicorn there are no definite legends attached to it. It is crossed by the Milky Way, and the general area is decidedly rich. There are two stars above the fourth magnitude; of these, β is a double, and the magnitude is combined. The stars are:

β 3·7 (combined)   30 3·95

See chart for Orion

BRIGHTEST STARS

Star	Mag.	R.A. h	m	s	Dec. °	′	″	Abs. Mag.	Spectrum	Dist. l/y	
5 γ	4·09	06	13	47	−06	16	01	−0·3	K2	250	
8 ε	4·48	06	22	36	+04	36	18	+1·5	A5	130	
11 β	3·7	06	27	45	−07	01	09	−1·0, −1·1	B3+B3	470	
13	4·50	06	31	43	+07	21	00	−5·4	A0p	3300	
15 S	4·68	06	39	46	+09	55	02	−5·5	O7	3300	
22 δ	4·09	07	10	44	−00	27	19	+0·4	A0	180	
26 α	4·07	07	40	12	−09	29	56	+0·3	K0	180	
29 ζ	4·41	08	07	29	−02	55	08	−1·1, −4·9	B3+B3	1900	?
30	3·95	08	24	34	−03	50	02	+0·6	A0	160	
(31	4·70	08	42	36	−07	09	13	−4·1	G0	1400)	?

31 MONOCEROTIS IS ALSO KNOWN AS F HYDRÆ.

IN MONOCEROS THERE ARE FOUR MORE STARS ABOVE THE FIFTH MAGNITUDE:

Star	Mag.	Abs. Mag.	Spectrum
17	5·00	−0·2	K5
18	4·80	+0·1	K0
19	4·89	−3·5	B1
28	4·88	−0·1	K5

DOUBLES

Star	R.A. h	m	Dec. °	′	Mags.	P.A.	Dist.″	
ε	06	22·6	+04	36	4·5, 6·5	027	13·2	Fixed
β	06	27·8	−07	02	4·6, 4·7	132	7·4	A is double (separation 2″·8)

VARIABLES

Star	R.A. h	m	Dec. °	′	Range	Period, d	Spectrum	Type
V	06	20·2	+02	10	6·0—13·7	335	M	Mira
T	06	22·5	+07	07	5·8—6·8	27·0	F-G	Cepheid
X	06	54·8	−09	00	6·9—10·0	156	M	Semi-regular
U	07	28·4	−09	40	6·0—8·0	92	G	RV Tauri
RY	07	04·5	−07	29	7·7—9·2	466	R	Semi-regular
SU	07	39·9	−10	46	7·7—9·0	—	S	Semi-regular

CLUSTERS AND NEBULÆ

Object	R.A. h	m	Dec. °	′	Type	Notes
NGC 2236	06	27·0	+06	53	Open cluster	Mag. 12. Fairly rich
NGC 2237-9	06	29·6	+04	00	Nebula	Rosette Nebula
NGC 2244 H.VII.2	06	30·0	+04	54	Open cluster	Naked-eye; round 12 Mon
NGC 2261	06	36·4	+08	46	Nebula	Hubble's variable nebula (R Mon)
NGC 2264	06	38·2	+09	57	Nebula	Cone nebula (S Mon)
NGC 2301	06	49·2	+00	31	Open cluster	Mag. 6. About 60 stars
NGC 2323 M.50	07	00·6	−08	16	Open cluster	Mag. 6·3
NGC 2506 H.VI.37	07	57·5	−10	27	Open cluster	Best seen with low power

MUSCA AUSTRALIS

(Abbreviation: Mus).

This small but fairly distinctive constellation, not far from Crux, is generally known simply as 'Musca'. There are four stars above the fourth magnitude:

α 2·66v   δ 3·63
β 3·06   λ 3·80

α is actually variable over a very small range (2·66–2·73).

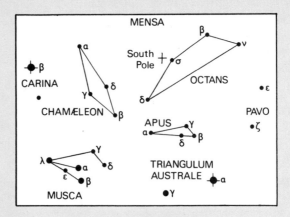

BRIGHTEST STARS

Star	Mag.	R.A.			Dec.			Abs. Mag.	Spectrum	Dist. l/y
		h	m	s	°	′	″			
λ	3·80	11	44	34	−66	36	24	?	A5	?
μ	4·71	11	47	10	−66	41	33	−0·8	K5	400
ε	4·16	12	16	22	−67	50	19	+2·2	M4	75
γ	4·04	12	31	08	−72	00	42	−0·6	B5	270
α	2·66v	12	35	51	−69	00	52	−2·9	B3	430
β	3·06	12	44	55	−67	59	17	−2·1	B3	470
δ	3·63	13	00	44	−71	25	51	+0·2	K2	155

THERE ARE FOUR MORE STARS ABOVE THE FIFTH MAGNITUDE:

Star	Mag.	Abs. Mag.	Spectrum
CP 16206	4·90	−2·1	K2
η	4·95	+1·0	B8
CP 17959	4·78	−3·7	K6
ι¹	4·96	+0·4	G7

DOUBLES

Star	R.A.		Dec.		Mags.	P.A.	Dist."	
	h	m	°	′				
β	12	44·9	−67	59	3·9, 4·2	007	1·6	Binary
θ	13	04·9	−65	02	5·6, 7·2	186	5·7	Fixed

VARIABLES

Star	R.A.		Dec.		Range	Period, d	Spectrum	Type
	h	m	°	′				
S	12	10·1	−69	52	6·5—7·3	9·66	F-G	Cepheid
R	12	39·0	−69	08	6·3—7·3	7·5	G	Cepheid
BO	12	32·0	−67	29	6·0—6·7	—	Mb	Irregular

CLUSTERS AND NEBULÆ

Object	R.A.		Dec.		Type	Notes
	h	m	°	′		
NGC 4372	12	23·0	−72	24	Globular	Mag. 8
H.6	12	45·0	−68	10	Open cluster	Mag. 10·6. About 75 stars
NGC 4833	12	56·0	−70	36	Globular	Mag. 7
I 4191	13	05·5	−67	22	Planetary	Mag. 12. Stellar appearance
H.8	13	15·0	−66	49	Open cluster	Mag. 10. About 25 stars

NORMA

(Abbreviation: Nor).

An obscure constellation, once known as Quadra Euclidis (Euclid's Quadrant); it was added to the sky by Lacaille. It contains no star above the fourth magnitude.

See charts for Ara and Lupus

BRIGHTEST STARS

Star	Mag.	R.A. h	m	s	Dec. °	′	″	Abs. Mag.	Spectrum	Dist, l/y
η	4·74	16	01	35	−49	10	11	+0·9	G5	190
γ²	4·14	16	18	11	−50	06	11	+2·1	K0	80

THERE ARE FOUR MORE STARS DOWN TO THE FIFTH MAGNITUDE:

Star	Mag.	Abs. Mag.	Spectrum
ι¹	4·87	+1·9	B0
δ	4·84	+0·6	A3p
γ¹	5·00	—	G4p
ε	4·80	−1·7	B5

VARIABLES

Star	R.A. h	m	Dec. °	′	Range	Period, d	Spectrum	Type
T	15	40·2	−54	50	6·2—13·4	293	M	Mira
R	15	32·3	−49	21	6·5—13·9	490	M	Mira
S	16	14·7	−57	47	6·8—7·8	9·75	F-G	Cepheid

CLUSTERS AND NEBULÆ

Object	R.A. h	m	Dec. °	′	Type	Notes
NGC 5946	15	42·8	−37	37	Globular	Mag. 9
—	15	47·4	−51	21	Planetary	Mag. 8·4. Annular
NGC 6067 △ 360	16	09·4	−54	05	Open cluster	Rich; about 120 stars
NGC 6087	16	14·7	−57	47	Open cluster	Mag. 6. About 35 stars
NGC 6152	16	28·8	−52	31	Open cluster	Mag. 8. Rather sparse

OCTANS

(Abbreviation: Oct).

The south polar constellation. It is very obscure, and contains only one star above the fourth magnitude: ν (3·74). The south polar star, within a degree of the pole, is σ; magnitude 5·48, spectral type A7.

See chart for Musca

BRIGHTEST STARS

Star	Mag.	R.A. h	m	s	Dec. °	′	″	Abs. Mag.	Spectrum	Dist. l/y
δ	4·14	14	23	13	−83	34	09	+0·2	K2	200
ν	3·74	21	39	05	−77	29	20	+1·7	K0	80
β	4·34	22	43	54	−81	29	52	—	F0	—
θ	4·73	00	00	29	−77	11	13	+0·3	K0	250

DOUBLE

Star	R.A. h	m	Dec. °	′	Mags.	P.A.	Dist."	
λ	21	43·5	−82	57	5·5, 7·6	069	3·1	Slow binary

VARIABLES

Star	R.A. h	m	Dec. °	′	Range	Period, d	Spectrum	Type
R	05	41·1	−86	26	6·4—13·2	450	M	Mira
S	17	46·0	−85	48	7·4—14·0	259	M	Mira

OPHIUCHUS

(Abbreviation: Oph).

This constellation is also sometimes known as Serpentarius. It commemorates Æsculapius, son of Apollo and Coronis, who became so skilled in medicine that he was even able to restore the dead to life. To avoid depopulation of the Underworld, Jupiter reluctantly disposed of Æsculapius with a thunderbolt, but relented sufficiently to place him in the sky!

Ophiuchus is a very large constellation, with twelve stars above the fourth magnitude:

α 2·09	β 2·77	72 3·73
η 2·46	ϰ 3·18	γ 3·74
ζ 2·57	ε 3·22	λ 3·85
δ 2·72	θ 3·29	67 3·92

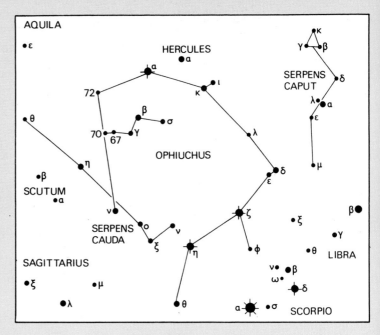

BRIGHTEST STARS

Star		Mag.	R.A. h m s	Dec. ° ′ ″	Abs. Mag.	Spectrum	Dist. l/y
1 δ	Yed Prior	2·72	16 13 11	−03 38 20	−0·5	M1	140
2 ε	Yed Post	3·22	16 17 09	−04 38 24	+1·0	K0	90
3 υ		4·68	16 26 37	−08 19 25	+1·4	A2	620
4 ψ		4·59	16 22 49	−19 59 14	0·0	K0	270
5 ρ		4·3	16 22 35	−23 20 02	−0·6, −1·3	B5+B5	400
7 χ		var.	16 24 07	−18 20 41	var.	B3p	390
8 φ		4·40	16 29 53	−16 33 57	0·0	A0	250
9 ω		4·57	15 30 50	−21 25 14	+1·7	F0	120
10 λ		3·85	16 29 48	+02 01 51	0·0	A0	190
13 ζ	Han	2·57	16 35 57	−10 31 26	−4·3	B0	520
20		4·73	16 48 37	−10 44 43	+2·7	F5	80
25 ι		4·29	16 52 58	+10 12 02	0·0	B8	230
27 ϰ		3·18	16 56 38	+09 24 29	−0·1	K0	150
35 η	Sabik	2·46	17 09 07	−15 41 56	+1·4	A2	69
40 ξ		4·46	17 19 41	−21 05 26	+3·4	F5	50
42 θ		3·29	17 20 39	−24 58 44	−3·4	B3	710
44 b		4·28	17 25 02	−24 09 23	+2·3	F0	80
47		4·61	17 25 28	−05 04 06	+1·7	F1	125
49 σ		4·44	17 25 55	+04 09 30	−0·8	K1	360
45 d		4·37	17 25 57	−29 50 55	+1·2	F5	140
55 α	Rasalhague	2·09	17 33 55	+12 34 30	+0·8	A5	58
57 μ		4·65	17 36 39	−08 06 24	−0·1	B8	300
60 β	Cheleb	2·77	17 42 23	+04 34 31	−0·1	K0	124
62 γ		3·74	17 46 47	+02 42 52	+1·3	A0	100
64 ν		3·32	17 57 49	−09 46 20	+0·2	K0	140
67		3·92	17 59 33	+02 55 53	−5·7	B5p	2300
68		4·44	18 00 38	+01 18 16	+1·0	A2	155
70 p		4·07	18 04 21	+02 30 13	+5·6	K0	16·7
71		4·73	18 06 15	+08 43 48	+0·7	G5	205
72		3·73	18 06 18	+09 33 35	+1·7	A3	80

THERE ARE SIX MORE STARS DOWN TO THE FIFTH MAGNITUDE:

Star		Mag.	Abs. Mag.	Spectrum
30		5·00	+1·1	K4
51 c		4·89	+2·7	A0
58		4·89	+3·6	F5
66		4·81	−2·4	B5
69 τ		4·88	+2·4	F3
74		4·92	+0·5	G5

DOUBLES

Star	R.A.		Dec.		Mags.	P.A.	Dist."	
	h	m	°	'				
ρ	16	22·6	−23	20	5·2, 5·9	347	3·5	Slow binary
τ	18	00·4	−08	11	5·3, 6·0	270	2·0	Binary
70	18	02·9	+02	31	4·3, 6·0		<7	Binary: 88 years

VARIABLES

Star	R.A.		Dec.		Range	Period, d	Spectrum	Type
	h	m	°	'				
V	16	23·9	−12	19	7·3—11·0	298	N	Mira
χ	16	24·1	−18	21	4·1—5·0	—	B	Novalike
R	17	04·9	−16	02	7·0—13·6	302	M	Mira
U	17	14·0	+01	16	5·8—6·5	1·7	B	Algol
Z	17	17·0	+01	34	7·6—13·2	349	M	Mira
RS	17	47·5	−06	42	5·2—12·2	—	Op	Recurrent nova (1898, 1933, 1958)
Y	17	50·0	−06	08	7·0—7·8	17·1	F-G	Cepheid
RY	18	14·1	+03	40	7·6—13·8	150	M	Mira
X	18	36·0	+08	47	5·9—9·2	334	M	Mira

CLUSTERS AND NEBULÆ

Object		R.A.		Dec.		Type	Notes
		h	m	°	'		
I 4603-4		16	22·3	−23	20	Nebula	Region of ρ Oph
NGC 6171	M.107	16	29·7	−12	57	Globular	Mag. 9·2. ('Extra' Messier object)
NGC 6218	M.12	16	44·6	−01	52	Globular	Mag. 6·6
NGC 6254	M.10	16	54·5	−04	02	Globular	Mag. 6·7
NGC 6266	M.62	16	58·1	−30	03	Globular	Mag. 6·6
NGC 6273	M.19	16	59·5	−26	11	Globular	Mag. 6·6
NGC 6287		17	02·1	−24	41	Globular	Mag. 9·7
NGC 6293		17	07·1	−26	30	Globular	Mag. 8·4
NGC 6333	M.9	17	16·2	−18	28	Globular	Mag. 7·3
NGC 6356		17	20·7	−17	46	Globular	Mag. 8·7
NGC 6369		17	26·3	−23	44	Planetary	Mag. 9·9; central star mag. 17
NGC 6402	M.14	17	35·0	−03	13	Globular	Mag. 8·0
I 4665		17	43·8	+05	44	Open cluster	Mag. 6. Rather sparse
NGC 6572	Σ6	18	10·2	+06	50	Planetary	Mag. 9·6; central star mag. 12
NGC 6633		18	25·1	+06	32	Open cluster	Mag. 5. About 65 stars

ORION

(Abbreviation: Ori).

One of the most magnificent constellations in the sky; it represents the mythological hunter who boasted that he could kill any creature on earth, but who was fatally stung by a scorpion. The two leading stars are Rigel, which is actually variable over a very small range (0·08 to 0·20) and the red variable Betelgeux – a name which may also be spelled Betelgeuse or Betelgeuze. The gaseous nebula M. 42, in the Sword, is the most famous example of its type, and is easily visible with the naked eye. Altogether there are fifteen stars above the fourth magnitude:

β 0·08v	κ 2·06	λ 3·40
α 0·4v	δ 2·20v	τ 3·68
γ 1·64	ι 2·76	π⁵ 3·7v
ε 1·70	π³ 3·17	π⁴ 3·78
ζ 1·79	η 3·32v	σ 3·78

(π^5 is an eclipsing binary with an extremely small range, and η is a β Lyræ eclipsing binary with a range of from 3·32 to 3·50).

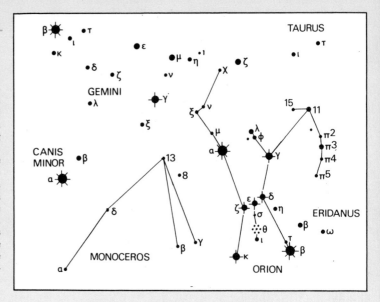

BRIGHTEST STARS

Star		Mag.	R.A. h	m	s	Dec. °	′	″	Abs. Mag.	Spectrum	Dist. l/y
1 π³		3·17	04	48	39	+06	55	26	+3·7	F8	26
2 π²		4·35	04	49	25	+08	51	49	+1·3	A0	130
3 π⁴		3·78	04	50	02	+05	34	07	−4·5	B3	1300
8 π⁵		3·7v	04	53	06	+02	24	20	var.	B3	1300
7 π¹		4·74	04	53	41	+10	07	01	+0·9	A0	270
9 o²		4·28	04	55	08	+13	28	51	+0·4	K2	190
10 π⁶		4·73	04	57	24	+01	40	53	−2·0	K0	650
11		4·65	05	03	19	+15	22	28	0·0	B9	270
17 ρ		4·64	05	12	08	+02	50	10	−0·8	K3	410
19 β Rigel		0·08v	05	13	29	−08	13	34	−7·1	B8p	900
20 τ		3·68	05	16	32	−06	52	02	−1·8	B8	410
22 o		4·65	05	20	38	−00	24	11	−3·7	B3	1300
29 e		4·21	05	22	53	−07	49	38	−0·4	G9	270
28 η Algjebbah		3·3v	05	23	22	−02	24	58	−3·7	B1	950
25		4·73	05	23	36	+01	49	38	−3·4	B3p	1300
24 γ Bellatrix		1·64	05	23	57	+06	19	51	−4·2	B2	470
30 ψ		4·66	05	25	41	+03	04	40	−3·5	B2	1300
32 A		4·32	05	29	36	+05	55	57	−0·9	B3	360
36 υ Thabit		4·64	05	30	52	−07	19	00	−3·6	B3	1300
34 δ Mintake		2·2v	05	30	53	−00	18	52	−6·1	O9	1500
37 φ¹		4·53	05	33	37	+09	28	33	−4·9	B0	1900
39 λ Heka		3·40	05	33	56	+09	55	14	−5·1	O8	1800
42 c		4·65	05	34	18	−04	51	06	−3·6	B3	1300
44 ι Hatysa		2·76	05	34	21	−05	55	24	−6·1	O9	2000
46 ε Alnilam		1·70	05	35	06	−01	12	54	−6·8	B0	1600
40 φ²		4·39	05	35	42	+09	16	47	+1·5	K0	125
48 σ		3·78	05	37	38	−02	36	42	−4·6	B0	1300
47 ω		4·54	05	38	01	+04	06	36	−4·1	B3p	1300
50 ζ Alnitak		1·79	05	39	39	−01	57	12	−6·6	O9	1600
54 κ Saiph		2·06	05	46	43	−09	40	35	−6·9	B0	2100
54 χ¹		4·62	05	53	05	+20	16	24	+4·7	F8	31
58 α Betelgeux		var.	05	53	59	+07	24	15	−5·6v	M2	520
Boss 7587		4·68	05	58	57	−03	04	26	+1·2	K2	160
61 μ		4·19	06	01	10	+09	38	55	+1·2	A2	130
62 χ²		4·71	06	02	37	+20	08	24	−7·0	B2p	4100
67 ν		4·40	06	08	19	+14	46	40	−1·4	B2	470
70 ξ		4·35	06	10	41	+14	12	54	−1·4	B3	470

THERE ARE FIVE MORE STARS ABOVE THE FIFTH MAGNITUDE:

Star	Mag.	Abs. Mag.	Spectrum
15	4·86	+0·1	F0
23 m	4·99	−3·3	B1
31 CI	5·00	−0·5	K5
49 d	4·88	+2·0	A3
69 f¹	4·92	−1·8	B3

DOUBLES

Star	R.A. h m	Dec. ° '	Mags.	P.A.	Dist."	
ρ	05 12·1	+02 50	4·7, 8·6	063	7·0	Fixed
β	05 13·5	−08 14	0·1, 7·0	206	9·2	Fixed
η	05 23·4	−02 25	3·7, 5·1	083	1·5	
λ	05 33·9	+09 55	3·7, 5·6	042	4·4	Fixed
δ	05 30·9	−00 19	2·2v, 6·9	000	52·8	Fixed
ζ	05 39·7	−01 57	2·0, 4·2	164	2·4	P.A. increasing

In addition there are the two famous multiples in the Sword θ (the 'Trapezium') and σ

VARIABLES

Star	R.A. h m	Dec. ° '	Range	Period, d	Spectrum	Type
W	05 02·8	+01 07	5·9—7·7	212	N	Semi-regular
S	05 26·5	−04 44	7·5—13·5	416	M	Mira
α	05 53·9	+07 24	0·1—0·9	±2070	M	Semi-regular
U	05 52·9	+20 10	5·3—12·6	372	M	Mira

The range and period of Betelgeux are uncertain. The star has been recorded as equalling or even surpassing Rigel, and at other times descending to equality with Aldebaran, but no precise limits are available. Generally, Betelgeux is comparable with Procyon, decidedly inferior to Rigel, and considerably brighter than Aldebaran. Mintake, in the Belt, has long been listed as a variable, but it is in fact an Algol system with a range of only 2·20 to 2·35.

CLUSTERS AND NEBULÆ

Object	R.A. h m	Dec. ° '	Type	Notes
NGC 1976 M.42 } NGC 1982 M.43 }	05 32·9	−05 25	Nebula	"Sword of Orion". Naked-eye; contains the multiple θ
NGC 2068 M.78	05 44·2	+00 02	Nebula	Mag. 8·3. Reflection nebula

PAVO

(Abbreviation: Pav).

One of the 'Southern Birds'. The brightest star, α, is somewhat isolated from the main pattern; the most celebrated object is κ, often called a Cepheid even though it is, strictly speaking, a W Virginis type variable.

Pavo includes four stars above the fourth magnitude:

α 1·95　　β 3·60
η 3·58　　δ 3·64

BRIGHTEST STARS

Star	Mag.	R.A.			Dec.			Abs. Mag.	Spectrum	Dist. l/y
		h	m	s	°	′	″			
η	3·58	17	43	34	−64	42	55	−0·1	K0	180
π	4·44	18	06	28	−63	40	17	+1·6	A5	120
ξ	4·25	18	21	12	−61	30	21	+0·1	K2	220
ζ	4·10	18	40	28	−71	26	58	+1·5	K0	105
λ	4·42	18	50	11	−62	12	54	−3·2	B2	1100
κ	var.	18	54	41	−67	15	48	var.	F5p	650
ε	4·10	19	58	04	−72	58	15	−0·5	A0	270
δ	3·64	20	06	35	−66	14	24	+4·7	G5	19·2
α	1·95	20	23	55	−56	48	25	−2·9	B3	310
β	3·60	20	42	59	−66	17	01	+0·9	A5	110
γ	4·30	21	24	39	−65	28	01	+4·6	F8	29

THERE ARE THREE MORE STARS ABOVE THE FIFTH MAGNITUDE:

Star	Mag.	Abs. Mag.	Spectrum
ν	4·81	−0·2	B8
Boss 25604	4·90	+2·3	A2
φ¹	4·84	+1·5	F0

DOUBLE

Star	R.A.		Dec.		Mags.	P.A.	Dist.″	
	h	m	°	′				
ξ	18	21·2	−61	30	4·2, 8·6	151	3·3	Fixed

VARIABLES

Star	R.A.		Dec.		Range	Period, d	Spectrum	Type
	h	m	°	′				
R	18	08·1	−63	38	7·5—13·8	230	M	Mira
κ	18	54·6	−67	16	4·0—5·5	9·1	F–G	W Virginis
T	19	45·1	−71	54	7·0—14·0	244	M	Mira
S	19	51·0	−59	20	6·6—10·4	387	M	Semi-regular
Y	21	19·8	−69	57	5·7—8·5	233	N	Semi-regular

CLUSTERS AND NEBULÆ

Object	R.A.		Dec.		Type	Notes
NGC 6684	18	44·1	−65	14	Galaxy	Mag. 12. Spiral
NGC 6752 △ 295	19	06·4	−60	04	Globular	Bright; diameter 18′

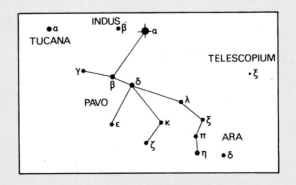

PEGASUS

(Abbreviation: Peg).

One of the most distinctive of the northern constellations. It commemorates the flying horse which the hero Bellerophon rode during an expedition to slay the fire-breathing Chimæra. The main stars of Pegasus make up a square; three of these are α, β and γ. The fourth is Alpheratz, which used to be included in Pegasus as δ Pegasi, but has been officially – and, frankly, illogically – transferred to Andromeda, as α Andromedæ. In Pegasus, excluding Alpheratz, there are nine stars above the fourth magnitude:

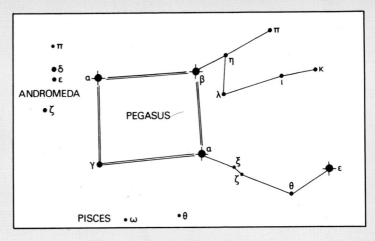

ε 2·31	γ 2·84v	μ 3·67
β 2·4 (max)	η 2·95	θ 3·70
α 2·49	ζ 3·40	ι 3·96

BRIGHTEST STARS

Star			Mag.	R.A. h m s			Dec. ° ′ ″			Abs. Mag.	Spectrum	Dist. l/y
1			4·24	21 21 04			+19 42 35			+0·4	K0	190
8	ε	Enif	2·31	21 43 06			+09 46 25			−4·6	K0	780
9			4·52	21 43 28			+17 14 54			−4·0	G5	1250
10	κ		4·27	21 43 39			+25 32 36			+1·3	F5	130
24	ι		3·96	22 05 39			+25 14 14			+3·4	F5	42
29	π²		4·38	22 09 00			+33 04 11			−0·2	F5	270
26	θ	Biham	3·70	22 09 05			+06 05 21			+1·7	A2	80
42	ζ	Homan	3·40	22 40 22			+10 42 58			−0·6	B8	210
44	η	Matar	2·95	22 41 58			+30 06 21			−2·2	G2	360
47	λ		4·14	22 45 28			+23 26 58			+0·5	K0	150
46	ξ		4·31	22 45 35			+12 03 35			+3·1	F3+M1	65
48	μ	Sadalbari	3·67	22 48 56			+24 29 06			+1·1	K0	105
53	β	Scheat	var.	23 02 42			+27 57 47			−1·5v	M0	210
54	α	Markab	2·49	23 03 40			+15 02 12			−0·1	A0	100
55			4·69	23 05 54			+09 17 25			+0·4	M0	230
62	τ		4·65	23 19 33			+23 37 11			+2·0	A5	110
68	υ		4·57	23 24 17			+23 16 58			+0·8	G0	140
70	q		4·67	23 28 02			+12 38 21			−0·3	K0	325
84	ψ		4·75	23 56 38			+25 01 09			0·0	F8p	300
88	γ	Algenib	2·84v	00 12 06			+15 03 41			−3·4	B2	570

THERE ARE NINE MORE STARS ABOVE THE FIFTH MAGNITUDE:

Star	Mag.	Abs. Mag.	Spectrum	Star	Mag.	Abs. Mag.	Spectrum
2	4·76	+0·2	M1	43 o	4·85	+1·0	A2
22 ν	4·90	+0·1	K5	50 ρ	4·95	−0·3	A0
32	4·88	−0·6	B8	56	4·98	−2·0	K0
31	4·93	−1·2	B3	89 χ	4·94	−0·5	M2
35	4·93	+1·6	K0				

DOUBLE

Star	R.A. h m	Dec. ° ′	Mags.	P.A.	Dist.″	
ξ	22 45·6	+12 04	4·3, 11·7	108	11·9	Slow binary

VARIABLES

Star	R.A. h m	Dec. ° ′	Range	Period, d	Spectrum	Type
TW	22 01·7	+28 07	7·0—9·2	956	M	Semi-regular
β	23 02·7	+27 58	2·4—2·8	±36	M	Semi-regular
R	23 04·1	+10 16	7·1—13·8	378	M	Mira
W	23 17·4	+26 00	7·9—13·0	344	M	Mira
S	23 18·0	+08 39	7·4—13·8	319	M	Mira

CLUSTERS AND NEBULÆ

Object	R.A. h m	Dec. ° ′	Type	Notes
NGC 7078 M.15	21 27·6	+11 57	Globular	Mag. 6. Bright and condensed
NGC 7217	22 05·6	+31 07	Galaxy	Type Sb. Mag. 11
NGC 7331	22 34·8	+34 10	Galaxy	Type Sb. Mag. 9·7
NGC 7332	22 35·0	+25 32	Galaxy	Type E7, Mag. 11·8
NGC 7479	23 02·4	+12 03	Galaxy	Type SBb. Mag. 11·3
NGC 7741	23 41·4	+25 48	Galaxy	Type SBc. Mag. 11·6

PERSEUS

(Abbreviation: Per).

A prominent constellation, containing the prototype eclipsing star Algol as well as the superb Sword-Handle cluster (H.VI.33–4). Mythologically, Perseus was the hero of one of the most famous of all legends; he killed the Gorgon, Medusa, and married Andromeda, daughter of Cepheus and Cassiopeia. The Gorgon's Head is marked by the winking 'Demon Star', Algol.

Perseus includes ten stars above the fourth magnitude:

α 1·80   ε 2·88   o 3·82
β 2·06 (max) γ 2·91   ν 3·93
ζ 2·83   δ 3·03   η 3·93
            ρ 3·2 (max)

BRIGHTEST STARS

Star		Mag.	R.A. h	m	s	Dec. °	'	"	Abs. Mag.	Spectrum	Dist. l/y
φ		4·02v	01	42	16	+50	34	42	var.	B1p	550
13 θ		4·22	02	42	41	+49	08	11	+3·7	F8	42
15 η		3·93	02	49	05	+55	48	19	−4·7	K0	1100
16		4·27	02	49	11	+38	13	44	+1·2	K5	140
17		4·67	02	50	09	+34	58	12	−0·6	K2	360
18 τ	Kerb	4·06	02	52	41	+52	40	25	+0·8	G1+A5	150
22 π		4·62	02	57	21	+39	34	32	0·0	A2	270
23 γ		2·91	03	03	12	+53	25	17	+0·3	F7+A3	110
25 ρ		var.	03	03	46	+38	45	22	−1·0v	M3	260
26 β	Algol	var.	03	06	44	+40	52	19	−0·5	B8	105
ι		4·17	03	07	28	+49	31	50	+3·8	G0	39
27 κ	Misam	4·00	03	08	00	+44	46	30	+1·0	K0	130
33 α	Mirphak	1·80	03	22	45	+49	47	03	−4·4	F5	570
34		4·67	03	27	47	+49	26	02	−0·6	B5	360
35 σ		4·55	03	29	01	+47	55	14	−0·7	K0	360
37 ψ		4·26	03	34	55	+48	07	15	−1·2	B5p	410
39 δ		3·03	03	41	21	+47	43	06	−3·3	B5	590
38 o	Ati	3·82	03	42	56	+32	13	11	−4·8	B1	1160
41 ν		3·93	03	43	42	+42	30	37	−2·0	F4	470
44 ζ	Atik	2·83	03	52	45	+31	49	10	−6·1	B1	1000
45 ε		2·88	03	56	22	+39	56	52	−3·7	B1	680
46 ξ	Menkib	4·05	03	57	22	+35	43	44	−5·4	O7	1900
47 λ		4·33	04	04	56	+50	17	35	+0·7	A0	170
48 υ	Nembus	4·03	04	07	04	+47	39	19	−2·0	B3	410
51 μ		4·28	04	13	17	+48	21	17	−4·6	G2	1250
b¹		4·6v	04	16	35	+50	14	34	var.	A2	180
58 e		4·46	04	35	10	+41	13	15	−2·5	K0+A3	650

THERE ARE EIGHT MORE STARS ABOVE THE FIFTH MAGNITUDE:

Star	Mag.	Abs. Mag.	Spectrum
4 g	4·99	−0·2	B8
12	4·99	+2·1	F9
24	4·97	+0·2	K2
28 ω	4·82	+1·0	G9
Boss 3948	4·92	−0·6	K4
32 l	4·98	+1·0	A1
52 f	4·89	−4·6	G3+A5
53 d	4·89	−1·2	B4

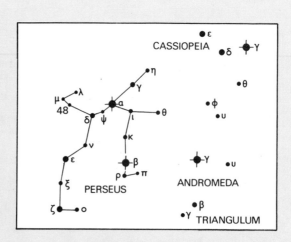

DOUBLES

Star	R.A. h m	Dec. ° '	Mags.	P.A.	Dist."	
η	02 49·1	+55 48	3·9, 8·6	301	28·4	Fixed
β	03 06·7	+40 52	var, 10·5	193	82·1	Optical
ζ	03 52·8	+31 49	2·9, 9·4	208	12·9	Fixed
ε	03 56·4	+39 57	2·9, 8·1	009	9·0	Fixed

VARIABLES

Star	R.A. h m	Dec. ° '	Range	Period, d	Spectrum	Type
U	01 56·2	+54 35	7·6—12·3	321	M	Mira
S	02 19·3	+56 23	7·9—11·1	?(slow)	M	Semi-regular
ρ	03 03·8	+38 45	3·2—4·2	33—55	M8	Semi-regular
β	03 06·7	+40 52	2·1—3·3	2·87	B8	Prototype Algol
X	03 52·3	+30 54	6·0—6·6	—	0p	Irregular

(X Persei is a very peculiar star, and appears to be a source of radio waves.)

CLUSTERS AND NEBULÆ

Object	R.A. h m	Dec. ° '	Type	Notes
NGC 650-1 M.76	01 38·8	+51 19	Planetary	Mag. 12·2 (faintest Messier object)
NGC 869/884 H.VI.33-4	02 18	+56 54	Double cluster	Sword-handle; naked-eye
NGC 957	02 28·9	+57 18	Open cluster	Mag. 7. About 40 stars
NGC 1039 M.34	02 38·8	+42 34	Open cluster	Mag. 5·5. About 80 stars
NGC 1245	03 11·2	+47 03	Open cluster	Mag. 7. About 40 stars
NGC 1342	03 28·4	+37 09	Open cluster	Mag. 7. About 40 stars
I 2003	03 53·2	+33 44	Planetary	Mag. 12·6. Very faint central star
NGC 1499	04 00·1	+36 17	Nebula	California Nebula
NGC 1528	04 11·4	+51 07	Open cluster	Mag. 6. About 80 stars

PHŒNIX

(Abbreviation: Phe).

One of the 'Southern Birds'. It is not very distinctive, and Ankaa is the only bright star; there are however seven stars above the fourth magnitude:

α 2·39   γ 3·44   ε 3·94
β 3·30   ζ 3·6 (max)   δ 3·96
          χ 3·90

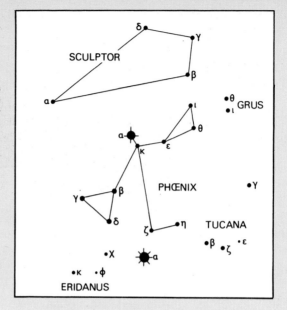

BRIGHTEST STARS

Star		Mag.	R.A.			Dec.			Abs. Mag.	Spectrum	Dist. l/y
			h	m	s	°	′	″			
ε		3·94	00	08	18	—45	52	08	+1·8	K0	88
χ		3·90	00	25	08	—43	48	07	+3·0	A3	50
α	Ankaa	2·39	00	25	12	—42	25	32	+0·1	K0	93
μ		4·65	00	40	17	—46	12	20	—0·3	K0	320
η		4·53	00	42	22	—57	35	01	+1·5	A0	130
β		3·30	01	05	06	—46	50	11	+0·3	K0	190
ζ		var.	01	07	28	—55	21	48	—1·1	B7	360
γ		3·44	01	27	25	—43	25	51	—4·6	K5	1300
δ		3·96	01	30	20	—49	11	12	+1·1	K0	120
ψ		4·41	01	52	46	—46	24	36	—	M3	—
—47°597		4·74	01	56	18	—47	29	32	+2·5	G4	90

THERE ARE SIX MORE STARS ABOVE THE FIFTH MAGNITUDE:

Star	Mag.	Abs. Mag.	Spectrum
ι	4·80	—5·2	A2p
Boss 32836	4·86	+0·1	A2
λ¹	4·88	+1·8	A2
ν	4·88	+4·4	G0
φ	5·00	—	B9
χ	4·96	—	M0

DOUBLES

Star	R.A.		Dec.		Mags.	P.A.	Dist.″	
	h	m	°	′				
θ	23	36·7	—46	55	6·7, 7·4	272	4·2	Slow binary
β	01	05·1	—46	50	4·1, 4·1	352	1·3	Slow binary

VARIABLES

Star	R.A.		Dec.		Range	Period, d	Spectrum	Type
	h	m	°	′				
R	23	53·9	—50	05	7·5—14·4	268	M	Mira
S	23	56·5	—56	51	7·4—8·2	141	M	Semi-regular
ζ	01	07·5	—55	22	3·6—4·1	1·67	B7	Eclipsing

PICTOR

(Abbreviation: Pic).

An unremarkable constellation near Canopus, known originally under the cumbersome name of Equuleus Pictoris (the Painter's Easel). There are two stars above the fourth magnitude:

α 3·27   β 3·94

See chart for Carina

BRIGHTEST STARS

Star	Mag.	R.A.			Dec.			Abs. Mag.	Spectrum	Dist. l/y
		h	m	s	°	'	"			
β	3·94	05	46	46	−51	04	26	+2·8	A3	55
γ	4·38	05	49	26	−56	10	18	+0·3	K1	220
α	3·27	06	47	58	−61	55	03	+2·1	A5	57

There are two more stars above the fifth magnitude. One is η² (mag. 4·92, absolute mag. −3·6, spectrum M2). The other is δ, which is a B-type star variable over a small range (4·66—4·89)

DOUBLES

Star	R.A.		Dec.		Mags.	P.A.	Dist."	
	h	m	°	'				
ι	04	49·8	−53	33	5·6, 6·4	058	12·0	Fixed
μ	06	31·2	−58	43	5·8, 9·3	231	2·4	Fixed

VARIABLES

Star	R.A.		Dec.		Range	Period, d	Spectrum	Type
	h	m	°	'				
R	04	44·8	−49	20	6·7—10.0	171	M	Semi-regular
S	05	09·6	−48	34	7·2—14·0	427	M	Mira
W	05	42·0	−46	29	11·8—13·3	—	N	Irregular

The bright nova of 1925 (RR Pictoris) was in this constellation.

PISCES

(Abbreviation: Psc).

A large but faint Zodiacal constellation; it now contains the Vernal Equinox. Its mythological associations are rather vague, but it may represent the fishes into which Venus and Cupid once changed themselves in order to escape from the monster Typhon. There are three stars above the fourth magnitude:

η 3·72   γ 3·85   α 3·94

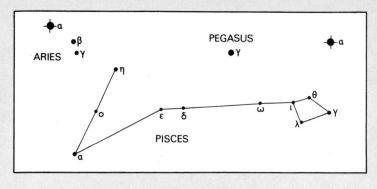

BRIGHTEST STARS

Star		Mag.	R.A. h m s			Dec. ° ′ ″			Abs. Mag.	Spectrum	Dist. l/y
4 β		4·58	23 02 45			+03 42 05			−0·4	B5	330
6 γ		3 85	23 16 01			+03 09 43			+0·9	K0	125
10 θ		4·45	23 26 51			+06 15 28			+0·3	G5	220
17 ι		4·28	23 38 49			+05 30 25			+3·4	F5	50
18 λ		4·61	23 40 55			+01 39 32			+1·5	A5	140
19 TX		var.	23 43 50			+03 12 34			var.	N	?
28 ω		4·03	23 58 11			+06 44 29			+0·8	F3	140
30		4·66	00 00 50			−06 08 11			+0·4	M3	230
33		4·68	00 04 13			−05 49 50			+1·0	K0	180
63 δ		4·55	00 47 32			+07 27 56			+0·3	K5	230
71 ε		4·45	01 01 48			+07 46 19			+1·4	K0	140
83 τ		4·70	01 10 27			+29 58 23			+1·2	K0	160
85 φ		4·64	01 12 33			+24 28 03			−0·1	K0	300
90 υ		4·67	01 18 15			+27 08 56			+0·9	A2	180
99 η		3·72	01 30 18			+15 13 58			−2·0	G5	410
106 ν		4·68	01 40 17			+05 22 36			+1·2	K0	160
110 o		4·50	01 44 17			+09 02 51			+0·6	K0	190
113 α Kaïtain		3·94	02 00 54			+02 39 29			+0·9	A2p	140

THERE ARE SIX MORE STARS ABOVE THE FIFTH MAGNITUDE:

Star	Mag.	Abs. Mag.	Spectrum
8 x	4·94	+2·2	A3p
47 TV	var.	var.	M3
74 ψ	4·7	+1·3, +1·5	A2+A0
84 χ	4·89	+0·5	G9
86 ζ	4·9	+2·1	A5+F
111 ξ	4·84	−0·2	G7

DOUBLES

Star	R.A. h m	Dec. ° ′	Mags.	P.A.	Dist."	
ζ	01 11·1	+07 19	5·6, 6·5	063	23·6	Fixed
α	02 00·9	+02 39	4·3, 5·2	297	2·1	Slow binary

VARIABLES

Star	R.A. h m	Dec. ° ′	Range	Period, d	Spectrum	Type
TX	23 43·8	+03 13	4·3—5·1	—	N	Irregular
TV	00 25·4	+17 37	4·8—5·2	±49	M3	Semi-regular
Z	01 13·4	+25 30	7·0—7·9	144	N	Semi-regular
R	01 28·1	+02 37	7·1—14·8	344	M	Mira

NEBULA

Object	R.A. h m	Dec. ° ′	Type	Notes
NGC 628 M.74	01 34·0	+15 32	Galaxy	Mag. 10·2. Type Sc

PISCIS AUSTRALIS

(Abbreviation: PsA).

Also known as Piscis Austrinus. No specific mythological legends have been associated with it. It contains Fomalhaut, which is, incidentally, the most southerly of the first-magnitude stars to be visible from England (mag. 1·16), but there are no other stars above the fourth magnitude.

BRIGHTEST STARS

Star		Mag.	R.A.			Dec.			Abs. Mag.	Spectrum	Dist. l/y
			h	m	s	°	′	″			
9	ι	4·35	21	43	38	−33	07	37	+2·1	A0	95
14	μ	4·62	22	07	06	−33	05	48	+1·7	A2	125
17	β	4·36	22	30	15	−32	27	33	+0·2	A0	220
18	ε	4·22	22	39	26	−27	09	32	−0·2	B8	250
22	γ	4·52	22	51	18	−32	59	34	+0·8	A0	180
23	δ	4·33	22	54	44	−32	39	27	+0·6	K0	180
24	α Fomalhaut	1·16	22	56	26	−29	44	21	+1·9	A3	22·6

There are no other stars above the fifth magnitude.

DOUBLES

Star	R.A.		Dec.		Mags.	P.A.	Dist.″	
	h	m	°	′				
β	22	30·3	−32	28	4·4, 7·9	172	30·4	Optical
γ	22	51·3	−33	00	4·5, 8·1	264	4·3	Slow binary
η	21	58·0	−28	42	5·8, 6·8	119	1·6	Binary

NEBULA

Object	R.A.		Dec.		Type	Notes
	h	m	°	′		
NGC 7314	22	33·0	−26	18	Galaxy	Mag. 12. Type Sc

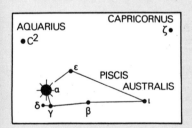

PUPPIS

(Abbreviation: Pup).

The poop of the dismembered ship, Argo Navis. There are ten stars above the fourth magnitude: an eleventh, k, has a combined magnitude of 3·8.

ζ 2·23	τ 2·97	L² 3·4
ρ 2·72v	ν 3·19	(max)
π 2·81	σ 3·28	c 3·72
	ξ 3·34	a 3·76

ρ is a variable with very small range (2·72–2·87).

See chart for Carina

BRIGHTEST STARS

Star		Mag.	R.A.			Dec.			Abs. Mag.	Spectrum	Dist. l/y
			h	m	s	°	′	″			
ν		3·19	06	37	05	−43	10	34	−3·2	B8	620
τ		2·97	06	49	23	−50	35	17	+0·1	K0	125
l		4·47	07	11	56	−46	13	49	+2·6	F0	75
L²		var.	07	12	52	−44	36	12	−3·1v	M3	650
π		2·81	07	16	22	−37	03	26	−0·3	K5	140
υ¹		4·68	07	17	31	−36	41	36	−1·1	B3	460
σ		3·28	07	28	28	−45	15	08	−0·5	K5	180
108 G		4·52	07	33	07	−22	14	52	−2·0	F8	65
p		4·55	07	34	30	−28	19	12	+0·1	B8	250
f		4·62	07	36	33	−34	55	06	+1·6	B8	130
m		4·64	07	37	23	−25	18	51	−0·1	B8	300
k		3·8	07	37	55	−26	45	04	−0·7, −0·6	B8+B5	360
3 l		4·10	07	42	55	−28	54	06	−6·2	A2p	3300
c		3·72	07	44	28	−37	54	53	−6·3	K5	3300
o		4·59	07	47	10	−25	52	55	−3·5	B2	1100
Q		4·64	07	47	41	−47	01	18	+1·0	K0	1100
7 ξ	Asmidiske	3·34	07	48	22	−24	48	14	−4·6	G0p	1200
P		4·25	07	48	34	−46	19	02	−4·8	B0	1900
a		3·76	07	51	28	−40	31	06	−0·5	G5	140
b		4·53	07	51	52	−38	48	19	−0·5	B3	330
J		4·32	07	52	39	−48	02	42	−6·0	B1	2500
11 e		4·35	07	55	55	−22	49	14	−1·8	F8	540
V		var.	07	57	36	−49	11	05	var.	B1p+B3p	1100
232 G		4·64	07	58	33	−18	20	17	+1·2	A2	160
ζ	Suhail Hadar	2·23	08	02	49	−39	56	26	−7·1	O5	2400
15 ρ	Turais	2·7v	08	06	56	−24	14	24	+0·3	F5	105
16		4·34	08	08	03	−19	10	47	−1·1	B5	410
19		4·68	08	10	14	−12	51	39	+1·6	K0	140
h¹		4·43	08	10	34	−39	33	09	—	K5	—
h²		4·43	08	13	16	−40	16	48	−0·6	K2	400
q		4·43	08	17	44	−36	35	26	+2·4	A5	85

THERE ARE ELEVEN MORE STARS ABOVE THE FIFTH MAGNITUDE:

Star	Mag.	Abs. Mag.	Spectrum
Y	5·00	−0·2	G7
Boss 9137	4·88	?	M3
A	4·85	−0·6	B3
Boss 9635	4·88	+1·3	B8
Boss 10043	4·80	−5·0	A5
d¹	4·91	−0·9	B3
1	4·82	0·0	K5
Boss 10686	4·83	−1·7	B3
Boss 10774	4·85	+1·6	A2
r	4·77	−1·3	B3
W	4·94	+1·8	M0

DOUBLES

Star	R.A.		Dec.		Mags.	P.A.	Dist.″	
	h	m	°	′				
Y	06	37·3	−48	10	5·0, 8·3	320	12·8	Slow binary
σ	07	28·5	−43	15	3·3, 8·5	074	22·4	Fixed

VARIABLES

Star	R.A.		Dec.		Range	Period, d	Spectrum	Type
	h	m	°	′				
L²	07	12·9	−44	36	3·4—6·2	141	M	Semi-regular
Z	07	30·5	−20	33	7·2—14·6	510	M	Mira
V	07	56·8	−49	07	4·3—5·1	1·45	B1+B3	β Lyræ
AP	07	56·0	−39	59	7·6—8·7	5·08	F	Cepheid
XZ	08	11·4	−23	48	8·0—10·7	2·10	A	Algol

CLUSTERS AND NEBULÆ

Object	R.A.		Dec.		Type	Notes
	h	m	°	'		
NGC 2298	06	47·2	−35	57	Globular	Mag. 12·5
NGC 2422 M.47	07	34·3	−14	22	Open cluster	Naked eye (mag. 5·2)
NGC 2423	07	34·8	−13	45	Open cluster	Mag. 7. About 60 stars
NGC 2437 M.46	07	39·5	−14	42	Open cluster	Mag. 6. Within it lies the faint planetary NGC 2438
NGC 2348	07	39·6	−14	36	Planetary	Mag. 11·3 (central star mag. 17)
NGC 2440 H.IV.64	07	39·6	−18	05	Planetary	Mag. 11 (central star mag. 17). Annular
NGC 2447 M.93	07	42·4	−23	45	Open cluster	Mag. 6. Loose
NGC 2452	07	45·6	−27	13	Planetary	Mag. 13 (central star mag. 19)
NGC 2477	07	50·5	−38	25	Open cluster	Mag. 5·7. Rich. Just possibly it is a globular
NGC 2489	07	56·2	−29	56	Open cluster	Mag. 9·4. Fairly rich
NGC 2539	08	08·4	−12	41	Open cluster	Mag. 8·2. About 150 stars
NGC 2567	08	16·6	−30	29	Open cluster	Mag. 8·3
NGC 2571	08	16·9	−29	35	Open cluster	Mag. 7·5. About 25 stars

PYXIS

(Abbreviation: Pyx).

Also originally part of Argo. The only star above the fourth magnitude is α (3·70).

See chart for Carina

BRIGHTEST STARS

Star	Mag.	R.A.			Dec.			Abs. Mag.	Spectrum	Dist. l/y
		h	m	s	°	'	"			
β	4·04	08	39	15	−35	13	46	−0·2	G5	230
α	3·70	08	42	42	−33	06	23	−2·1	B2	470
γ	4·19	08	49	36	−27	37	40	+0·3	K4	190

THERE ARE FOUR MORE STARS ABOVE THE FIFTH MAGNITUDE:

Star	Mag.	Abs. Mag.	Spectrum
δ	4·87	+2·4	A2
ϰ	4·82	−0·4	M0
θ	4·93	+0·1	M1
λ	4·90	+1·3	G7

VARIABLE

Star	R.A.		Dec.		Range	Period, d	Spectrum	Type
	h	m	°	'				
T	09	02·6	−32	11	7·0—14·0	—	Pec	Recurrent nova (1920, 1944)

CLUSTERS AND NEBULÆ

Object	R.A.		Dec.		Type	Notes
	h	m	°	'		
NGC 2613	08	31·1	−22	48	Galaxy	Mag. 11. Type Sb
NGC 2627	08	35·2	−29	46	Open cluster	Mag. 8·3. About 40 stars
NGC 2635	08	36·5	−34	35	Open cluster	Mag. 10. Sparse (about 20 stars)

RETICULUM

(Abbreviation: Ret).

Originally Reticulum Rhomboidalis (the Rhomboidal Net). A small but quite distinctive constellation of the far south. There are two stars above the fourth magnitude:

α 3·33   β 3·80

BRIGHTEST STARS

Star	Mag.	R.A.			Dec.			Abs. Mag.	Spectrum	Dist. l/y
	`	h	m	s	°	′	″			
β	3·80	03	43	55	−64	52	34	+2·0	K0	75
δ	4·41	03	58	24	−61	27	43	?	M0	?
γ	4·46	04	00	35	−62	13	14	?	M5	?
α	3·33	04	14	08	−62	31	43	−2·1	G5	390
ε	4·42	04	16	06	−59	21	17	+2·5	K2	80

THERE ARE TWO MORE STARS ABOVE THE FIFTH MAGNITUDE:

Star	Mag.	Abs. Mag.	Spectrum
χ	4·80	+3·6	F5
ι	4·81	−1·3	M0

DOUBLE

Star	R.A.		Dec.		Mags.	P.A.	Dist."
	h	m	°	′			
θ	04	17·1	−62	23	6·2, 8·3	004	3·9 Fixed

VARIABLE

Star	R.A.		Dec.		Range	Period, d	Spectrum	Type
	h	m	°	′				
R	04	33·0	−63	08	6·8—14·0	278	M	Mira

NEBULA

Object	R.A.		Dec.		Type	Notes
	h	m	°	′		
NGC 1313	03	17·6	−66	40	Galaxy	Mag. 11

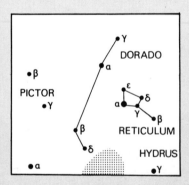

SAGITTA
(Abbreviation: Sge).

An original constellation; small though it is, it is quite distinctive. It has been identified with Cupid's bow, and also with the arrow used by Apollo against the one-eyed Cyclops. There are two stars above the fourth magnitude:

γ 3·71   δ 3·78

See chart for Aquila

BRIGHTEST STARS

Star		Mag.	R.A. h	m	s	Dec. °	′	″	Abs. Mag.	Spectrum	Dist. l/y
5	α	4·37	19	39	07	+17	57	44	−1·7	G0	550
6	β	4·45	19	40	04	+17	25	27	0·0	K0	250
7	δ	3·78	19	46	24	+18	28	45	−1·7	M0+A0	410
12	γ	3·71	19	57	47	+19	25	53	−0·1	K5	190

The only other star above the fifth magnitude is 8 (ζ); mag. 4·95, absolute mag +0·5, spectrum A2.

DOUBLE

Star	R.A. h	m	Dec. °	′	Mags.	P.A.	Dist.″	
θ	20	07·7	+20	46	6·3, 8·7	328	11·6	Slow binary

VARIABLE

Star	R.A. h	m	Dec. °	′	Range	Period, d	Spectrum	Type
U	19	16·6	+19	31	6·4—9·0	3·4	B+G	Algol

CLUSTERS AND NEBULÆ

Object	R.A. h	m	Dec. °	′	Type	Notes
H.20	19	50·9	+18	13	Open cluster	Mag. 9. Has been suspected of being a globular
NGC 6838 M.71	19	51·5	+18	39	Open cluster	Mag. 9. Very distant (18,000 l/y)
NGC 6886	20	10·5	+19	50	Planetary	Mag. 12·2 (central star, mag. 17)
I 4997	20	17·9	+16	35	Planetary	Mag. 11·4. (central star, mag. 14)

SAGITTARIUS

(Abbreviation: Sgr).

The southernmost of the Zodiacal constellations, and not wholly visible from England. Mythologically it has been associated with Chiron, the wise centaur who was tutor to Jason and many others; but it would certainly be more logical to associate Chiron with Centaurus, and another version states that Chiron merely invented the constellation Sagittarius to help in guiding the Argonauts in their quest of the Golden Fleece. The centre of the Milky Way lies behind the star-clouds here, and the whole area is exceptionally rich; it abounds in Messier objects. It is worth commenting that the stars lettered α and β are relatively faint. There are fourteen stars above the fourth magnitude:

ε 1·81	π 2·89	ξ² 3·51
σ 2·12	γ 2·97	μ 3·8v
ζ 2·61	η 3·17	o 3·90
δ 2·71	φ 3·20	ρ¹ 3·95
λ 2·80	τ 3·30	

μ is an Algol binary with a very small range (3·8 to 3·9).

Star		Mag.	R.A.			Dec.			Abs. Mag.	Spectrum	Dist. l/y
			h	m	s	°	′	″			
3 X		var.	17	46	10	−27	49	26	var.	F5—G0	1100
W		var.	18	03	37	−29	34	57	var.	F—G	1300
10 γ	Alnasr	2·97	18	04	24	−30	25	33	+0·1	K0	125
−28° 14174		4·66	18	06	42	−28	27	39	+1·0	K0	170
13 μ		3·8v	18	12	27	−21	03	57	var.	B8p	3300
η		3·17	18	16	08	−36	46	11	+1·1	M3	86
−27° 12684		4·69	18	16	41	−27	03	07	+0·3	K5	250
19 δ	Kaus Meridionalis	2·71	18	19	35	−29	50	20	+0·7	K0	84
20 ε	Kaus Australis	1·81	18	22	43	−34	23	48	−1·1	A0	124
22 λ	Kaus Borealis	2·80	18	26	37	−25	26	07	+1·1	K0	70
27 φ		3·20	18	44	17	−27	00	53	−3·1	B8	590
34 σ	Nunki	2·12	18	53	54	−26	19	32	−2·7	B3	300
37 ξ²		3·51	18	56	25	−21	08	13	0·0	K0	160
38 ζ	Ascella	2·61	19	01	13	−29	54	47	+0·1	A2	140
39 o		3·90	19	03	22	−21	46	30	+1·3	K0	105
40 τ		3·30	19	05	34	−27	42	14	+1·2	K0	86
41 π	Albaldah	2·89	19	08	27	−21	03	36	−0·7	F2	250
44 ρ¹		3·95	19	20	24	−17	53	23	+2·0	A5	65
46 υ		4·58	19	20	28	−15	59	51	−3·0	B8+F2	1100
β¹	Arkab	4·24	19	21	03	−44	30	07	+0·9	B8	270
β²		4·51	19	21	38	−44	50	33	+1·5	F0	130
α	Rukbat	4·11	19	22	22	−40	39	41	−0·3	B8	250
52 h²		4·66	19	35	22	−24	56	00	+0·9	B9	180
ι		4·21	19	53	45	−41	55	39	+0·9	K0	150
θ¹		4·39	19	58	18	−35	20	14	−1·1	B3	400
62 c		4·60	20	01	18	−27	46	20	+0·3	M3	230

THERE ARE SEVEN MORE STARS ABOVE THE FIFTH MAGNITUDE:

Star	Mag.	Abs. Mag.	Spectrum
4	4·76	+0·6	A0
21	4·96	0·0	K1+A0
Boss 25613	4·82	−0·9	B5
32 v¹	4·96	−0·8	K2
42 ψ	4·93	+1·0	F5
58 ω	4·81	+3·4	G5
60 A	4·95	0·0	G5

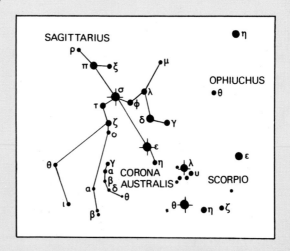

DOUBLE						
Star	R.A.	Dec.	Mag s.	P.A.	Dist."	
	h m	° ′				
χ²	20 20·5	−42 35	5·9, 7·3	229	0·9	Closing

VARIABLES

Star	R.A.		Dec.		Range	Period, d	Spectrum	Type
	h	m	°	'				
X	17	46·2	—27	50	4·1—5·1	7·01	F—G	Cepheid
W	18	03·6	—29	35	4·0—5·2	7·59	F—G	Cepheid
U	18	28·9	—19	10	7·0—8·2	6·7	F—G	Cepheid. In M.25
ST	18	58·7	—12	50	7·6—15	395	S	Mira
RY	19	13·3	—33	37	6·5—14·0	—	G	R Coronæ
R	19	13·8	—19	24	6·7—12·8	269	M	Mira
S	19	16·5	—19	07	9·5—16·0	231	M	Mira

CLUSTERS AND NEBULÆ

Object		R.A.		Dec.		Type	Notes
		h	m	°	'		
NGC 6440		17	45·9	—20	21	Globular	Mag. 10·4
NGC 6494	M.23	17	54·0	—19	01	Open cluster	Mag. 6·9. About 120 stars
NGC 6514	M.20	17	58·9	—23	02	Nebula	Trifid Nebula. Mag. 9
NGC 6520		18	00·3	—27	54	Open cluster	Mag. 8. Rich
NGC 6523	M.8	18	01·6	—24	20	Nebula	Lagoon Nebula. Mag. 6
NGC 6531	M.21	18	01·8	—22	30	Open cluster	Mag. 6·5. About 50 stars
NGC 6567		18	10·8	—19	05	Planetary	Mag. 11·7 (central star mag. 15)
NGC 6603	M.24	18	15·5	—18	27	Open cluster	Mag. 4·6. Rich; some 50 stars
NGC 6613	M.18	18	17·0	—17	09	Open cluster	Mag. 7·5. Sparse
NGC 6618	M.17	18	17·9	—16	12	Nebula	Omega Nebula. Mag. 7
NGC 6626	M.28	18	21·5	—24	54	Globular	Mag. 7·3
NGC 6629		18	22·7	—23	14	Planetary	Mag. 10·6 (central star mag. 14)
NGC 6638		18	27·9	—25	32	Globular	Mag. 9·8
NGC 6637	M.69	18	28·1	—32	33	Globular	Mag. 8·9
I 4725	M.25	18	28·8	—19	17	Open cluster	Mag. 6·5. Contains U Sgr
NGC 6644		18	29·5	—25	11	Planetary	Mag. 12·2
NGC 6656	M.22	18	33·3	—23	58	Globular	Mag. 5·9 (First globular to be discovered)
NGC 6681	M.70	18	40·0	—32	21	Globular	Mag. 9·6
NGC 6715	M.54	18	52·0	—30	32	Globular	Mag. 7·3
NGC 6723		18	56·2	—36	42	Globular	Mag. 6
NGC 6809	M.55	19	36·9	—31	03	Globular	Mag. 7·6
NGC 6818		19	41·1	—14	17	Planetary	Mag. 10 (central star mag. 15)
NGC 6822		19	42·1	—14	53	Galaxy	Mag. 9. Elliptical
NGC 6864	M.75	20	03·2	—22	04	Globular	Mag. 8·0

SCORPIO

(Abbreviation: Sco).

Alternatively, and probably more correctly, known as Scorpius. Mythologically it is usually associated with the scorpion which Juno caused to attack and kill the great hunter Orion. Note that Orion and Scorpio are now on opposite sides of the sky – placed there, it is said, so that the creature could do Orion no further damage!

Scorpio is one of the most magnificent of all constellations, and one of the few which gives at least a vague impression of the creature it is meant to represent. It is dominated by Antares, but the whole area is exceptionally rich. The 'sting', which includes Shaula – only just below the first magnitude – is to all intents and purposes invisible from England.

There are seventeen stars above the fourth magnitude. Antares itself is very slightly variable (range 0·86—1·02). The leaders are:

α 0·86v   β 2·65   μ¹ 2·99
λ 1·60   υ 2·71   G 3·21
θ 1·86   σ 2·82v   η 3·33
ε 2·28   τ 2·85   μ² 3·64
δ 2·34   π 2·92   ζ² 3·75
χ 2·39   ι¹ 2·99

σ is very slightly variable (2·82-2·90)

Star			Mag.	R.A.			Dec.			Abs. Mag.	Spectrum	Type
				h	m	s	°	′	″			
2	A		4·66	15	52	17	−25	15	46	−1·6	B3	580
5	ρ		4·02	15	55	31	−29	09	03	−2·4	B3	630
6	π		2·92	15	57	31	−26	03	07	−3·3	B2	570
7	δ	Dschubba	2·34	15	59	02	−22	33	37	−4·0	B0	590
	ξ		4·16	16	03	09	−11	18	49	+2·1	F8	85
8	β	Graffias	2·65	16	04	09	−19	44	47	−3·7, −1·2	B1+B3	650
9	ω¹	Jabhat al Akrab	4·13	16	05	31	−20	36	39	−3·5	B3	960
10	ω²		4·58	16	06	07	−20	48	37	+0·5	G0	220
14	ν	Jabbah	4·29	16	10	43	−19	24	17	−1·6	B3	360
13	c²		4·70	16	10	57	−27	52	14	−1·6	B3	600
20	σ	Alniyat	2·82v	16	19	51	−25	32	29	−4·4	B1	570
21	α	Antares	0·86v	16	28	03	−26	23	04	−5·1	M1+A3	520
	N		4·33	16	29	56	−34	39	28	−2·4	B3	680
23	τ		2·85	16	34	31	−28	10	18	−4·0	B0	750
	H		4·30	16	34	55	−35	12	42	+0·6	M2	180
26	ε	Wei	2·28	16	48	44	−34	15	17	+0·7	K0	65
	μ¹		2·99	16	50	23	−38	00	40	−3·0	B3	520
	μ²		3·64	16	50	51	−37	58	53	−3·2	B2	740
	ζ²		3·75	16	53	02	−42	19	31	+0·4	K5	160
	η		3·33	17	10	34	−43	12	42	+2·3	F2	52
34	υ	Lesath	2·71	17	29	16	−37	16	47	−3·4	B3	540
35	λ	Shaula	1·60	17	32	07	−37	05	21	−3·3	B2	610
	Q		4·34	17	35	02	−38	37	17	+0·6	K0	180
	θ	Sargas	1·86	17	35	44	−42	59	07	−4·6	F0	650
	κ	Girtab	2·39	17	40	58	−39	01	13	−3·4	B2	470
	ι¹		2·99	17	46	03	−40	07	12	−7·1	F5p	3400
	G		3·21	17	48	22	−37	02	16	+0·7	K2	100

THERE ARE ELEVEN MORE STARS ABOVE THE FIFTH MAGNITUDE:

Star		Mag.	Abs. Mag.	Spectrum
1	b	4·77	−1·6	B4
15	ψ	4·91	+1·1	A3
	d	4·87	+1·9	A0
19	o	4·76	−0·2	A3
22	i	4·87	−1·6	B3
	ζ¹	4·88	−3·6	B1p
	k	4·87	−4·8	B1
Boss 24176		4·83	−0·2	B8
	ι²	4·88	−0·6	A3p
Boss 24374		4·98	+0·4	K5
Boss 24402		4·89	?	M1

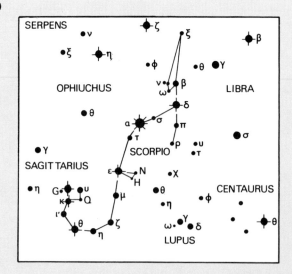

DOUBLES

Star	R.A. h m	Dec. ° '	Mags.	P.A.	Dist."
ξ	16 03·2	−11 19	4·9, 4·9	—	<2. Binary, 45·7 years
β	16 04·2	−19 05	2·9, 5·1	023	13·7 A is double; mag. 9·7, dist. 0"·8
ν	16 10·7	−19 24	4·3, 6·5	337	41·4 Both double
α	16 28·1	−26 23	0·9, 6·5	274	2·9 Fixed. Red and green. B is a radio source

VARIABLES

Star	R.A. h m	Dec. ° '	Range	Period, d	Spectrum	Type
SS	16 52·0	−32 33	7·5—9·5	—	K	Irregular
RS	16 52·0	−45 01	6·2—13·0	320	M	Mira
RV	16 55·1	−33 32	7·2—8·4	6·06	F—G	Cepheid
RT	17 00·2	−36 51	7·0—14·6	448	M	Mira

CLUSTERS AND NEBULÆ

Object		R.A. h m	Dec. ° '	Type	Notes
NGC 6093	M.80	16 14·1	−22 52	Globular	Mag. 7·7
NGC 6121	M.4	16 20·6	−26 24	Globular	Mag. 6·4
NGC 6124		16 22·2	−40 35	Open cluster	Mag. 6. About 120 stars
NGC 6153		16 28·0	−40 08	Planetary	Mag. 11·5
NGC 6242		16 52·2	−39 25	Open cluster	Mag. 8. About 40 stars
NGC 6259		16 57·1	−44 36	Open cluster	Mag. 8·6. About 100 stars
NGC 6268		16 58·6	−39 39	Open cluster	Mag. 9·5. Rather sparse
NGC 6302		17 10·4	−37 03	Planetary	Mag. 11·4
NGC 6302		17 10·5	−37 03	Nebula	'Bug' Nebula
NGC 6318		17 14·3	−39 24	Open cluster	Mag. 12, but concentrated
NGC 6322		17 15·2	−42 50	Open cluster	Mag. 7·0. About 20 stars
NGC 6357		17 21·3	−34 07	Nebula	
NGC 6405	M.6	17 36·8	−32 11	Open cluster	Mag. 5·3. Naked-eye; superb cluster
NGC 6451		17 47·4	−30 11	Open cluster	Mag. 8·3. About 50 stars
NGC 6475	M.7	17 50·7	−34 48	Open cluster	Mag. 3·2. Glorious brilliant naked-eye cluster

SCULPTOR

(Abbreviation: Scl).

Originally Apparatus Sculptoris. There is no star above the fourth magnitude.

See chart for Phoenix

BRIGHTEST STARS

Star	Mag.	R.A. h m s	Dec. ° ' "	Abs. Mag.	Spectrum	Dist. l/y
γ	4·51	23 17 38	−32 39 07	+1·1	K0	160
β	4·46	23 31 48	−37 56 25	0·0	B9	250
δ	4·64	23 47 47	−28 15 07	+1·0	A0	165
α	4·39	00 57 33	−29 28 34	−0·2	B5	270

THERE ARE TWO MORE STARS ABOVE THE FIFTH MAGNITUDE:

Star	Mag.	Abs. Mag.	Spectrum
ζ	4·99	−2·5	B7
η	4·96	—	M5

DOUBLE

Star	R.A. h m	Dec. ° '	Mags.	P.A.	Dist."
ε	01 43·3	−25 18	5·4, 9·4	048	4·7 Slow binary

VARIABLES

Star	R.A. h m	Dec. ° '	Range	Period, d	Spectrum	Type
S	00 12·9	−32 19	6·1—13·6	366	M	Mira
R	01 24·7	−32 48	5·8—7·7	363	N	Semi-regular

CLUSTERS AND NEBULÆ

Object	R.A. h m	Dec. ° '	Type	Notes
I 5332	23 31·1	−36 22	Galaxy	Mag. 11·9. Spiral
NGC 7713	23 38·8	−38 13	Galaxy	Mag. 11·8 Spiral
NGC 7793	23 55·3	−32 57	Galaxy	Mag. 10. Spiral
NGC 288	00 50·2	−26 52	Globular	Mag. 7
NGC 55	00 12·5	−39 30	Galaxy	Mag. 8. Spiral
NGC 134	00 27·9	−33 32	Galaxy	Mag. 12. Spiral
NGC 253	00 45·1	−25 34	Galaxy	Mag. 9. Spiral (type Sc)
NGC 300	00 52·6	−37 58	Galaxy	Mag. 11·3. Spiral
NGC 613	01 32·0	−29 40	Galaxy	Mag. 10·2. Spiral (type SBc)

SCUTUM

(Abbreviation: Sct).

Originally Scutum Sobieskii or Clypeus Sobieskii (Sobieski's Shield) It has no star brighter than the fourth magnitude, but it is a rich area bordering Aquila, and contains the glorious 'Wild Duck' cluster M.11.

See chart for Aquila

BRIGHTEST STARS

Star	Mag.	R.A. h m s	Dec. ° ' "	Abs. Mag.	Spectrum	Dist. l/y
γ	4·73	18 27 57	−14 34 52	+1·4	K0	150
α	4·06	18 34 01	−08 15 38	+0·1	K0	205
δ	4·7v	18 41 04	−09 04 29	var.	F0	190
β	4·47	18 46 00	−04 46 22	−4·3	G0	1300
R	var.	18 46 18·4	−05 43 48	var.	G+K	1600

δ is a variable with a very small range (4·7—4·8) and a period of 0·19 day.

α was formerly known as 1 Aquilæ, δ as 2 Aquilæ and β as 6 Aquilæ.

There is one other star above the fifth magnitude: ζ (mag. 4·83, absolute mag. +0·6, spectrum K0).

VARIABLES

Star	R.A. h m	Dec. ° '	Range	Period, d	Spectrum	Type
R	18 46·2	−05 44	5·7—8·6	±144	G+K	RV Tauri
S	18 47·6	−07 58	7·0—8·0	148	N	Semi-regular

CLUSTERS

Object	R.A. h m	Dec. ° '	Type	Notes
NGC 6649	18 30·7	−10 26	Open cluster	Mag. 8·8. Rich
NGC 6694 M.26	18 42·5	−09 27	Open cluster	Mag. 9. Not comparable with M.11
NGC 6705 M.11	18 48·4	−06 20	Open cluster	Mag. 6·3. Lovely fan-shaped cluster

SERPENS

(Abbreviation: Ser).

A curious constellation inasmuch as it is divided into two parts: Caput (the Head) and Cauda (the Body). It evidently represents the serpent with which Ophiuchus is struggling – and has had the worst of the encounter, since it has been pulled in half! Caput contains six stars above the fourth magnitude, and Cauda three:

Caput	Cauda
α 2·65	η 3·23
μ 3·63	θ 3·4 (combined)
β 3·74	ξ 3·64
ε 3·75	
δ 3·8 (combined)	
γ 3·86	

See chart for Ophiuchus

BRIGHTEST STARS

CAPUT

Star		Mag.	R.A. h m s	Dec. ° ′ ″	Abs. Mag.	Spectrum	Dist. l/y
13 δ		3·8	15 33 45	+10 36 42	+0·6, +1·5	A7+A9	170
21 ι		4·49	15 40 34	+19 44 25	+0·1	A2	250
24 α	Unukalhai	2·65	15 43 11	+06 29 38	+1·0	K2	71
28 β		3·74	15 45 10	+15 29 23	+0·9	A2	120
27 λ		4·42	15 45 22	+07 25 16	+4·3	G0	34
35 κ		4·28	15 47 45	+18 12 31	+0·8	K5	165
32 μ		3·63	15 48 28	−03 21 50	−0·2	A0	190
37 ε		3·75	15 49 43	+04 32 35	+1·6	A2	85
41 γ		3·86	15 55 26	+15 43 57	+3·4	F5	40

CAUDA

Star		Mag.	R.A. h m s	Dec. ° ′ ″	Abs. Mag.	Spectrum	Dist. l/y
53 ν		4·35	17 19 35	−12 49 33	+1·1	A0	150
55 ξ		3·64	17 36 20	−15 23 10	+1·1	A5	105
56 o		4·39	17 40 11	−12 51 53	−0·4	A2	300
57 ζ		4·60	17 59 19	−03 41 24	+2·6	F0	80
58 η		3·23	18 20 10	−02 54 20	+1·9	K0	60
63 θ	Alya	3·4	18 55 08	+04 10 26	+1·3, +1·3	A5+A5	140

THERE ARE TWO OTHER STARS ABOVE THE FIFTH MAGNITUDE, BOTH IN CAPUT:

Star	Mag.	Abs. Mag.	Spectrum
38 ρ	4·88	+0·1	K5
44 π	4·82	+0·8	A2

DOUBLES

Star	R.A. h m	Dec. ° ′	Mags.	P.A.	Dist.″	
δ	15 33·8	+10 42	4·2, 5·2	179	3·9	Widening
θ	18 55·1	+04 08	4·5, 4·5	103	22·6	Fixed, Lovely, easy pair

VARIABLES

Star	R.A. h m	Dec. ° ′	Range	Period, d	Spectrum	Type
S	15 19·3	+14 30	7·7—14·1	367	M	Mira
R	15 48·4	+15 17	5·7—14·4	357	M	Mira
U	16 04·9	+10 04	7·8—14·0	238	M	Mira

CLUSTERS AND NEBULÆ

Object	R.A. h m	Dec. ° ′	Type	Notes
NGC 5904 M.5	15 16·0	+02 16	Globular	Mag. 6·2. Fine globular
NGC 5921	15 19·5	+05 15	Galaxy	Mag. 12. Type SBb
I 4593	16 09·5	+12 12	Planetary	Mag. 10·2. Concentrated
H.19	18 14·5	−13 18	Open cluster	Mag. 12. Concentrated
NGC 6611 M.16	18 16·0	−13 48	Nebula+cluster	Mag. 7. Easy object
I 4756	18 36·6	+05 26	Open cluster	Mag. 5. Rich cluster

SEXTANS

(Abbreviation: Sxt).

A very obscure constellation, with no star as bright as the fourth magnitude.

See chart for Hydra

BRIGHTEST STAR

Star	Mag.	R.A. h m s	Dec. ° ′ ″	Abs. Mag.	Spectrum	Dist. m/l/y
15 α	4·50	10 06 49	−00 15 49	−0·1	A0	270

The only other star above the fifth magnitude is 30 (β); mag. 4·95, absolute magnitude −0·3, spectrum B5.

NEBULÆ

Object	R.A. h m	Dec. ° ′	Type	Notes
NGC 3115	10 02·8	−07 28	Galaxy	Mag. 10·2. Elliptical (E6)
NGC 3166	10 11·2	+03 40	Galaxy	Mag. 11·4. Spiral
NGC 3169	10 11·7	+03 43	Galaxy	Mag. 11·7. Spiral (Sb)

TAURUS

(Abbreviation: Tau).

One of the brightest of the Zodiacal constellations. Mythologically it has been said to represent the bull into which Jupiter transformed himself when he wished to carry off Europa, daughter of the King of Crete.

Taurus includes the reddish first-magnitude star Aldebaran, and also the two most famous open clusters in the sky: the Pleiades and the Hyades. Altogether there are four-teen stars above the fourth magni-tude:

α 0·78v	θ²	3·42	17	3·81	
β 1·65	ε	3·54	γ	3·86	
η 2·86	ξ	3·75	δ	3·93	
ζ 3·07	o	3·80	ν	3·94	
λ 3·31 (max)	27	3·80			

Aldebaran has a very small range (0·78–0·93). β was formerly included in Auriga, as γ Aurigæ.

BRIGHTEST STARS

Star			Mag.	R.A. h m s	Dec. ° ′ ″	Abs. Mag.	Spectrum	Dist. l/y
1	o		3·80	03 23 38	+08 57 08	−0·5	G5	230
2	ξ		3·75	03 25 59	+09 39 25	−0·1	B8	190
5	f		4·28	03 29 39	+12 51 44	+0·4	K0	190
10			4·40	03 35 45	+00 19 58	+3·2	G5	57
17		Electra	3·81	03 43 34	+24 02 43	−0·2	B5p	(Pleiades)
19	q	Taygate	4·37	03 43 54	+24 23 57	−0·4	B5	(Pleiades)
20		Maia	4·02	03 44 31	+24 18 00	−0·4	B5	(Pleiades)
23		Merope	4·25	03 45 01	+23 52 51	−0·5	B5	(Pleiades)
25	η	Alcyone	2·86	03 46 11	+24 02 17	−3·2	B5p	(Pleiades)
27		Atlas	3·80	03 47 51	+23 59 14	−1·0	B8	(Pleiades)
35	λ		var.	03 59 28	+12 25 45	var.	B3+A4	470
38	ν		3·94	04 01 59	+05 55 45	+1·0	A0	125
37	A¹		4·50	04 03 24	+22 01 22	+0·4	K0	220
49	μ		4·32	04 14 20	+08 59 18	−1·2	B3	400
54	γ	Hyadum Primus	3·86	04 18 21	+15 34 42	+0·7	K0	(Hyades)
61	δ		3·93	04 21 40	+17 29 31	+0·3	K0	(Hyades)
65	χ		4·06	04 24 03	+22 14 41	+1·3	A3	140
68			4·24	04 24 13	+17 52 44	+1·2	A2	(Hyades)
69	υ		4·40	04 24 59	+22 45 53	+1·5	A5	125
71			4·60	04 25 05	+15 34 10	+1·5	A5	(Hyades)
74	ε	Ain	3·54	04 27 20	+19 07 58	+0·1	K0	(Hyades)
77	θ¹		4·04	04 27 21	+15 43 10	+1·1	G8	(Hyades)
78	θ²		3·42	04 27 24	+15 49 23	+0·2	A7	(Hyades)
86	ρ		4·75	04 32 36	+14 47 57	+1·6	A5	140
88	d		4·38	04 34 27	+10 07 00	+1·6	A3	115
87	α	Aldebaran	0·78v	04 34 39	+16 27 58	−0·7	K5	68
90	c¹		4·30	04 36 56	+12 28 04	+1·2	A3	140
94	τ		4·33	04 40 55	+22 54 57	−3·4	B5	105
102	ι		4·70	05 01 47	+21 33 35	+1·0	A5	180
112	β	Al Nath	1·65	05 24 54	+28 35 25	−3·2	B7	300
119	CE		var.	05 30 55	+18 34 45	var.	M2	180
123	ζ		3·07	05 37· 20	+21 07 49	−4·2	B2	940
136			4·54	05 51 57	+27 36 30	+0·6	A0	205

THERE ARE ELEVEN MORE STARS DOWN TO THE FIFTH MAGNITUDE:

Star	Mag.	Abs. Mag.	Spectrum
28 BU Pleione	var.	var.	B8p. (Shell star)
47	4·98	−0·5	G5
50 ω	4·80	+2·3	A5
64	4·84	+1·7	A6
73 π	4·94	+0·3	G6
92 σ²	4·84	+1·3	A5
114 o	4·83	−3·5	B3
125	5·00	−1·5	B3
126	4·87	−1·6	B3
134	4·92	+0·1	B9
139	4·90	−6·0	B1

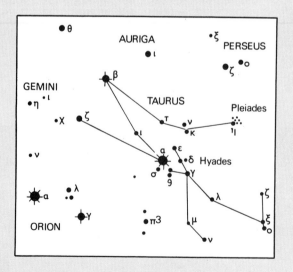

DOUBLES

Star	R.A.		Dec.		Mags.	P.A.	Dist."	
	h	m	°	'				
φ	04	17·3	+27	14	5·1, 8·7	250	52·1	Optical
χ	04	19·5	+25	31	5·4, 8·2	025	19·9	Fixed
α	04	34·7	+16	28	0·8, 11·0	112	31·4	Optical. Widening

θ¹ and θ² form a very wide naked-eye pair—too wide, indeed, to be classed as a true double even though the components are at the same distance from us.

VARIABLES

Star	R.A.		Dec.		Range	Period, d	Spectrum	Type
	h	m	°	'				
BU (Pleione)	03	46·2	+23	59	5·0—5·5	—	B8p	Shell star
λ	03	59·5	+12	26	3·3—4·2	3·95	B+A	Algol
CE	05	30·9	+18	35	4·3—4·7	165	M2	Semi-regular
Y	05	42·7	+20	40	6·8—9·2	241	N	Semi-regular

CLUSTERS AND NEBULÆ

Object	R.A.		Dec.		Type	Notes
	h	m	°	'		
Mel 22 M.45	03	44·1	+23	58	Open cluster	Pleiades. Distance, 410 l/y
Mel. 25	04	17	+15	30	Open cluster	Hyades. Distance 130 l·y. Very scattered
NGC 1554-5	04	19·9	+19	25	Nebula	Hind's variable nebula (T Tauri)
NGC 1647	04	43·2	+18	59	Open cluster	Mag. 6·3. About 30 stars
NGC 1952 M.1	05	31·5	+21	59	Supernova remnant	Mag. 8·4. Crab Nebula
NGC 1807	06	07·8	+16	28	Open cluster	Mag. 7·8. About 15 stars
NGC 1817	06	09·2	+16	38	Open cluster	Mag. 7·9. Only about 10 stars

---

TELESCOPIUM

(Abbreviation: Tel).

A small constellation with only one star above the fourth magnitude: α (3·76).

See chart for Ara

BRIGHTEST STARS

Star	Mag.	R.A.			Dec.			Abs. Mag.	Spectrum	Dist. l/y
		h	m	s	°	'	"			
ε	4·60	18	09	36	−45	57	35	−0·2	K0	90
α	3·76	18	25	21	−45	58	55	−2·8	B3	200
ζ	4·14	18	27	08	−49	05	04	+0·8	K0	45

The only other star above the fifth magnitude is ξ ; mag. 4·86, absolute magnitude −0·4, type M2.

VARIABLES

Star	R.A.		Dec.		Range	Period, d	Spectrum	Type
	h	m	°	'				
BL	19	02·7	−51	30	7·5—9·6	778	F+M	Algol
RR	20	00·3	−55	52	6·5—16·5	—	Fp	Novalike

CLUSTERS AND NEBULÆ

Object	R.A.		Dec.		Type	Notes
	h	m	°	'		
NGC 6584	18	14·6	−52	14	Globular	Mag. 8·3
I 4699	18	14·8	−46	01	Planetary	Mag. 12

TRIANGULUM

(Abbreviation: Tri).

A small but original constellation – and its main stars really do form a triangle! There are two stars above the fourth magnitude:

β 3·00   α 3·45.

See chart for Andromeda

BRIGHTEST STARS

Star	Mag.	R.A. h m s	Dec. ° ' "	Abs. Mag.	Spectrum	Dist./ly
2 α Rasalmothallah	3·45	01 51 49	+29 28 20	+2·0	F5	65
4 β	3·00	02 08 14	+34 53 02	−0·1	A5	140
9 γ	4·07	02 16 00	+33 44 46	+1·5	A0	105

There are no other stars above the fifth magnitude.

DOUBLE

Star	R.A. h m	Dec. ° '	Mags.	P.A.	Dist."	
ι	02 09·5	+30 04	5·4, 7·0	071	3·6	Slow binary

VARIABLE

Star	R.A. h m	Dec. ° '	Range	Period, d	Spectrum	Type
R	02 34·0	+34 03	5·4—12·0	266	M	Mira

NEBULÆ

Object		R.A. h m	Dec. ° '	Type	Notes
NGC 598	M.33	01 31·0	+30 24	Galaxy	Mag. 6·7. Type Sc. Member of the Local Group
NGC 925		02 24·3	+33 22	Galaxy	Mag. 10·5. Type SBc

TRIANGULUM AUSTRALE

(Abbreviation: TRA).

This 'triangle' also merits its name. There are three stars above the fourth magnitude:

α 1·93   β 2·87   γ 2·94

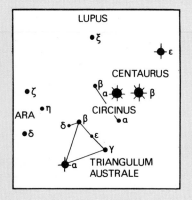

BRIGHTEST STARS

Star	Mag.	R.A. h m s	Dec. ° ' "	Abs. Mag.	Spectrum	Dist. l/y
γ	2·94	15 16 50	−68 35 59	+0·2	A0	113
ε	4·11	15 34 41	−66 14 42	+1·4	K0	110
β	2·87	15 53 11	−63 21 52	+2·3	F0	42
δ	4·03	16 13 25	−63 37 53	+1·0	G0	130
α Atria	1·93	16 46 19	−68 59 22	−0·1	K2	82

There is one other star above the fifth magnitude: ζ, mag. 4·93, absolute magnitude +4·6, distance 40 light-years.

DOUBLE

Star	R.A. h m	Dec. ° '	Mags.	P.A.	Dist."	
ι	16 23·3	−63 57	5·3, 9·7	061	19·7	Optical

VARIABLES

Star	R.A. h m	Dec. ° '	Range	Period, d	Spectrum	Type
R	15 15·3	−66 19	6·8—7·7	3·39	F—G	Cepheid
S	15 56·7	−63 38	6·4—7·6	6·3	F—G	Cepheid
U	16 02·9	−62 47	7·9—8·7	2·6	F—G	Cepheid

CLUSTER

Object		R.A. h m	Dec. ° '	Type	Notes
NGC 6025	Δ 304	15 59·4	−60 21	Open cluster	Mag. 5·8. About 30 stars

TUCANA

(Abbreviation: Tuc).

The dimmest of the 'Southern Birds', but graced by the presence of the glorious globular cluster 47 Tucanæ – inferior only to ω Centauri. The only star above the fourth magnitude is α (2·87), but the conbined magnitude of β¹ and β² is 3·7.

BRIGHTEST STARS

Star	Mag.	R.A. h m s	Dec. ° ′ ″	Abs. Mag.	Spectrum	Dist. l/y
α	2·87	22 17 00	−60 22 12	+1·5	K2	62
γ	4·10	23 16 09	−58 21 24	+2·0	F0	85
ε	4·71	23 58 47	−65 41 59	+0·4	B9	230
ζ	4·34	00 18 56	−65 00 15	+4·9	F8	23
β	3·7	00 30 33	−63 05 13	+1·2, +1·2	B9 + A2	150

THERE ARE TWO MORE STARS ABOVE THE FIFTH MAGNITUDE:

Star	Mag.	Abs. Mag.	Spectrum
δ	4·80	+0·7	B9
ν	4·92	−2·1	M5

DOUBLES

Star	R.A. h m	Dec. ° ′	Mags.	P.A.	Dist."	
δ	22 23·8	−65 13	4·8, 9·3	283	6·8	Fixed
β	00 29·6	−63 24	4·5, 4·5	170	27·1	Both double
{ β¹	00 29·6	−63 24	4·5, 14·0	149	2·2	
{ β²	00 29·6	−63 24	4·9, 5·7	—		Binary; 43·1 years
λ	00 53·1	−59 48	5·3, 7·3	080	20·8	Optical
κ	01 14·1	−69 08	5·1, 7·3	341	5·7	Slow binary

CLUSTERS

Star	R.A. h m	Dec. ° ′	Type	Notes
NGC 362	62 01 00·7	−71 06	Globular	Mag. 6. Fringe of naked-eye visibility
NGC 104	47 00 21·9	−72 22	Globular	Mag. 5. Magnificent globular

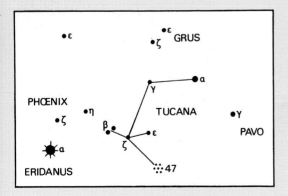

BRIGHTEST STAR

Star		Mag.	R.A. h	m	s	Dec. °	'	"	Abs. Mag.	Spectrum	Dist, l/y
1 o	Muscida	3·37	08	28	27	+60	47	35	+0·1	G1	150
9 ι	Talita	3·12	08	59	42	+48	07	44	+2·2	A5+M1	49
10		4·09	08	59	13	+41	52	15	+3·4	F5	45
12 ϰ		3·68	09	02	08	+47	14	40	−0·6	A0	230
15 f		4·54	09	07	20	+51	41	39	+2·1	A3p	95
14 τ		4·74	09	09	07	+63	36	15	+0·5	F6+A5	230
23 h		3·75	09	29	49	+63	09	32	+1·6	F0	85
25 θ		3·19	09	31	24	+51	46	42	+1·8	F8	63
24 d		4·57	09	32	34	+69	55	41	+3·2	F9	80
26		4·65	09	33	20	+52	09	00	+1·5	A0	140
29 υ		3·89	09	49	26	+59	08	35	+1·6	A0	95
30 φ		4·54	09	50	37	+54	10	04	+0·9	A2	170
33 λ	Tania Borealis	3·45	10	15	47	+43	01	29	+0·1	A2	150
34 μ	Tania Australis	3·05	10	21	01	+41	36	38	+0·5	K5	105
48 β	Merak	2·37	11	01	31	+56	30	02	+0·5	A0	78
50 α	Dubhe	1·81	11	02	23	+62	51	12	−0·7	K0	105
52 ψ		3·00	11	08	26	+44	37	05	0·0	K0	130
53 ξ	Alula Australis	3·88	11	17	01	+31	39	11	+4·5	G0	25
54 ν	Alula Borealis	3·71	11	17	18	+33	12	52	−0·9	K0	270
63 χ	Alkafzah	3·85	11	44	54	+47	54	05	+0·1	K0	180
64 γ	Phad	2·44	11	52	41	+53	49	01	+0·2	A0	90
69 δ	Megrez	3·30	12	14	21	+57	09	17	+1·9	A2	63
77 ε	Alioth	1·79	12	53	04	+56	04	44	+0·2	A0p	68
79 ζ	Mizar	2·09	13	23	03	+55	02	23	+0·1, +2·1	A2+A6	88
80 g	Alcor	4·02	13	24	21	+55	06	08	+1·9	A5	84
83		4·75	13	39	54	+54	47	33	+0·1	M2	270
85 η	Alkaid	1·87	13	46	41	+49	25	21	−2·1	B3	210

URSA MAJOR

(Abbreviation: UMa).

The most famous of all northern constellations; circumpolar in England and the northern United States. Mythologically it represents Callisto, daughter of King Lycaon of Arcadia. Her beauty surpassed that of Juno, which so infuriated the goddess that she ill-naturedly changed Callisto into a bear. Years later Arcas, Callisto's son, found the bear while out hunting, and was about to shoot it when Jupiter intervened, swinging both Callisto and Arcas – also transformed into a bear – up to the sky: Callisto as Ursa Major, Arcas as Ursa Minor. The sudden jolt explains why both bears have tails stretched out to decidedly un-ursine length!

The seven main stars of Ursa Major are often called the Plough; sometimes King Charles' Wain, and, in America, the Big Dipper. One of the Plough stars is Mizar, the most celebrated naked-eye double in the sky since it makes a pair with Alcor; Mizar is itself a compound system, and the two main components are easily separable with a small telescope.

Ursa Major contains nineteen stars above the fourth magnitude:

ε	1·79	ψ 3·00	ϰ 3·68
α	1·81	μ 3·05	ν 3·71
η	1·87	ι 3·12	h 3·75
ζ	2·09	θ 3·19	χ 3·85
β	2·37	δ 3·30	ξ 3·88
γ	2·44	o 3·37	υ 3·89
		λ 3·45	

Several of the Plough stars have alternative proper names; thus η may also be called Benetnasch, while γ may be Phekda or Phecda. However, Mizar is the only star whose proper name is generally used.

THERE ARE EIGHT MORE STARS ABOVE THE FIFTH MAGNITUDE:

Star		Mag.	Abs. Mag.	Spectrum
4 π²	Ta Tsun	4·76	+0·6	K2
8 ρ		4·99	−0·2	M3
13 σ²		4·87	+3·5	F4
18 e		4·89	+1·6	A0
36		4·84	+4·4	F8
45 ω		4·84	+0·6	A0
55		4·78	+1·2	A0
78		4·89	+2·6	A6

DOUBLES

Star	R.A. h	m	Dec. °	'	Mags.	P.A.	Dist."	
ι	08	42·7	+48	08	3·1, 10·8	0·4	5·0	Closing
σ²	09	06·0	+67	20	4·9, 8·5	024	2·2	Binary. Widening
ξ	11	17·0	+31	39	4·4, 4·8	—	2·9 (in 1980). Binary, 60 years	
ν	11	17·3	+33	13	3·7, 9·7	147	7·2	Fixed
ζ	13	23·1	+55	02	2·4, 3·9	150	14·5	Naked-eye pair with Alcor

VARIABLES

Star	R.A. h	m	Dec. °	'	Range	Period, d	Spectrum	Type
R	10	41·1	+69	02	6·7—13·4	302	M	Mira
VY	10	41·6	+67	40	6·0—6·6	—	N	Irregular
TX	10	42·4	+45	50	6·8—8·9	3·06	B+F	Algol
Z	11	53·9	+58	09	6·6—9·1	198	M	Semi-regular
T	12	34·1	+59	46	6·6—13·4	257	M	Mira
S	12	41·8	+61	22	7·4—12·3	226	S	Mira

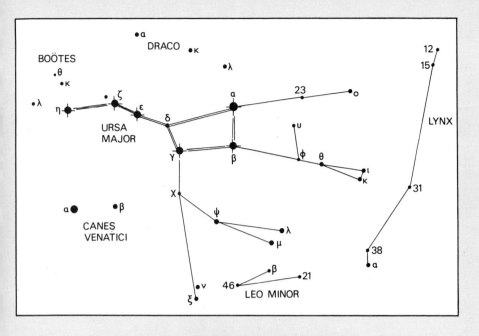

CLUSTERS AND NEBULÆ

Object		R.A. h m	Dec. ° ′	Type	Notes
NGC 2681		08 50·0	+51 31	Galaxy	Mag. 10·4. Type Sa
NGC 2768		09 07·8	+60 16	Galaxy	Mag. 10·5. Type E5
NGC 2841		09 18·6	+51 12	Galaxy	Mag. 9·3. Type Sb
NGC 2950		09 39·1	+59 05	Galaxy	Mag. 10·9. Type Sa(p)
NGC 2976		09 43·2	+68 08	Galaxy	Mag. 11·4. Type S
NGC 3031	M.81	09 51·5	+69 18	Galaxy	Mag. 7·9. Type Sb
NGC 3034	M.82	09 51·9	+69 56	Galaxy	Mag. 8·8. Irregular. Radio source
NGC 3079		09 58·6	+55 57	Galaxy	Mag. 11·2. Type Sb
NGC 3077		09 59·4	+68 58	Galaxy	Mag. 10·9. Type E2(p)
NGC 3184		10 15·2	+41 40	Galaxy	Mag. 12. Type Sc
NGC 3198		10 16·7	+45 49	Galaxy	Mag. 12·4. Type Sc
NGC 3310		10 35·7	+53 46	Galaxy	Mag. 10·1. Irregular
NGC 3359		10 43·4	+63 30	Galaxy	Mag. 11. Type Sc
NGC 3348		10 43·5	+73 07	Galaxy	Mag. 11·2. Type E
NGC 3516		11 03·4	+72 50	Galaxy	Mag. 11·6. Type Sa
NGC 3556	M.108	11 08·7	+55 57	Galaxy	Mag. 10·7. Type Sc. "Extra" M object
NGC 3587	M.97	11 12·0	+55 18	Planetary	Mag. 12. Owl Nebula. Central star (mag. 14·3)
NGC 3613		11 15·7	+58 17	Galaxy	Mag. 11·2. Type E5
NGC 3619		11 16·5	+58 02	Galaxy	Mag. 11·7. Type S
NGC 3631		11 18·3	+53 28	Galaxy	Mag. 11·2. Type Sc
NGC 3642		11 19·6	+59 21	Galaxy	Mag. 11·4. Type Sc
NGC 3665		11 22·1	+39 02	Galaxy	Mag. 11·4. Type E2
NGC 3675		11 23·5	+43 52	Galaxy	Mag. 10·6. Type Sb
NGC 3718		11 29·9	+53 21	Galaxy	Mag. 11·2. Type S
NGC 3738		11 33·1	+54 48	Galaxy	Mag. 11·8. Irregular
NGC 3893		11 46·1	+49 00	Galaxy	Mag. 11·3. Type Sc
NGC 3898		11 46·7	+56 22	Galaxy	Mag. 11·4. Type Sb
NGC 3938		11 50·2	+44 24	Galaxy	Mag. 11·5. Type Sc
NGC 3941		11 50·3	+37 16	Galaxy	Mag. 9·8. Type Sa
NGC 3949		11 51·1	+48 08	Galaxy	Mag. 11·0. Type Sb
NGC 3982		11 53·9	+55 24	Galaxy	Mag. 11·3. Type Sb
NGC 3992	M.109	11 55·0	+53 39	Galaxy	Mag. 10·8. Type Sb. "Extra" M object
NGC 3998		11 55·3	+55 44	Galaxy	Mag. 11·3. Type E2
NGC 4026		11 56·9	+51 14	Galaxy	Mag. 10·7. Type E8
NGC 4036		11 58·9	+62 10	Galaxy	Mag. 10·7. Type E6
NGC 4041		11 59·7	+62 25	Galaxy	Mag. 11·0. Type Sc
NGC 4088		12 03·0	+50 49	Galaxy	Mag. 10·9. Type Sc
NGC 4111		12 04·5	+43 21	Galaxy	Mag. 9·7. Type E8

URSA MINOR

(Abbreviation: UMi).

The north polar constellation. Mythologically it represents Arcas, son of Callisto (see Ursa Major). There are three stars above the fourth magnitude:

α 1·99v   β 2·04   γ 3·08

Polaris is actually variable over a very small range – too slight for its fluctuations to be detected with the naked eye.

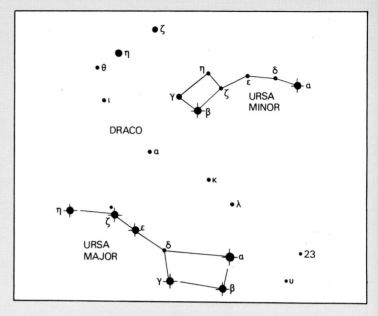

BRIGHTEST STARS

Star		Mag.	R.A. h	m	s	Dec. °	′	″	Abs. Mag.	Spectrum	Dist. l/y
2		4·52	01	01	31	+85	59	24	0·0	K0	250
1 α	Polaris	1·99v	02	10	01	+89	09	51	−4·6	F8	680
5		4·37	14	27	33	+75	47	37	−0·2	K5	270
7 β	Kocab	2·04	14	50	45	+74	14	43	−0·5	K5	105
13 γ	Pherkad Major	3·08	15	20	45	+71	54	44	−1·5	A2	270
16 ζ	Alifa	4·34	15	44	49	+77	51	46	+0·2	A2	220
22 ε		4·4v	16	48	10	+82	04	32	var.	G5	300
23 δ	Yildun	4·44	17	39	16	+86	35	56	−0·6	A0	330

ε is an eclipsing binary with a very small range (4·40—4·44) and a period of 39·5 days.

The only other star above the fifth magnitude is RR (variable; spectrum M5). 21 (η), magnitude 5·04, has a proper name; Alasco.

DOUBLES

Star	R.A. h	m	Dec. °	′	Mags.	P.A.	Dist.″	
α	02	10·0	+89	10	2·0, 9·0	217	18·3	Optical
π¹	15	32·0	+80	37	6·5, 7·2	081	31·0	Slow binary

VARIABLES

Star	R.A. h	m	Dec. °	′	Range	Period, d	Spectrum	Type
V	13	37·8	+74	34	7·4—8·8	72	M	Semi-regular
U	14	16·2	+67	01	7·4—12·7	327	M	Mira
RR	14	56·8	+66	08	4·6—5·0	±40	M	Semi-regular
SU	15	31·4	+78	48	8·0—12·9	327	M	Mira

VELA

(Abbreviation: Vel).

The Sails of the dismembered ship Argo. There are ten stars above the fourth magnitude:

γ² 1·88		ϰ	2·45	o	3·68
δ 1·95		μ	2·67	c	3·69
λ 2·24		N	3·19	φ	3·70
		ψ	3·64		

δ and ϰ make up the 'False Cross' with ε and ι Carinæ.   See chart for Carina

BRIGHTEST STARS

Star		Mag.	R.A. h	m	s	Dec. °	′	″	Abs. Mag.	Spectrum	Dist. l/y
γ²		1·88	08	08	51	−47	16	16	−4·1	Oap	520
e		4·13	08	36	52	−42	52	42	−4·5	A5	1630
o		3·68	08	39	40	−52	50	36	−2·8	B3	650
b		4·06	08	39	54	−46	34	12	−7·0	F5	3300
Boss 11997		4·30	08	40	59	−52	56	02	+0·2, +0·4	B5+B9	360
d		4·12	08	43	37	−42	34	08	+0·6	G5	160
δ	Koo She	1·95	08	44	06	−54	37	39	+0·2	A0	76
a		4·09	08	45	17	−45	57	38	?	A0	?
W		4·42	08	59	16	−41	10	03	+1·5	F8	120
c		3·69	09	03	24	−47	00	35	0·0	K0	180
λ	Al Suhail al Wazn	2·24	09	07	11	−43	20	36	−4·6	K5	750
k²		4·70	09	14	52	−37	19	15	+3·5	F5	55
ϰ	Markeb	2·45	09	21	26	−54	54	59	−3·4	B3	470
ψ		3·64	09	29	50	−40	22	12	+2·7	A7	50
N		3·19	09	30	33	−56	56	13	−0·4	K5	170
M		4·49	09	36	02	−49	15	22	+1·0	A5	160
m		4·56	09	50	50	−46	26	39	−5·4	G5	3300
φ		3·70	09	56	05	−54	27	45	−5·1	B5	1800
q		4·09	10	13	49	−42	00	46	+1·5	A2	105
Boss 14185		4·58	10	18	47	−54	55	07	−1·2	K0	470
l		4·65	10	20	06	−55	55	55	−2·4	B5p	780
p		4·06	10	36	22	−48	06	40	+1·5	G0+A3	105
x		4·25	10	38	26	−55	29	19	+0·4	G0	190
μ		2·67	10	45	49	−49	18	13	+0·1	G5	108
i		4·56	10	59	08	−42	06	27	?	A2	?

THERE ARE FIVE MORE STARS ABOVE THE FIFTH MAGNITUDE:

Star	Mag.	Abs. Mag.	Spectrum
B	4·90	−3·5	B1
C	4·87	−2·1	K0
n	4·85	?	A3
l	4·98	+0·7	K2
r	4·99	+1·4	K1

DOUBLE

Star	R.A. h	m	Dec. °	′	Mags.	P.A.	Dist."	
γ	08	08·9	−47	16	2·2, 4·8	220	41·0	Fixed

VARIABLES

Star	R.A. h	m	Dec. °	′	Range	Period, d	Spectrum	Type
AH	08	10·4	−46	30	5·8—6·4	4·23	F—G	Cepheid
AI	08	12·4	−44	25	6·4—7·1	0·11	A—F	RR Lyræ
RZ	08	35·3	−43	56	7·0—8·5	20·4	F—G	Cepheid
CV	08	59·0	−51	21	6·6—7·3	6·89	B+B	Algol
V	09	20·8	−55	45	7·6—8·7	4·37	F—G	Cepheid

CLUSTERS AND NEBULÆ

Object	R.A. h	m	Dec. °	′	Type	Notes
NGC 2547	08	08·9	−49	07	Open cluster	Mag. 5·1. About 50 stars
NGC 2626	08	33·8	−40	28	Nebula	
I 2391	08	38·8	−52	53	Open cluster	Mag. 2·6. 0 Velorum
I 2488	09	25·7	−56	45	Open cluster	Mag. 7·2. Fairly rich
NGC 3105	09	58·9	−54	32	Open cluster	Mag. 11. Faint but rich
H.4	10	36·9	−53	54	Open cluster	Mag. 9. Rich

VIRGO

(Abbreviation: Vir).

A very large constellation, representing Astræa – the goddess of
justice, daughter of Jupiter and
Themis. The 'bowl' of Virgo
abounds in faint galaxies. Of the
eight stars down to the fourth
magnitude, Spica is an eclipsing
variable over a small range (0·91–
1·01), and γ (Arich) used to be one of
the most spectacular binary pairs in
the sky, though it is closing up and
will have become very difficult to
separate by the end of the 20th
century. The leading star sare:

α 0·91v	ζ 3·40	β 3·80
γ 2·76	δ 3·66	η 4·00
ε 2·86	109 3·76	

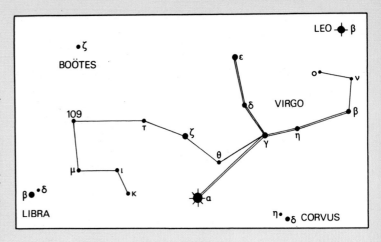

BRIGHTEST STARS

Star			Mag.	R.A.			Dec.			Abs. Mag.	Spectrum	Dist. l/y
				h	m	s	°	′	″			
3	ν		4·20	11	44	44	+06	39	10	−0·1	M1	230
5	β	Zawijah	3·80	11	49	33	+01	53	19	+3·8	F8	32
8	π		4·57	11	59	45	+06	44	13	+1·1	A3	160
9	o		4·24	12	02	06	+08	51	18	+1·1	A0	125
15	η	Zaniah	4·00	12	48	47	−00	32	41	+1·1	A0	125
29	γ	Arich	2·76	12	40	33	−01	19	45	+3·5	F0+F0	32
43	δ	Minelauva	3·66	12	54	30	+03	31	00	−0·1	M0	180
47	ε	Vindemiatrix	2·86	13	01	05	+11	04	37	+0·6	K0	90
51	θ		4·44	13	08	49	−05	25	19	+0·5	A0	205
67	α	Spica	0·91v	13	24	02	−11	02	49	−3·3	B2	220
79	ζ	Heze	3·40	13	33	34	−00	29	03	+1·1	A2	93
93	τ		4·34	14	00	32	+01	39	01	+0·6	A1	180
98	ϰ		4·31	14	11	43	−10	10	19	+0·8	K0	160
99	ι	Syrma	4·16	14	14	52	−05	53	46	+2·4	F5	75
100	λ		4·60	14	17	55	−13	16	14	+1·3	A2	150
107	μ		3·95	14	41	54	−05	33	48	+2·3	F5	70
109			3·76	14	45	08	+01	59	06	+1·1	A0	105
110			4·62	15	01	47	+02	10	36	+0·9	K0	180

THERE ARE EIGHT MORE STARS ABOVE THE FIFTH MAGNITUDE:

Star		Mag.	Abs. Mag.	Spectrum
26	χ	4·78	+0·5	K3
30	ρ	4·95	+0·5	B9
40	ψ	4·91	−0·3	M3
61		4·80	+5·0	G6
69		4·89	−0·6	K3
74	I²	4·83	+0·6	M3
78	o	4·93	+0·9	A2p
105	φ	4·97	+2·8	F8

DOUBLES

Star	R.A.		Dec.		Mags.	P.A.	Dist."	
	h	m	°	′				
γ	12	40·6	−01	20	3·6, 3·6	Binary: 180 years. Closing		
							(> 5″)	
θ	13	08·8	−05	25	4·4, 8·6	345	7·2	Fixed
τ	14	00·6	+01	39	4·3, 9·5	290	80·1	Optical
φ	14	25·6	−02	00	5·0, 9·2	110	4·7	Fixed

VARIABLES

Star	R.A.		Dec.		Range	Period, d	Spectrum	Type
	h	m	°	′				
X	11	59·4	+09	22	7·3—11·2	?	Fp	?
SS	12	22·7	+01	03	6·0—9·6	355	N	Mira
R	12	36·0	+07	16	6·2—12·1	146	M	Mira
U	12	48·6	+05	50	7·5—13·5	207	M	Mira
SW	13	11·5	−02	53	6·8—8·1	150	M	Semi-regular
S	13	30·4	−06	56	6·3—13·2	378	M	Mira

Object		R.A. h m	Dec. °	,	Type	Notes
NGC 4179		12 10·3	+01	35	Galaxy	Mag. 11·6. Type E8
NGC 4216		12 13·4	+13	25	Galaxy	Mag. 10·4. Type Sb
NGC 4261		12 16·8	+06	06	Galaxy	Mag. 10·3. Type E2
NGC 4270		12 17·3	+05	44	Galaxy	Mag. 11·9. Type Sa
NGC 4303	M.61	12 19·4	+04	45	Galaxy	Mag. 10·1. Type Sc
NGC 4365		12 22·0	+07	36	Galaxy	Mag. 11·1. Type E2
NGC 4371		12 22·4	+11	59	Galaxy	Mag. 11·6. Type SBa
NGC 4374	M.84	12 22·6	+13	10	Galaxy	Mag. 9·3. Type E1
NGC 4378		12 22·8	+05	12	Galaxy	Mag. 11·7. Type E
NGC 4388		12 23·3	+12	56	Galaxy	Mag. 12. Type Sb
NGC 4406	M.86	12 23·7	+13	13	Galaxy	Mag. 9·7. Type E3
NGC 4429		12 24·9	+11	23	Galaxy	Mag. 11·2. Type Sa
NGC 4435		12 25·2	+13	21	Galaxy	Mag. 10·3. Type E4
NGC 4438		12 25·3	+13	17	Galaxy	Mag. 10·8. Type S
NGC 4442		12 25·6	+10	05	Galaxy	Mag. 10·8. Type E5
NGC 4457		12 26·4	+03	51	Galaxy	Mag. 12. Type Sa
NGC 4472	M.49	12 27·3	+08	16	Galaxy	Mag. 8·6. Type E4
NGC 4486	M.87	12 28·3	+12	40	Galaxy	Mag. 9·2. Type E0. Radio source
NGC 4517		12 29·0	+00	21	Galaxy	Mag. 12. Type Sc
NGC 4504		12 29·7	−07	17	Galaxy	Mag. 11·7. Type Sc
NGC 4526		12 31·6	+07	58	Galaxy	Mag. 10·9. Type E7
NGC 4546		12 32·9	−03	21	Galaxy	Mag. 10·0. Type E6
NGC 4552	M.89	12 33·1	+12	50	Galaxy	Mag. 9·5. Type E0
NGC 4569	M.90	12 34·3	+13	26	Galaxy	Mag. 10·0. Type Sc
NGC 4570		12 34·4	+07	31	Galaxy	Mag. 10·9. Type E8
NGC 4579	M.58	12 35·1	+12	05	Galaxy	Mag. 9·2. Type SB
NGC 4594	M.104	12 37·3	−11	21	Galaxy	Mag. 8·7. Type Sb
NGC 4596		12 37·4	+10	27	Galaxy	Mag. 11·4. Type SBa
NGC 4621	M.59	12 39·5	+11	55	Galaxy	Mag. 9·6. Type E3
NGC 4636		12 40·3	+02	57	Galaxy	Mag. 10·4. Type E1
NGC 4643		12 40·8	+02	15	Galaxy	Mag. 10·6. Type SBa
NGC 4649	M.60	12 41·1	+11	49	Galaxy	Mag. 8·9. Type E1
NGC 4660		12 42·0	+11	26	Galaxy	Mag. 10·9. Type E5
NGC 4665		12 42·6	+03	19	Galaxy	Mag. 11·1. Type SBa
NGC 4666		12 42·6	−00	12	Galaxy	Mag. 11·4. Type Sc
NGC 4684		12 44·7	−02	28	Galaxy	Mag. 11·6. Type Sa
NGC 4698		12 45·8	+08	45	Galaxy	Mag. 11·3. Type Sb
NGC 4697		12 46·0	−05	32	Galaxy	Mag. 9·6. Type E4
NGC 4699		12 46·5	−08	24	Galaxy	Mag. 9·3. Type Sa
NGC 4713		12 47·5	+05	35	Galaxy	Mag. 11·7. Type Sc
NGC 4742		12 49·2	−10	12	Galaxy	Mag. 11·7. Type E3
NGC 4754		12 49·7	+11	35	Galaxy	Mag. 10·5. Type E
NGC 4753		12 49·8	−00	55	Galaxy	Mag. 10·8. Type S
NGC 4762		12 50·4	+11	31	Galaxy	Mag. 11·0. Type S
NGC 4781		12 51·8	−10	16	Galaxy	Mag. 11·2. Type Sc
NGC 4866		12 57·0	+14	27	Galaxy	Mag. 11·4. Type Sb
NGC 4902		12 58·3	−14	15	Galaxy	Mag. 11·7. Type SBb
NGC 4958		13 03·1	−07	45	Galaxy	Mag. 10·9. Type E6
NGC 4995		13 07·0	−07	34	Galaxy	Mag. 11·2. Type Sb
NGC 5363		13 53·6	+05	29	Galaxy	Mag. 10·7. Type E(p)
NGC 5566		14 17·8	+04	11	Galaxy	Mag. 10·4. Type Sb
NGC 5634		14 27·0	−05	45	Globular	Mag. 10·4
NGC 5746		14 42·3	+02	10	Galaxy	Mag. 10·1. Type Sb
NGC 5846		15 04·0	+01	48	Galaxy	Mag. 10·5. Type E0

VOLANS

(Abbreviation: Vol).

Originally Piscis Volans. A small constellation, intruding into Carina. There are three stars above the fourth magnitude:

γ 3·6 (combined)

β 3·65   ζ 3·89

See chart for Carina

BRIGHTEST STARS

Star	Mag.	R.A. h m s	Dec. ° ′ ″	Abs. Mag.	Spectrum	Dist. l/y
γ	3·6	07 08 56	−70 27 48	+0·9, +2·8	K0 + G0	130
δ	4·02	07 16 51	−67 55 01	−4·4	F5	1100
ζ	3·89	07 42 06	−72 33 13	+1·1	K0	120
ε	4·46	08 07 52	−68 33 09	−2·1	B5	650
β	3·65	08 25 30	−66 03 48	+1·0	K0	110
α	4·18	09 02 06	−66 18 29	+0·5	A5	70

There is one other star above the fifth magnitude: the double κ, which is really too wide to be considered a true pair. The two components are of magnitudes 5·4 and 5·7, giving a combined magnitude of 4·7 (spectra B9 and A0).

DOUBLES

Star	R.A. h m	Dec. ° ′	Mags.	P.A.	Dist. ″	
γ	07 08·9	−70 28	3·9, 5·8	299	13·8	Little change
ε	08 07·9	−68 33	4·5, 8·0	022	6·1	Fixed
θ	08 38·9	−70 13	5·3, 9·8	108	45·0	Optical

VARIABLE

Star	R.A. h m	Dec. ° ′	Range	Period, d	Spectrum	Type
R	07 06·6	−72 56	8·8–13·9	448	M	Mira

VULPECULA

(Abbreviation: Vul).

Originally Vulpecula et Anser, the Fox and Goose – nowadays the goose has disappeared (possibly the fox has eaten it). Vulpecula contains no star above magnitude 4·5, and is notable only because of the presence of the Dumbbell Nebula and the fact that several novæ have appeared within the boundaries of the constellation.

See chart for Cygnus

BRIGHTEST STARS

Star		Mags.	R.A. h m s	Dec. ° ′ ″	Abs. Mag.	Spectrum	Dist. l/y
1		4·60	19 15 16	+21 21 02	−0·4	B3	330
6	α	4·63	19 27 47	+24 37 10	0·0	M1	270
13		4·50	19 52 32	+24 01 17	−0·1	A0	270
15		4·74	20 00 12	+27 41 31	+1·7	A5	130
23		4·73	20 14 51	+27 44 46	−0·5	K5	360

THERE ARE FIVE OTHER STARS ABOVE THE FIFTH MAGNITUDE:

Star	Mag.	Abs. Mag.	Spectrum
9	4·88	−0·1	B9
12	4·91	−0·9	B5
Boss 28140	4·82	−1·3	B3
29	4·78	+0·2	B9
31	4·76	+1·0	G2

VARIABLES

Star	R.A. h m	Dec. ° ′	Range	Period, d	Spectrum	Type
Z	18 19·6	+25 29	7·0—8·6	2·5	B + A	Algol
RS	19 15·5	+22 21	6·0—7·6	4·48	B + A	Algol
U	19 34·4	+20 13	7·8—9·0	7·99	F—K	Cepheid
SV	19 49·5	+27 19	7·8—9·6	45	F—K	Cepheid
R	21 02·2	+23 37	7·4—13·4	137	M	Mira

NEBULA

Object	R.A. h m	Dec. ° ′	Type	Notes
NGC 6853 M.27	19 57·4	+22 35	Planetary	Mag. 7·6. Dumb-bell Nebula

TELESCOPES & OBSERVATORIES

Dome of the 98 in Isaac Newton telescope at Herstmonceaux (Patrick Moore)

Historical notes about the development of optical and radio telescopes are given elsewhere in this book. By now the emphasis has started to shift to space telescopes, and no doubt the first lunar optical and radio astronomy observatory will have been set up well before the end of the century.

The largest optical telescope in the world is the 236·2 in (600 cm) reflector at Mount Semirodriki, Caucasus, USSR.

The largest optical telescope in the western hemisphere is the 200 in (508 cm) Hale reflector at Palomar Mountain, California, USA.

The largest optical telescope ever set up in Britain is the 98 in (249 cm) Isaac Newton reflector at the Royal Greenwich Observatory, Herstmonceux, Sussex. At present (1978) plans for moving it to La Palma, in the Canary Islands, are well advanced.

The largest refractor in the world is the 40 in (101·6 cm) telescope at the Yerkes Observatory, USA. (The largest object-glass ever made was 49 in (124·5 cm) in diameter, and was shown at the Paris Exhibition of 1900, but it was never used seriously.)

Herstmonceux Castle, East Sussex, to which the Royal Observatory, established at Greenwich by Charles II in 1675, moved during the 1950s (Patrick Moore)

The present condition of Lord Rosse's 72 in telescope (Patrick Moore)

The largest speculum metal mirror ever made for an optical telescope was 72 in (182·9 cm) in diameter. This was the mirror made by Lord Rosse for the 'Leviathan of Parsonstown', the Birr Castle reflector, in 1845 (he later made a duplicate mirror for it).

The largest infra-red telescope in the world is 150 in (381 cm) in diameter. It was completed at Sheffield (England) in 1977, and is to be set up in Hawaii. Though a thin mirror designed for infra-red, it has proved to be so good that it will also be usable for optical observations.

The largest balloon-borne telescope in the world is the 40 in (101·6 cm) reflector, designed and used by Dr. R. Joseph, Imperial College, University of London.

The largest 'dish' radio telescope in the world is the 1000 ft (300 m) at Arecibo, Puerto Rico, built in a natural hollow in the ground.

The largest fully-steerable 'dish' is the 100 m at Bonn (Germany).

The radio telescope at Jodrell Bank, Cheshire (Patrick Moore)

SOME OF THE WORLD'S GREAT OBSERVATORIES

Location	Name	Latitude °	′	″		Longitude °	′	″	Altitude metres
Aarhus, Denmark	Ole Rømer Obs.	+56	07	40·0	E	10	11·8		50
Alma Ata, USSR	Mountain Obs. of Academy of Sciences	+43	11	16·9	E	76	57·4		1450
Ann Arbor, Michigan	Univ. of Michigan	+42	16	48·7	W	83	43·8		282
Arcetri (Florence), Italy	Astrophysical Obs.	+43	45	44·7	E	11	15·3		184
Armagh, N. Ireland	Armagh Obs.	+54	21	11·1	W	06	38·9		64
Athens, Greece	National Obs.	+37	58	19·7	E	23	43·0		110
Auckland, N. Zealand	Auckland Public Obs.	−36	54	28·0	E	174	46·6		80
Barcelona, Spain	Fabra Obs.	+41	24	59·3	E	02	07·6		415
Beirut Lebanon	American Univ. Obs.	+33	54	22·0	E	35	28·2		38
Belgrade, Jugoslavia	Obs. of Academy of Science	+44	48	13·2	E	20	30·8		253
Berlin, E. Germany	Wilhelm Förster Obs.	+52	28	30·0	E	13	25·5		40
Berlin-Treptow, E. Germany	Archenhold Obs.	+52	29	07·0	E	13	28·6		38
Berne, Switzerland	Astr. Inst. of Univ.	+46	57	12·7	E	07	25·7		563
Bloemfontein, S. Africa	Boyden Obs.	−29	02	18·0	E	26	24·3		1387
Bochum, Germany	Astronomical Station	+51	27	54·8	E	07	13·4		132
Bogota, Colombia	National Obs.	+04	35	55·2	W	74	04·9		2640
Bologna, Italy	University Obs.	+44	29	52·8	E	11	21·1		84
Bombay, India	Government Obs.	+18	53	36·2	E	72	48·9		14
Bonn, W. Germany	University Obs.	+50	43	45·0	E	07	05·8		62
Bordeaux, France	Obs. of Univ. of Bordeaux	+44	50	07	E	00	31·6		73
Boulder, Colorado	Sommers-Bausch Obs.	+40	00	13·0	W	105	15·7		1648
Bucharest, Roumania	National Obs.	+44	24	49·4	E	26	05·8		83
Budapest, Hungary	Konkoly Obs.	+47	29	58·6	E	18	57·9		474
Buenos Aires, Argentina	Naval Obs.	−34	37	18·3	W	58	21·3		6
Byurakan, Armenia	Astronomical Obs. of Academy of Sciences	+40	20	07·0	E	44	17·5		1500
Cambridge, England	University Obs.	+52	12	51·6	E	00	05·7		28
Cambridge, Mass.	Harvard College Obs.	+42	22	47·6	W	71	07·8		24
Canberra, Australia	Mount Stromlo Obs.	−35	19	16·0	E	149	00·3		768
Cape, South Africa	S. African Astr. Obs.	−33	56	02·5	E	18	28·6		10
Caracas, Venezuela	Cagigal Obs.	+10	30	24·3	W	66	55·7		1042
Castel Gandolfo, Italy	Vatican Obs.	+41	44	47·4	E	12	39·1		450
Catania, Sicily	Astrophysical Obs.	+37	31	42·0	E	15	04·7		193
Cincinnati, Ohio	Cincinnati Obs.	+39	08	19·8	W	84	25·3		247
Copenhagen, Denmark	Urania Obs.	+55	41	19·2	E	12	32·3		10
Cordoba, Argentina	National Obs.	−31	25	16·4	W	64	11·8		434
Cracow, Poland	University Obs.	+50	03	52·0	E	19	57·6		221
Crimea, USSR	Crimean Astro. Obs.	+44	43	42·0	E	34	01·0		550

Location	Name	Latitude °	"	Longitude °	"	Altitude metres
Dublin, Eire	Dunsink Obs.	+53 23	13·1	W 06	20·3	86
Edinburgh, Scotland	Royal Obs.	+55 55	30·0	W 03	11·0	146
Flagstaff, Arizona	US Naval Obs	+35 11	00·0	W 111	44·4	2310
Flagstaff, Arizona	Lowell Obs.	+35 12	06·0	W 111	39·8	2210
Fort Davis, Texas	McDonald Obs.	+30 40	17·7	W 104	0·13	2081
Göttingen, Germany	University Obs.	+51 31	48·2	E 09	56·6	161
Greenbelt, Maryland	Goddard Research Obs.	+39 01	11·5	W 76	49·6	49
Groningen, Holland	Kapteyn Astron. Lab.	+53 13	13·8	E 06	33·8	4
Haleakala, Hawaii	Univ. of Hawaii	+20 42	22·0	W 156	15·4	3054
Hamburg, W. Germany	Bergedorf Obs.	+53 28	46·9	E 10	14·4	41
Hartebeespoort, S. Africa	Republic Obs. Annexe	−25 46	22·4	E 27	52·6	1220
Helsinki, Finland	University Obs.	+60 09	42·3	E 24	57·3	33
Helwan, Egypt	Helwan Obs.	+29 51	31·1	E 31	20·5	115
Hurstmonceux, England	Royal Greenwich Obs.	+50 52	18·0	E 00	20·3	34
Jena, E. Germany	Karl Schwarzschild Obs.	+50 58	51·0	E 11	42·8	331
Johannesburg, S. Africa	Republic Obs.	−26 10	55·3	E 28	04·5	1806
Juvisy, France	Flammarion Obs.	+48 41	37·0	E 02	22·3	92
Kodaikanal, India	Astrophysical Obs.	+10 13	50·0	E 77	28·1	2343
Kyoto, Japan	Kwasan Obs.	+34 59	40·8	E 135	47·6	234
Leyden, Holland	University Obs.	+52 09	19·8	E 04	29·0	6
Los Angeles, Calif.	Griffith Obs.	+34 06	46·8	W 118	18·1	357
Lund, Sweden	Royal University Obs.	+55 41	51·6	E 13	11·2	34
Madrid, Spain	Astronomical Obs.	+40 24	30·0	W 03	41·3	655
Mauna Kea, Hawaii	Univ. of Hawaii	+19 49	34·0	W 155	28·3	4205
Meudon, France	Obs. of Physical Astr.	+48 48	18·0	E 02	13·9	162
Milan, Italy	Brera Obs.	+45 27	59·2	E 09	11·5	120
Mill Hill, London, Eng.	Univ. of London Obs.	+51 36	46·3	W 00	14·4	82
Minneapolis, Minnesota	Univ. of Minnesota Obs.	+44 58	40·0	W 93	14·3	260
Montevideo, Uruguay	National Obs.	−34 54	33·0	W 56	12·7	24
Montreal, Quebec. Canada	McGill Univ. Obs.	+45 30	20·0	W 73	34·7	57
Moscow, USSR	Sternberg Inst. Obs.	+55 45	19·8	E 37	34·2	166
Mount Hamilton, Calif.	Lick Obs.	+37 20	25·3	W 121	38·7	1283
Mount Semirodriki, Caucasus, USSR	Zelenchukskaya	+43 49	32·0	E 41	35·4	973
Mount Wilson, Calif.	Hale Obs.	+34 12	59·5	W 118	03·6	1742
Nanking, China	Purple Mountain Obs.	+32 03	59·9	E 118	49·3	367
Naples, Italy	Capodimonte Obs.	+40 51	45·7	E 14	15·4	164
Nice, France	Nice Obs.	+43 43	17·0	E 07	18·0	376
Ondřejov, Czecho-slovakia	Astrophysical Obs.	+49 54	38·1	E 14	47·0	533
Ottawa, Ontario, Canada	Dominion Obs.	+45 23	38·1	W 75	43·0	87
Palermo, Sicily	University Astron. Obs.	+38 06	43·6	E 13	21·5	72
Palomar, California	Hale Obs.	+33 21	22·4	W 116	51·8	1706
Pic du Midi, France	Obs. of University of Toulouse	+42 56	12·0	E 00	08·5	2862
Pittsburgh, Penn.	Allegheny Obs.	+40 20	58·1	W 80	01·3	370
Potsdam, E. Germany	Astrophysical Obs.	+52 22	56·0	E 13	04·0	107
Pulkovo, USSR	Astron. Obs. of Academy of Sciences	+59 46	18·5	E 30	19·6	75
Quito, Ecuador	National Obs.	−00 14	00·0	W 78	29·6	2908
Richmond, Hill, Ontario, Canada	David Dunlap Obs.	+43 51	46·0	W 79	25·3	244
Rio de Janeiro, Brazil	National Obs.	−22 53	42·2	W 43	13·4	33
St. Andrews, Scotland	University Obs.	+56 20	12·0	W 02	48·9	30
Santiago, Chile	Cerro Tololo Obs.	−33 23	50·0	W 70	32·9	860
São Paulo, Brazil	Astronomical and Geophysical Inst.	−23 39	06·9	W 46	37·4	800
Siding Spring, NSW, Australia	Siding Spring Obs.	−31 16	37·3	E 149	04·0	1165
Stockholm, Sweden	Saltsjöbaden Obs.	+59 16	18·0	E 18	18·5	55
Sutherland, CP, South Africa	S. African Astro-nomical Observatories	−32 22	46·5	E 20	48·6	1830
Sydney, Australia	Government Obs.	−33 51	41·1	E 151	12·3	44
Tokyo, Japan	Tokyo Obs. at Mitaka	+35 40	21·4	E 139	32·5	59
Tonantzintla, Mexico	National Astro. Obs.	+19 01	57·9	W 98	18·8	2150
Toruń, Poland	Copernicus Univ. Obs.	+53 05	47·7	E 18	33·3	90
Tucson, Arizona	Catalina Obs.	+32 25	00·7	W 110	43·9	2510
Tucson, Arizona	Kitt Peak Nat. Obs.	+31 57	30·3	W 111	35·7	2064
Uccle, Belgium	Royal Obs.	+50 47	55·0	E 04	21·5	105
Uppsala, Sweden	Univ. Astron. Obs.	+59 51	29·4	E 17	37·5	21
Utrecht, Holland	Sonnenborgh Obs.	+52 05	09·6	E 05	07·8	14
Victoria, BC, Canada	Dominion Astro. Obs.	+48 31	15·7	W 123	25·0	229
Vienna, Austria	University Obs.	+48 13	55·1	E 16	20·3	240
Washington, DC, USA	US Naval Obs.	+38 55	14·0	W 77	03·9	86
Wellington, N. Zealand	Carter Obs.	−41 17	03·9	E 174	45·9	129
Williams Bay, Wisconsin USA	Yerkes Obs.	+42 34	13·4	W 88	33·4	334

RADIO OBSERVATORIES

Arcetri (Florence) Italy	Astrophysical Obs.	+43	45	14·4	E 11	15·3	184
Arecibo, Puerto Rico	Arecibo Obs., Cornell University	+18	20	36·6	W 66	45·2	496
Bochum, Germany	Radio Telescope Stn.	+51	25	43·0	E 07	11·5	160
Boulder, Colorado	High Altitude Obs.	+40	04	42·0	W 105	16·5	1692
Cambridge, England	Mullard Radio Astron. Obs.	+52	09	45·0	E 00	02·4	26
Crimea, USSR	Crimean Astrophys. Obs.	+44	43	42·0	E 34	01·0	550
Delaware, Ohio	Ohio State Obs.	+40	15	04·7	W 83	02·9	282
Eschweiler, W. Germany	Stockert (Bonn Univ)	+50	34	14·0	E 06	43·4	435
Goldstone, California	Jet Propulsion Lab.	+35	23	34·2	W 116	50·9	1038
Green Bank, W. Virginia	Nat. Radio Astron. Obs.	+38	26	17·0	W 79	50·2	823
Harestua, Norway	Univ. of Oslo Obs.	+60	12	30·0	E 10	45·5	585
Jodrell Bank, England	Nuffield Rad. Ast. Lab.	+53	14	11·0	W 02	18·4	70
Kitt Peak, Arizona	Nat. Radio Ast. Obs.	+31	57	11·0	W 111	36·8	1920
Nançay, France	Rad. Obs. of Nançay	+47	22	48·0	E 02	11·8	150
Nederhorst den Berg, Holland	Rad. Astr. Obs.	+52	14	03·0	E 05	04·6	0
Parkes, NSW, Australia	Australian Nat. Rad. Obs.	−33	00	00·4	E 148	15·7	392
Pulkovo, USSR	Astron. Obs. Acad. Sciences	+59	46	05·5	E 30	19·4	70
Richmond Hill, Ontario, Canada	David Dunlap Obs.	+43	51	44·0	W 79	25·2	244
St. Michel, France	Nat. Centre of Scientific Research	+43	55	00·0	E 05	42·5	614
Tokyo, Japan	Tokyo Obs. at Mitaka	+35	40	18·2	E 139	32·4	70
Washington, DC, USA	Radio Astron. Obs. National Lab.	+38	49	16·6	W 77	01·6	30

THE WORLD'S LARGEST TELESCOPES

The first really giant telescope was built by William Herschel in 1789; it had a 49 in (124·5 cm) mirror. This was surpassed in 1845 by Lord Rosse's 72 in (182·9 cm), which was actually the largest in the world until 1917, when the Hooker reflector at Mount Wilson was completed. Of course, these two early giants are now dismantled,

The 60 in Robert McMath solar telescope at Kitt Peak, Arizona, the largest of its kind in the world

but both were used for outstanding work. The 1978 list of giant telescopes is as follows:

REFLECTORS

Aperture in	cm	Observatory	Date of completion
236·2	600·0	Mount Semirodriki, USSR	1976
200	508	Palomar, California (Hale Reflector)	1948
158	401	Cerro Tololo, Chile	1970
158	401	Kitt Peak, Arizona	1970
153	389	Siding Spring, Australia	1974
150	381	Mount Stromlo, Canberra, Australia	1972
150	381	La Cilla, Chile	1975
120	305	Lick, California	1959
107	272	McDonald, Fort Davis, Texas	1968
104	264	Crimea, USSR	1960
102	260	Byurakan (Armenia)	1976
100	254	Mount Wilson, California (Hooker Reflector)	1917
98	249	*Herstmonceux (Royal Greenwich Observatory)	1967
88	224	Mauna Kea, Hawaii	1970

Dome of the 200 in Hale reflector on Palomar Mountain (Patrick Moore)

REFRACTORS

40·0	101·6	Yerkes, Williams Bay, USA	1897
36·0	91·4	Lick, California	1888
32·7	83·1	Meudon, France	1893
32·0	81·3	Potsdam, East Germany	1899
30·0	76·2	Allegheny, Pittsburgh	1914
30·0	76·2	Nice, France	1880

*The 98-inch 'Isaac Newton' reflector is being moved to a new site in La Palma, Canary Islands, and will be given a slightly larger mirror. All the other telescopes listed here are in full use.

X-RAY ASTRONOMY

X-ray astronomy began in 1962, with rocket-borne equipment. By now many X-ray sources are known, some permanent (such as Cygnus X–1, Hercules X–1, and of course the Crab Nebula) and others temporary; 'X-ray novæ' seem to be common. One famous example was detected in December 1974, near the known source Centaurus A (though certainly unconnected with it). It reached its peak on Christmas Day, and was inevitably nicknamed CenXmas. Another was detected in May 1975, near the Crab Nebula (though similarly unconnected with it) and was nicknamed Fresh Crab! The nature of X-ray novæ is still obscure.

The first X-ray source detected (1962) was in Scorpio, and is known as Sco-X1. **The first X-ray source to be optically identified** was the Crab Nebula (also 1962).

The first X-ray artificial satellite was UHURU, launched in 1971, which detected some 160 sources – many of them in our Galaxy, but some extragalactic (such as M.87 and NGC 4151).

The most successful X-ray satellite to date has been the British-built Ariel 5, launched in October 1974.

On 14 June 1978, equipment in the Satellite HEAO 1 detected X-rays from SS Cygni.

GAMMA-RAY ASTRONOMY

The ultra-short, highly-penetrating gamma-rays are difficult to detect. Spark chambers carried in balloons and artificial satellites are used, though care must be taken to distinguish gamma-rays from the much more plentiful cosmic rays.

The first discrete gamma-ray source identified was in 1969 (in Sagittarius).

The most powerful gamma-ray source in the sky is the Vela pulsar. This is also the faintest optical object ever detected! The second most powerful source is the Crab Nebula, which was also the first gamma-ray source to be optically identified.

Gamma-rays come from the Milky Way, and may be due to collisions between cosmic rays and hydrogen gas, though discrete sources have also been identified. Gamma-rays are not abundant. Altogether, about a million gamma-ray photons have been collected – the same number as the photons of visible light received from a bright star, such as Vega, in one second.

Very high energy gamma-rays (wavelengths a million million times shorter than that of light) may be studied from Earth, because of the luminous effects caused by the polarization of atoms in the atmosphere by gamma-rays. **The main instrument used for this work** is the 10-metre reflector at Mount Hopkins, in Arizona.

The first interplanetary gamma-ray detectors were carried on Apollos 15 and 16; they were designed to study the background gamma-ray emission from the Milky Way.

Gamma-ray bursts have also been identified. They come from neither the Sun nor the Earth, but their origin is still obscure. Explanations range from flare stars to neutron stars, White Dwarfs, and even (less plausibly) comets hitting neutron stars! At present it must be admitted that gamma-ray astronomy is still in a very early stage.

THE HISTORY OF ASTRONOMY

To give every date of importance in the history of astronomy would be a mammoth undertaking. What I have therefore tried to do is to make a judicious selection, separating out purely space-research advances and discoveries.

It is impossible to say just when astronomy began, but even the earliest men capable of coherent thought must have paid attention to the various objects to be seen in the sky, so that it may be fair to say that astronomy is as old as *Homo sapiens*. Among the earliest peoples to make systematic studies of the stars were the Mesopotamians, the Egyptians and the Chinese, all of whom drew up constellation patterns. (There have also been suggestions that the constellations we use as a basis today were first worked out in Crete, but this is speculation only.) It seems that some constellation-systems date back to 3000 BC, probably earlier, but of course all dates in these very ancient times are uncertain.

The first essential among ancient civilizations was the compilation of a good calendar. Probably the first reasonably accurate value of the length of the year (365 days) was given by the Egyptians. (The first recorded monarch of all Egypt was Menes, who seems to have reigned around 3100 BC; he was eventually killed by a hippopotamus – possibly the only sovereign ever to have met with such a fate!) They paid great attention to the star Sirius (Sothis), because its 'heliacal rising', or date when it could first be seen in the dawn sky, gave a reliable clue to the time of the annual flooding of the Nile, upon which the Egyptian economy depended. The Pyramids are, of course, astronomically aligned, and arguments about the methods by which they were constructed still rage as fiercely as ever.

Obviously the Egyptians had no idea of the scale of the universe, and they believed the flat Earth to be all-important. So too did the Chinese, who also made observations. It has been maintained that a conjunction of the five naked-eye planets recorded during the reign of the Emperor Chuan Hsü refers to either 2449 or 2446 BC. There is also the legend of the Court Astronomers, Hsi and Ho, who were executed in 2136 BC (or, according to some authorities, 2159 BC) for their failure to predict a total solar eclipse; since the Chinese believed eclipses to be due to attacks on the Sun by a hungry dragon, this was clearly a matter of extreme importance! The story may be nothing more than myth; if true, it would show that the Chinese had learned how to predict eclipses by using the Saros cycle, but we cannot be sure.

The earliest data-collectors were the Assyrians; all students of ancient history know of the Library of Ashurbanipal (668–626 BC). This included the 'Venus Tablet', discovered by Sir Henry Layard and deciphered in 1911 by F. X. Kugler. It claims that when Venus appears, 'rains will be in the heavens'; when it returns after an absence of three months 'hostility will be in the land; the crops will prosper'. Early attempts at drawing up tables of the movements of the Moon and planets may well date from pre-Greek times, largely for astrological reasons; until relatively modern times astrology was regarded as a true science, and all the ancient astronomers (even Ptolemy) were also astrologers.

Babylonian astronomy continued well into Greek times, and some of the astronomers, such as Naburiannu (about 500 BC) and Kidinnu (about 380 BC) may have made great advances; but we know relatively little about them, and reliable dating begins only with the rise of Greek science. Among the great Greek philosophers were:

Thales of Miletus (624–547 BC). To him is attributed the prediction of the solar eclipse of 585 BC which ended the war between the Lydians and the Medes. He gave the length of the year as

365 days. Probably he owed much to the teachings of the Babylonians and the Egyptians, and his ideas were not greatly in advance of theirs; he believed that the Earth floated on water like a cork.

Anaximander, (611–547 BC), an associate of Thales, is credited with the declaration that the Earth is suspended freely in space, remaining fixed because it is at an equal distance from all the other celestial bodies. He believed the Earth to be cylindrical in shape, and thought that the Sun was as large as the Earth.

Anaximenes of Miletus (585–528 BC) held that the Sun, Moon and stars are made of fire, and that the stars are fastened on to a crystal sphere.

Pythagoras of Samos, the great geometer (born about 572 BC) was one of the first to maintain that the Earth is spherical, and he may well have believed that the universe also was spherical in form.

Xenophanes of Colophon (about 570–478 BC) was contemporary with Pythagoras, but his views were much more primitive. The Sun, Moon and stars were clouds set on fire; there were many suns and moons according to the regions of the Earth; the Earth itself was flat, its upper side touching the air and its lower side extending without limit.

Heraclitus of Ephesus (born about 544 BC) took fire to be the principal element, and maintained that the diameter of the Sun was about one foot!

Parmenides of Elea (second half of the 6th century BC) believed the stars to be of compressed fire, but agreed that the Earth was spherical, and in equilibrium because it was equidistant from all points on the sphere representing the universe.

Empedocles of Agrigentum (about 490–450 BC) believed the Sun to be a reflection of fire, but is credited with being the first to maintain that light has a finite velocity.

Anaxagoras of Clazomenæ (born about 500 BC) maintained that the Moon shines by reflected sunlight, and gave correct explanations for solar and lunar eclipses. His opinion that the Sun was a red-hot stone of considerable size led to his being accused of impiety, but he was saved from a death sentence by his friendship with Pericles, the most powerful man in Athens. He believed the Earth to be flat, and the Moon 'earthy', with plains, mountains and valleys.

Eudoxus of Cnidus (about 408–355 BC) was the first to attempt a mathematical basis for astronomy; he developed a theory of concentric spheres to explain the movements of the planets. This theory was further developed by **Calippus** (about 370–300 BC).

Heraclides of Pontus (about 388–315 BC) declared that the apparent daily rotation of the sky is due to the real rotation of the Earth. He also discovered that Mercury and Venus revolve round the Sun, not round the Earth.

Aristotle (384–322 BC) believed in a finite, spherical universe. He further developed the theory of concentric spheres, and gave the first practical proofs that the Earth cannot be flat.

Aristarchus of Samos (about 310–230 BC) maintained that the Earth moves round the Sun – thus anticipating Copernicus by about 19 centuries! He also wrote a treatise about the relative distances and sizes of the Sun and Moon which was a great advance on any previous work.

Eratosthenes of Cyrene (about 276–196 BC), Librarian at Alexandria, was the first to make an accurate measurement of the size of the Earth.

Apollonius of Perga (about 265–190 BC) discussed the motions of the planets in terms of epicycles and 'eccentric circles'. It seems that he believed the planets to move round the Sun, though the Sun itself moved round the Earth. This was therefore the original form of the system worked out by Tycho Brahe in the 16th century AD.

Hipparchus of Nicæa (lived between 161 and 126 BC, though the precise dates of his birth and death are unknown) discovered precession, compiled the first good star catalogue, and made many other notable advances, including the systematic use of trigonometry.

Ptolemy of Alexandria (more correctly, Claudius Ptolemæus) lived between AD 120 and 180. His great work, the *Syntaxis*, has come down to us by way of its Arab translation, the *Almagest*. He compiled a star catalogue, using that of Hipparchus as a basis, and he perfected the geocentric theory of the universe, which is always known as the Ptolemaic. With his death, Greek astronomy came to virtual end.

Little progress was made in the following centuries, though there were some interesting Indian writings (Aryabhāta, 5th century AD), and in AD 570 Isidorus, Bishop of Seville, was the first to draw a definite distinction between astronomy and astrology. The revival of astronomy was due to the Arabs. In 813 Al-Ma'mūn founded the Baghdad school of astronomy, and various star catalogues were drawn up, the most notable being that of Al-Sūfī (about 903). During this period two supernovæ were observed by Chinese astronomers; the star of 1006 (in Lupus) and 1054 (in Taurus, the remnant of which is today seen as the Crab Nebula).

The improved 'Alphonsine Tables' of planetary motions were published in 1270 by order of Alphonso X of Castile. In 1433 Ulūgh Beigh set up an elaborate observatory at Samarkand, but unfortunately he was a firm believer in astrology,

and was told that his eldest son Abdallatif was destined to kill him. He therefore banished his son, who duly returned at the head of an army and had Ulūgh Beigh murdered. This marked the end of the Arab school of astronomy, and subsequent developments were mainly European. Some of the important dates in the history of astronomy are as follows:

1543 Publication of Copernicus' book *De Revolutionibus Orbium Cœlestium*. This sparked off the 'Copernican revolution' which was not really complete until the publication of Newton's *Principia* in 1687.

1572 Tycho Brahe observed the supernova in Cassiopeia.

1576–96 Tycho worked at Hven, drawing up the best star catalogue of pre-telescopic times.

1600 Giordano Bruno burned at the stake in Rome, partly because of his defence of the theory that the Earth revolves round the Sun.

1603 Publication of Johann Bayer's star catalogue, *Uranometria*.

1604 Appearance of the last supernova to be observed in our Galaxy (Kepler's Star, in Ophiuchus).

1608 Invention of the telescope, by Lippershey in Holland. (Telescopes may well have been invented earlier, but we have no definite proof.)

1609 First telescopic lunar map, drawn by Thomas Harriot.

The earliest known drawing by Galileo of the Moon, 1609

1609 Serious telescopic work begun by Galileo, who made a series of spectacular discoveries in 1609–10 (phases of Venus, satellites of

Jupiter, stellar nature of the Milky Way).

1609 Publication of Kepler's first two Laws of Planetary Motion.

1618 Publication of Kepler's third Law of Planetary Motion.

1627 Publication by Kepler of improved planetary tables (the Rudolphine Tables).

1631 First transit of Mercury observed, by Gassendi (following Kepler's prediction of it).

1632 Publication of Galileo's *Dialogue*, which amounted to a defence of the Copernican system. In 1633 he was condemned by the Inquisition in Rome, and forced into a completely hollow recantation.

1632 Founding of the first official observatory (the tower observatory at Leiden in Holland).

1637 Founding of the first national observatory (Copenhagen, Denmark).

1638 Identification of the first variable star (Mira Ceti, by Phocylides Holwarda in Holland).

1639 First transit of Venus observed (by Horrocks and Crabtree, in England, following Horrocks' prediction of it).

1647 Publication of Hevelius' map of the Moon.

1651 Publication of Riccioli's map of the Moon, introducing the modern-type lunar nomenclature.

1655 Discovery of Saturn's main satellite, Titan, by C. Huygens. Huygens announced the correct explanation of Saturn's ring system in the same year.

1656 Founding of the second Copenhagen Observatory.

1659 Markings on Mars seen for the first time (by Huygens).

1663 First description of the principle of the reflecting telescope, by the Scottish mathematician James Gregory.

Isaac Newton's own sketch of the first reflecting telescope, which he built in 1668 (Royal Society of London)

1665 Newton's pioneer experiments on light and gravitation, carried out at Woolsthorpe in Lincolnshire while Cambridge University was temporarily closed because of the Plague.

1666 First observation of the Martian polar caps, by G. D. Cassini.

1667 Founding of the Paris Observatory, with Cassini as Director. (It was virtually in action by 1671.)

1668 First reflector made, by Newton. (This is the probable date. It was presented to the Royal Society in 1671, and still exists.)

1675 Founding of the Royal Greenwich Observatory

1675 Velocity of light measured, by O. Rømer (Denmark).

1676 First serious attempt at cataloguing the southern stars, by Edmond Halley from St Helena.

1685 First astronomical observations made from South Africa (Father Guy Tachard, at the Cape).

1687 Publication of Newton's *Principia*, finally proving the truth of the theory that the Sun is the centre of the Solar System.
1704 Publication of Newton's other major work, *Opticks*.
1705 Prediction of the return of a comet, by Halley (for 1758).
1723 Construction of the first really good reflecting telescope (a 6-inch, by Hadley).
1725 Publication of the final version of the star catalogue by Flamsteed, drawn up at Greenwich. (Publication was posthumous.)
1728 Discovery of the aberration of light, by James Bradley.
1750 First extensive catalogue of the southern stars, by Lacaille at the Cape. (His observations extended from 1750 to 1752.)
1750 Wright's theory of the origin of the Solar System.
1758 First observation of a comet at a predicted return (Halley's Comet, discovered on 25 December by Palitzsch. Perihelion occurred in 1759.)
1758 Principle of the achromatic refractor discovered by Dollond. (It had previously been described, by Chester More Hall in 1729, but More Hall's basic theory was erroneous, and his discovery had been forgotten.)
1761 Discovery of the atmosphere of Venus, during the transit of that year, by M. V. Lomonosov in Russia.
1762 Completion of a new star catalogue by James Bradley; it contained the measured positions of 60 000 stars.
1767 Founding of the *Nautical Almanac*, by Nevil Maskelyne.
1769 Observations of the transit of Venus made from many stations all over the world, including Tahiti (the expedition commanded by James Cook).
1774 First recorded astronomical observation by William Herschel.
1779 Founding of Johann Schröter's private observatory at Lilienthal, near Bremen.
1781 Publication of Charles Messier's catalogue of clusters and nebulæ.
1781 Discovery of the planet Uranus, by William Herschel.
1783 First explanation of the variations of Algol, by Goodricke.

William Herschel's 40 ft reflector, completed in 1789 (Royal Society of London)

(Algol's variability had been discovered by Montanari in 1669.)
1784 First Cepheid variable discovered; δ Cephei itself, by Goodricke.
1786 First reasonably correct description of the shape of the Galaxy given, by William Herschel.
1789 Completion of Herschel's great reflector, with a mirror 49 in (124·5 cm) in diameter and a focal length of 40 feet (12·2 m).
1796 Publication of Laplace's 'Nebular Hypothesis' of the origin of the Solar System.
1799 Great Leonid meteor shower, observed by W. Humboldt.
1800 Infra-red radiation from the Sun detected by W. Herschel.
1801 First asteroid discovered (Ceres, by Piazzi at Palermo).
1802 Second asteroid discovered (Pallas, by Olbers).
1802 Existence of binary star systems established by W. Herschel.
1802 Dark lines in the solar spectrum observed by W. H. Wollaston.
1804 Third asteroid discovered (Juno, by Harding).
1807 Fourth asteroid discovered (Vesta, by Olbers).
1814–18 Founding of the Calton Hill Observatory in Edinburgh.
1815 Fraunhofer's first detailed map of the solar spectrum (324 lines).

1820 Foundation of the Royal Astronomical Society.
1821 Arrival of F. Fallows at the Cape, as Director of the first observatory in South Africa.
1821 Founding of the Paramatta Observatory by Sir Thomas Brisbane, Governor of New South Wales. (This was the first Australian observatory. It was dismantled in 1847.)
1822 First calculated return of a short-period comet (Encke's, recovered by Rumker at Paramatta).
1824 First telescope to be mounted equatorially, with clock drive (the Dorpat refractor, made by Fraunhofer).
1827 First calculation of the orbit of a binary star (ξ Ursæ Majoris, by Savary).
1829 Completion of the Royal Observatory at the Cape.
1834–8 First really exhaustive survey of the southern stars, carried out by John Herschel at Feldhausen (Cape).
1835 Second predicted return of Halley's Comet.
1837 Publication of the famous lunar map by Beer and Mädler. (It remained the best for several decades.)
1837 Publication of the first good catalogue of double stars (W. Struve's *Mensuræ Micrometricæ*).
1838 First announcement of the

distance of a star (61 Cygni, by F. W. Bessel).

1839 Pulkova Observatory completed.

1840 First attempt to photograph the Moon (by J. W. Draper).

1842 Important total solar eclipse, from which it was inferred that the corona and prominences are solar rather than lunar. First attempt to photograph totality (by Majocci). though he recorded only the partial phase.

1843 First daguerreotype of the solar spectrum obtained (by Draper).

1844 Founding of the Harvard College Observatory (first official observatory in the United States). The 15-in. refractor was installed there in 1847.

1845 Completion of Lord Rosse's 72-in reflector at Birr Castle, and the discovery with it of the spiral forms of galaxies ('spiral nebulæ').

1845 Daguerreotype of the Sun taken by Fizeau and Foucault, in France.

1845 Discovery of the fifth asteroid (Astræa, by Hencke).

1851 First photograph of a total solar eclipse (by Berkowski).

1851 Schwabe's discovery of the solar cycle established by W. Humboldt.

1857 Clerk Maxwell proved that Saturn's rings must be composed of discrete particles.

1857 Founding of the Sydney Observatory.

1857 First good photograph of a double star obtained (Mizar, with Alcor, by Bond, Whipple and Black).

1858 First photograph of a comet (Donati's, photographed by Usherwood).

1859 Explanation of the absorption lines in the solar spectrum given by Kirchhoff and Bunsen.

1859 Discovery of the Sun's differential rotation (by Carrington).

1860 Total solar eclipse. Final demonstration that the corona and prominences are solar rather than lunar.

1861–2 Publication of Kirchhoff's map of the solar spectrum.

1862 Construction of the first great refractors, including the New-

1864 First spectroscopic proof that 'nebulæ' are gaseous (by Huggins).

1864 Founding of the Melbourne Observatory. (The 'Great Melbourne Reflector' was completed in 1869.)

1866 Association between comets and meteors established (by G. V. Schiaparelli).

1866 Great Leonid meteor shower.

1866 Announcement by J. Schmidt of an alteration in the lunar crater Linné. (Though the reality of change is now discounted, regular lunar observation dates from this time.)

1867 Studies of 'Wolf-Rayet' stars by Wolf and Rayet, at Paris.

1868 First description of the method of observing the solar prominences at times of non-eclipse (independently by Janssen and Lockyer).

1868 Publication of a detailed map of the solar spectrum, by A. Ångström.

1870 First photograph of a solar prominence (by C. Young.)

1872 First photograph of the spectrum of a star (Vega, by H. Draper).

1874 Transit of Venus; solar parallax re-determined. (Another transit occurred in 1882, but the overall results were disappointing.)

1874 Founding of observatories at Meudon (France) and Adelaide (Australia).

1876 First use of dry gelatine plates in stellar photography; spectrum of Vega photographed by Huggins.

1877 Discovery of the two satellites of Mars (by Hall, at Washington). Observations of the 'canals' of Mars (by Schiaparelli, at Milan).

1878 Publication of the elaborate lunar map by J. Schmidt (from Athens).

1878 Completion of the Potsdam Astrophysical Observatory.

1879 Founding of the Brisbane Observatory.

1880 First good photograph of a gaseous nebula (M.42, by Draper).

1882 Gill's classic photograph of the Great Comet of 1882, showing so many stars that the idea of stellar cataloguing by photography was born.

Schiaparelli showing the 'canals' he claimed to have seen on Mars and the co-ordinate system still used by areographers

1846 Discovery of Neptune, by Galle and D'Arrest at Berlin, from the prediction by Le Verrier. The large satellite of Neptune (Triton) was discovered by W. Lassell in the same year.

1850 First photograph of a star (Vega, from Harvard College Observatory). Castor was also photographed, and the image was extended, though the two components were not shown separately.

1850 Discovery of Saturn's Crêpe Ring (Bond, at Harvard).

all 25-in. made by Cooke. (It was for many years at Cambridge, and is now in Athens.)

1862 Discovery of the Companion of Sirius (by Clark, at Washington).

1862 Completion of the Bonner Durchmusterung.

1863 Secchi's classification of stellar spectra published.

1864 Huggins' first results in his studies of stellar spectra.

1864 First spectroscopic examination of a comet (Tempel's, by Donati).

1885 Founding of the Tokyo Observatory. (An earlier naval observatory in Tokyo had been established in 1874.)

1885 Supernova in M.31, the Andromeda Galaxy (S Andromedæ). This was the only recorded extragalactic supernova to reach the fringe of naked-eye visibility.

1886 Photograph of M.31 (the Andromeda Galaxy) by Roberts, showing spiral structure. (A better photograph was obtained by him in 1888.)

1887 Completion of the Lick 36-in refractor.

1888 Publication of J. L. E. Dreyer's *New General Catalogue* of clusters and nebulæ.

1888 Vogel's first spectrographic measurements of the radial velocities of stars.

1889 Spectrum of M.31 photographed by J. Scheiner, from Potsdam.

1889 Discovery at Harvard of the first spectroscopic binaries (ζ Ursæ Majoris and β Aurigæ).

1889 First photographs of the Milky Way taken (by E. E. Barnard).

1890 Foundation of the British Astronomical Association.

1890 Unsuccessful attempts to detect radio waves from the Sun, by Edison. (Sir Oliver Lodge was equally unsuccessful in 1896.)

1890 Publication of the Draper Catalogue of stellar spectra.

1891 Completion of the Arequipa southern station of Harvard College Observatory.

1891 Spectroheliograph invented by G. E. Hale.

1891 First photographic discovery of an asteroid (by Max Wolf, from Heidelberg).

1892 First photographic discovery of a comet (by E. E. Barnard).

1893 Completion of the 28-in. Greenwich refractor.

1894 Founding of the Lowell Observatory at Flagstaff, in Arizona.

1896 Publication of the first lunar photographic atlas (Lick).

1896 Founding of the Perth observatory.

1896 Completion of the Meudon 33-in (83-cm) refractor.

1896 Completion of the new Royal Observatory at Blackford Hill, Edinburgh.

1896 Discovery of the predicted Companion to Procyon (by Schaeberle).

1897 Completion of the Yerkes Observatory.

1898 Discovery of the first asteroid to come within the orbit of Mars (433 Eros, discovered by Witt at Berlin).

1899 Spectrum of the Andromeda Galaxy (M.31) photographed by Scheiner.

1900 Publication of Burnham's catalogue of 1290 double stars.

1900 Horizontal refractor, of 49-in. aperture, focal length 197 feet (60 m), shown at the Paris Exhibition. (It was never used for astronomical research.)

1905 Founding of the Mount Wilson Observatory (California).

1908 Giant and dwarf stellar divisions described by E. Hertzsprung (Denmark).

1908 Completion of the Mount Wilson 60-in. reflector.

1908 Fall of the Siberian meteorite.

1912 Studies of short-period variables in the Small Magellanic Cloud, by Miss H. Leavitt, leading on to the period-luminosity law of Cepheids.

1913 Founding of the Dominion Astrophysical Observatory, Victoria (British Columbia).

1913 H. N. Russell's theory of stellar evolution announced.

1915 W. S. Adams' studies of Sirius B, leading to the identification of White Dwarf stars.

1917 Completion of the 100-in Hooker reflector at Mount Wilson (the largest until 1948).

1918 Studies by H. Shapley leading him to the first accurate estimate of the size of the Galaxy.

1919 Publication of Barnard's catalogue of dark nebulæ.

1920 The Red Shifts in the spectra of galaxies announced by V. M. Slipher.

1923 Proof given (by E. Hubble) that the galaxies are true independent systems rather than parts of our Milky Way sytsem.

1923 Invention of the spectrohelioscope, by Hale.

1925 Establishment of the Yale observatory at Johannesburg. (It was finally dismantled in 1952, its work done.)

1927 Completion of the Boyden Observatory at Bloemfontein, South Africa.

1930 Discovery of Pluto, by Clyde Tombaugh at Flagstaff.

1930 Invention of the Schmidt camera, by Bernhard Schmidt (Estonia).

1931 First experiments by K. Jansky at Holmdel, New Jersey, with an improvised aerial, leading on to the founding of radio astronomy. Jansky published his first results in 1932, and in 1933 found that the radio emission definitely came from the Milky Way.

1932 Discovery of carbon dioxide in the atmosphere of Venus (by T. Dunham).

1933–5 Completion of the David Dunlap Observatory near Toronto (Canada).

1937 First intentional radio telescope built (by Grote Reber); it was a 'dish' 31 ft (9·4 m) in diameter.

1938 New (and correct) theory of stellar energy proposed by H. Bethe and, independently, by G. von Weizsäcker.

1942 Solar radio emission detected by M. H. Hey and his colleagues (27–28 February). The emission had previously been attributed to intentional jamming by the Germans!

1944 Suggestion, by H. C. van de Hulst, that interstellar hydrogen must emit radio waves at a wavelength of 21·1 cm.

1945 Thermal radiation from the Moon detected at radio wavelengths (by R. H. Dicke).

1945–6 First radar contact with the Moon, by Z. Bay (Hungary) and independently by the US Army Signal Corps Laboratory.

1946 Work at Jodrell Bank begun (radar reflections from the Giacobinid meteor trails, 10 October).

1946 Beginning of radio astronomy in Australia (solar work by a team led by E. G. Bowen).

1946 Identification of the radio source Cygnus A by Hey, Parsons and Phillips.

1947–8 Photoelectric observations of variable stars in the infrared carried out by Lenouvel, using a Lallemand electronic telescope.

1948 Completion of the 200-in Hale reflector at Palomar (California).

1948 Identification of the radio source Cassiopeia A, by M. Ryle and F. G. Smith.

1949 Identification of further radio sources; Taurus A (the Crab

Nebula), Virgo A (M.87), and Centaurus A (NGC 5128). These were the first radio sources beyond the Solar System to be identified with optical objects.

1950 M. 31 (the Andromeda Galaxy) detected at radio wavelengths by M. Ryle, F. G. Smith and B. Elsmore.

1950 Funds for the building of the great Jodrell Bank radio telescope obtained by Sir Bernard Lovell.

1951 Discovery by H. Ewen and E. Purcell of the 21 cm. emission from interstellar hydrogen, thus confirming van de Hulst's prediction.

1951 Optical identification of Cygnus A and Cassiopeia A (by Baade and Minkowski, using the Palomar reflector, from the positions given by Smith).

1952 W. Baade's announcement of an error in the Cepheid luminosity scale, showing that the galaxies are about twice as remote as had been previously thought.

1952 Electronic images of Saturn and θ Orionis obtained by Lallemand and Duchesne (Paris).

1952 Tycho's supernova of 1572 identified at radio wavelengths by Hanbury Brown and Hazard.

1953 I. Shklovskii explains the radio emission from the Crab Nebula as being due to synchroton radiation.

1955 Completion of the 250-ft. radio 'dish' at Jodrell Bank.

1955 First detection of radio emissions from Jupiter (by Burke and Franklin).

1955 Construction of a radio interferometer by M. Ryle, and also the completion of the second Cambridge catalogue of radio sources. (The third Cambridge catalogue was completed in 1959.)

1958 Observation of a red event in the lunar crater Alphonsus, by N. A. Kozyrev (Crimean Astrophysical Observatory, USSR).

1958 Venus detected at radio wavelengths (by Mayer).

1959 Radar contact with the Sun (Eshleman, at the Stanford Research Institute in the United States).

1960 Aperture synthesis method developed by M. Ryle and A. Hewish.

1961 Completion of the Parkes radio telescope, 330 km west of Sydney.

1962 Thermal radio emission detected from Mercury, by Howard, Barrett and Haddock, using the 85-ft. radio telescope at Michigan.

1962 First radar contact with Mercury (Kotelnikov, USSR).

1962 First X-ray source detected (in Scorpio).

1962 Sugar Grove fiasco; the US attempt to build a 600-ft fully steerable radio 'dish'. Work had been begun in 1959, and when discontinued had cost 96 000 000 dollars.

1963 Announcement, by P. van de Kamp, of a planet attending Barnard's Star.

1963 Identification of quasars (M. Schmidt, Palomar).

1965–6 Identification of the 3°K microwave radiation, as a result of theoretical work by Dicke and experiments by Penzias and Wilson.

1967 Completion of the 98-in. Isaac Newton reflector at Herstmonceux.

1967 Identification of the first pulsar, CP 1919, by Miss Jocelyn Bell at Cambridge.

1968 Identification of the Vela pulsar (Large, Vaughan and Mills).

1969 First optical identification of a pulsar; the pulsar in the Crab Nebula by Cocke, Taylor and Disney at the Steward Observatory, USA.

1970 Completion of the 100 m radio 'dish' at Bonn (Germany).

1970 Completion of the large reflectors for Kitt Peak (Arizona) and Cerro Tololo (Chile); each 158 in (401 cm) aperture. First large reflector to be erected on Mauna Kea, Hawaii; an 88-in. (224-cm).

1973 Opening of the Sutherland station of the South African Astronomical Observatories.

1974 Completion of the 153-in. (389 cm) reflector at the Siding Spring Observatory, Australia.

1976 Completion of the 236-in. (600 cm) reflector at Mount Semirodriki (USSR).

1977 Optical identification of the Vela pulsar (at Siding Spring).

1977 Discovery of Chiron (by C. Kowal, USA).

1977 Discovery of the rings of Uranus.

1978 Completion of the new Russian underground neutrino telescope.

1978 Reported discovery of a satellite of Pluto (J. Christy, USA).

1978 Discovery of the first satellite of an asteroid (Herculina).

ASTRONOMY AND SPACE RESEARCH

It is no longer possible to separate what may be called 'pure' astronomy from space research. The Space Age began on 4 October 1957, with the launching of Russia's first artificial satellite, Sputnik 1; less than a decade later Neil Armstrong and Edwin Aldrin stepped out on to the surface of the Moon, and there seems little doubt that elaborate space-stations will be in orbit during the 1980s. To talk of a lunar base before AD 2000 is no longer in the least far-fetched.

The following list has been restricted to the more 'astronomical' events. Full lists of all lunar and planetary probes are not given, because they will be found in the appropriate sections elsewhere in this book.

Pre-1957

about 150 Lucian of Samosata's *True History* about a journey to the Moon – possibly the first of all science-fiction stories.

1232 Military rockets used in a battle between the Chinese and the Mongols.

1865 Publication of Jules Verne's novel *From the Earth to the Moon*.

1881 Early rocket design by N. I. Kibaltchitch. (Unwisely, he made the bomb used to kill the Czar of Russia, and was predictably executed.)

1891 Public lecture about space-flight delivered by the eccentric German inventor Hermann Ganswindt.

1895 First scientific papers about space-flight by K. E. Tsiolkovskii. He published important papers in 1903 and subsequent years, and it is quite reasonable for the Russians to refer to him as 'the father of space-flight'.

1919 Monograph, *A Method of Reaching Extreme Altitudes*, published by R. H. Goddard in America. This included a suggestion of sending a small vehicle to the Moon, and adverse Press comments made Goddard disinclined to expose himself to further ridicule.

1924 Publication of *The Rocket into Interplanetary Space*, by H. Oberth. This was the first truly scientific account of space-research techniques.

1926 First liquid-propelled rocket launched, by Goddard.

1927 Formation of the German rocket group, *Verein für Raumschiffart*.

1931 First European firing of a liquid-propelled rocket (Winkler, in Germany).

1937 First rocket tests at the Baltic research station at Peenemünde. One of the leaders of the team was Wernher von Braun, who died in 1977.

1942 First firing of the A4 rocket (better known as the V2) from Peenemünde. Between 1944 and 1945 many V2s fell upon Southern England.

1945 White Sands proving ground established in New Mexico.

1945 Idea of synchronous artificial satellites for communications purposes proposed by Arthur C. Clarke.

1949 First step-rocket fired from White Sands; it reached an altitude of almost 400 km.

1949 Rocket testing ground established at Cape Canaveral, in Florida.

1955 Announcement of the United States 'Vanguard' project for launching artificial satellites.

The Space Age

1957 4 October; launching of the first artificial satellite, *Sputnik 1* (USSR).

1958 First successful American artificial satellite (*Explorer 1*). Instruments carried in it were responsible for the detection of the Van Allen radiation zones surrounding the Earth.

1959 First lunar probes; *Lunas 1, 2* and *3* (all Russian). *Luna 1* by-passed the Moon, *Luna 2* crash-landed there, and *Luna 3* went on a round trip, sending back pictures of the Moon's far side.

1960 First television weather satellite (*Tiros 1*, USA).

1961 First attempted Venus probe (Russian); contact with it lost.

1961 First manned space-flight (Yuri Gagarin, USSR.)

1961 First manned American space-flight (A. Shepard; sub-orbital).

1962 First American to orbit the Earth (J. Glenn).

1962 First British-built satellite (*Ariel 1*, launched from Cape Canaveral).

1962 First Transatlantic television pictures relayed by satellite (*Telstar*).

1962 First attempted Mars probe (Russian; contact lost).

1962 First successful planetary probe: *Mariner 2* to Venus.

1963 First occasion when two manned space-craft were in orbit simultaneously (Nikolayev and Popovich, USSR).

1963 First (and so far, only) space-woman; Valentina Tereshkova-Nikolayeva (USSR).

1964 First good close-range photographs of the Moon (*Ranger 7*, USA).

1965 First 'space-walk' (A. Leonov, USSR.).

1965 First successful Mars probe (*Mariner 4*, USA).

1966 First soft landing on the Moon by an automatic probe (*Luna 9*, USSR).

1966 First landing of a probe on Venus (*Venera 3*, USSR), though contact with it was lost.

1966 First soft-landing of an American probe on the Moon (*Surveyor 1*).

1966 First circum-lunar probe (*Luna 10*, USSR).

1966 First really good close-range lunar pictures (*Orbiter 1*, USA).

1967 First soft landing of an unmanned probe on Venus (*Venera 4*, USSR).

1968 First recovery of a circum-lunar probe (*Zond 5*, USSR).

1968 First manned Apollo orbital flight (*Apollo 7*; Schirra, Cunningham, Eisele).

1968 First manned flight round the Moon; *Apollo 8* (Borman, Lovell, Anders).

1969 First testing of the lunar module in orbit round the Moon (*Apollo 10*; Stafford, Cernan, Young).

1969 21 July; first lunar landing (*Apollo 11*; N. Armstrong, E. Aldrin).

1970 First Chinese and Japanese artificial satellites.

1971 First capsule landed on Mars, from the USSR probe *Mars-2*.

1971–2 First detailed pictures of Martian volcanoes, obtained from the probe *Mariner 9*, which entered orbit round the planet.

1972 End of the Apollo programme, with *Apollo 17* (Cernan, Schmitt, Evans).

1973–4 Operational 'life' of the United States space-station Skylab, manned by three successive three-man crews, and from which much pioneer astronomical work was carried out.

1973 First close-range information from Jupiter (including pictures) obtained from the fly-by probe *Pioneer 10*. *Pioneer 11* repeated the experiments in 1974. *Pioneer 10* was also the first probe to escape from the Solar System, while *Pioneer 11* was destined to be the first probe to by-pass Saturn (in 1979).

1974 First pictures of the cloud-tops of Venus from close range, from the two-planet probe *Mariner 10*; the probe then encountered Mercury, and sent back the first pictures of the cratered surface.

1975 First pictures received from the surface of Venus, from the Russian probes *Venera 9* and *Venera 10*.

1976 First successful soft landings on Mars (*Vikings 1* and *2*) sending back direct pictures and information from the surface of the planet.

1977 Launching of *Voyagers 1* and *2* to the outer planets.

1978 Launch of two Pioneer probes to Venus—the first American attempts to put a vehicle into orbit round the planet and to land capsules there.

ASTRONOMERS

Selecting a limited number of astronomers for short biographical notes may be somewhat invidious. However, the list given here includes most of the great pioneers and researchers. No astronomers still living at the time of writing (August 1978) are included. All dates are A.D. unless otherwise stated.

Abul Wafa, Mohammed. 959-988. Last of the famous Baghdad school of astronomers. He wrote a book called *Almagest*, really a summary of Ptolemy's great work which also is called the *Almagest* in Arabic.

Adams, John Couch. 1819-1892. English astronomer, born at Lidcot in Cornwall. He graduated brilliantly from Cambridge in 1843, but had already formulated a plan to search for a new planet by studying the perturbations of Uranus. By 1845 his results were ready, but no quick search was made, and the actual discovery was due to calculations by U. Le Verrier. Later he became Director of the Cambridge Observatory, and worked upon lunar acceleration, the orbit of the Leonid meteor shower, and upon various other investigations.

Airy, George Biddell. 1801-1892. English astronomer (Astronomer Royal, 1835-1881). Airy was born in Northumberland, and graduated from Cambridge in 1823; from 1826 to 1835 he was Professor of Astronomy there. On becoming Astronomer Royal he totally reorganized Greenwich Observatory (which had been at a low ebb in its career) and raised it to its present eminence. He was responsible for re-equipping the Observatory and ensuring that the best use was made of its instruments; it is rather ironical that he is probably best remembered for his failure to insti-

gate a prompt search for Neptune when receiving Adams' calculations.

Aitken, Robert Grant. 1864-1951. American astronomer, born at Jackson in California. In 1895 he joined the staff of Lick Observatory, and specialized in double star work; he discovered 31 000 new pairs, and wrote a standard book on the subject. From 1930 until his retirement in 1935 he was Director of the Lick Observatory.

Albategnius. *c.* 850-929. This is the Latinized form of the name of the Arab prince Al Battani. Born at Batan, Mesopotamia; he drew up improved tables of the Sun and Moon, and found a more accurate value for the precession of the equinoxes. His *Movements of the Stars* enabled Hevelius, in the 17th century, to discover the secular variation in the Moon's motion. Albategnius was also a pioneer mathematician; in trigonometry, he introduced the use of sines.

Alhazen, (Abu Ali al Hassan). 987-1038. Arab mathematician, born in Basra. He went to Cairo, where he made his observations and also wrote the first important book on optics since the time of Ptolemy.

Al-Ma'mūn, Abdalla. ?-833. Often referred to as Almanon. He was Caliph of Baghdad, son of Harun al Raschid; he collected and translated many Greek and Persian works, and built a major observatory in 829.

Alfonso X. 1223-1284. King of Castile. At Toledo he assembled many of the leading astronomers of the world, and drew up the famous Alphonsine Tables, which remained the standard for three centuries.

Al-Sūfi. 903-986. A Persian nobleman, who compiled an invaluable catalogue of 1018 stars, giving their approximate positions, magnitudes and colours.

Anaxagoras. 500-428 BC. Born at Clazomenæ, in Ionia. In Athens he became a friend of Pericles, and it was because of this friendship that he was merely banished, rather than being condemned to death, for teaching that the Moon contains plains, valleys and mountains, while the Sun is a blazing stone larger than the Peloponnesus (the peninsula upon which Athens stands).

Anaximander. *c.* 611-547 BC. Greek philosopher, born in Miletus. He believed the Earth to be a cylinder, suspended freely in the centre of a spherical universe. He attempted to draw up a map of the world, and introduced the gnomon into Greece.

Anaximenes. *c.* 585-525 BC. Greek philosopher, also born in Miletus. He believed the Sun to be hot because of its quick motion round the Earth, that the stars were too remote to send us detectable heat, and that the stars were fastened on to a crystal sphere.

Ångström, Anders. 1814-1874. Swedish physicist, who graduated from Uppsala. He mapped the solar spectrum, and was the first to examine the spectra of auroræ. The Ångström unit (one hundred-millionth part of a centimetre) is named in his honour.

Antoniadi, Eugene Michael. 1870-1944. Greek astronomer, who spent most of his life in France and became a naturalized Frenchman. He worked mainly at the Juvisy Observatory (with Camille Flammarion) and at Meudon, near Paris, where he used the 83-cm refractor to make classic observations of the planets. Before the Space Age, his maps of Mars and Mercury were regarded as the standard works. He died in Occupied France during the war.

Apian, Peter Bienewitz. 1495–1552. Born at Leisnig in Saxony; became professor of mathematics at Ingolstädt. He observed five comets, and was the first to note that their trails always point away from the Sun. The 1531 comet is known to be Halley's, and his observations of it enabled Edmond Halley to identify it with the comets of 1607 and 1682.

Apollonius. *c.* 250–200 BC. Born in Perga in Asia Minor, but lived in Alexandria. He was an expert mathematician, and was one of the first to develop the theory of epicycles to represent the movements of the Sun, Moon and planets.

Arago, François Jean Dominique. 1786–1853. Director of the Paris Observatory from 1830. He made many important contributions, including a recognition of the importance of photography in astronomy. He made an exhaustive study of the great total solar eclipse of 1842, and maintained (correctly!) that the Sun is wholly gaseous.

Argelander, Friedrich Wilhelm August. 1799 – 1875. German astronomer, who became Director of the Bonn Observatory in 1836. Here he drew up his atlas of the northern heavens (the *Bonn Durchmusterung*), containing the positions of 324 198 stars down to the ninth magnitude. This standard work was published in 1863.

Aristarchus. *c.* 310–250 BC. Greek astronomer, born in Samos. He was one of the first (quite possibly the very first) to maintain that the Earth moves round the Sun, and he also tried to measure the relative distances of the Sun and Moon by a method which was sound in theory, though inaccurate in practice.

Baade, Wilhelm Heinrich Walter. 1893–1959. German astronomer. In 1920, while assistant at the Hamburg Observatory, he discovered the unique asteroid 944 Hidalgo. In 1931 he went America, and joined the staff of Mount Wilson. In 1952 his work on the two classes of 'Cepheid' short-period variables enabled him to show that the galaxies are approximately twice as remote as had previously been thought.

Bailey, Solon Irving. 1854–1931. American astronomer, born in New Hampshire. He joined the Harvard staff in 1879, and was for many years in charge of the Harvard southern station at Arequipa, Peru. His studies of globular clusters led to the discovery of 'cluster variables', now known as RR Lyræ stars.

Barnard, Edward Emerson. 1857–1923. American astronomer, born at Nashville, Tennesee. He was self-taught, but joined the staff at Lick Observatory in 1888, moving to Yerkes in 1897. He was a renowned comet-hunter; he discovered the fifth satellite of Jupiter (Amalthea) and the swift-moving star in Ophiuchus now called Barnard's Star. He also specialized in studies of dark nebulæ.

Barrow, Isaac. 1630–1677. English mathematician. He made various important contributions to science, but is perhaps best known because in 1669 he resigned his post as Lucasian Professor at Cambridge so that his pupil Isaac Newton could succeed him.

Bayer, Johann. 1572–1625. German astronomer; a lawyer by profession and an amateur in science. He is remembered for his 1603 star catalogue, in which he introduced the system of allotting Greek letters to the stars in each constellation – the system still in use today.

Beer, Wilhelm. 1797–1850. A Berlin banker, who set up a private observatory and collaborated with Mädler in the great map of the Moon published in 1837–8. This map remained the standard for many years. Beer was the brother of Meyerbeer, the famous composer.

Belopolsky, Aristarch. 1854–1934. Russian astronomer, who went from Moscow to the Pulkova Observatory in 1888. He became Director of the Observatory in 1916, but resigned in 1918. He specialized in spectroscopic astronomy and in studies of variable stars.

Bessel, Friedrich Wilhelm, 1784–1846. German astronomer. He went to Lilienthal as assistant to Schröter, but in 1810 became Director of the Königsberg Observatory, retaining the post until his death. He determined the position of 75 000 stars by reducing Bradley's observations; was the first to obtain a parallax value for a star (61 Cygni, in 1838), and predicted the positions of the then-unknown companions of Sirius and Procyon.

FriedrichWilhelm Bessel, painted in 1844 by Johann Wolff (Radio Times Hulton Picture Library)

Biela, Wilhelm von. 1872–1856. Austrian army officer. He was an amateur astronomer, who is remembered for his discovery (in 1826) of the now-defunct periodical comet which bears his name.

Bode, Johann Elert. 1747–1826. German astronomer; born in Hamburg, and appointed to the directorship of Berlin Observatory in 1772. In the same year he drew attention to the 'law' of planetary distances which had been discovered by Titius of Wittenberg; rather unfairly, perhaps, this is known as Bode's Law. He published a star catalogue, did much to popularize astronomy, and fifty years edited the *Berlin Astronomisches Jahrbuch*.

Bond, George Phillips. 1825–1865. Son of W. C. Bond. He was born in Massachusetts, and in 1859 succeeded his father as Director of the Harvard Observatory. He was a pioneer of planetary and cometary photography, and was the first to assert upon truly scientific principles that Saturn's rings could not be solid.

Bond, William Cranch. 1789–1859. American astronomer, born in Maine. He began his career as a watchmaker, but his fame as an amateur astronomer led to his appointment as Director of the newly-founded Harvard Observatory. In 1848 he discovered Saturn's satellite Hyperion, and in 1850 he discovered Saturn's Crêpe Ring. He was also a pioneer of astronomical photography.

Bouvard, Alexis. 1767–1843. A shepherd boy, born in a hut at Chamonix. He went to Paris, taught himself mathematics, and was appointed assistant to Laplace. He made contributions to lunar theory and drew up tables of the motions of the outer planets, as well as discovering several comets.

Bradley, James. 1692–1762. English astronomer (Astronomer Royal, 1742–1762). He was educated in Gloucestershire, and entered the Ministry, becoming Vicar of Bridstow in 1719; in 1721 he went to Oxford as Professor of Astronomy, and remained there until his appointment to Greenwich, mainly on the recommendation of his close friend Halley. He discovered the aberration of light and the nutation of the Earth's axis, but his greatest work was his catalogue of the positions of 60 000 stars.

Brorsen, Theodor. 1819–1895. Danish astronomer, who discovered several comets, and in 1854 made the first scientific observations of the Gegenschein.

Brown, Ernest William. 1866–1938. English astronomer, who graduated from Cambridge and then went to the United States. His chief work was on lunar theory, and his tables of the Moon's motion are still recognized as the standard.

Burnham, Sherburne Wesley. 1838–1921. American astronomer, who began as an amateur and then went successively to Lick (1888) and Yerkes (1897). He specialized in double star work, and discovered over 1300 new pairs. His *General Catalogue of Double Stars* remains a standard reference work.

Campbell, William Wallace. 1862–1938. American astronomer, born in Ohio. He joined the staff at Lick Observatory in 1891, and was Director from 1900 until his retirement in 1930. His main work was in spectroscopy; he discovered 339 spectroscopic binaries (among them Capella), and determined the radial velocities of stars and of 125 nebulæ as well as carrying out spectroscopic observations of the planets.

Cannon, Annie Jump. 1863–1941. Outstanding American woman astronomer, born in Delaware. In 1896 she joined the staff at Harvard College Observatory, where she worked unceasingly on the classification of stellar spectra; the present system is due largely to her. She also discovered five novæ and over 300 variable stars. From 1938 she was William Cranch Bond Astronomer.

Annie Jump Cannon (Radio Times Hulton Picture Library)

Carrington, Richard Christopher. 1826–1875. English amateur astronomer, who had his observatory at Redhill (Surrey). He concentrated upon the Sun, and made many contributions, including the first observation of a solar flare and the independent discovery of Spörer's Law concerning the distribution of sunspots throughout a cycle.

Cassini, Giovanni Domenico. 1625–1712. Italian astronomer; Professor of Astronomy at Bologna from 1650 to 1669, when he went to Paris as the first Director of the observatory there. He discovered four of Saturn's satellites as well as the main division in the rings; he drew up new tables of Jupiter's satellites, made pioneer observations of Mars, and made the first reasonably good measurement of the distance of the Sun.

Cassini, Jacques J. 1677–1756. Son of G. D. Cassini; he was born in Paris and succeeded his father as Director of the Paris Observatory. He confirmed Halley's discovery of the proper motions of certain stars, and played an important part in measuring an arc of meridian from Dunkirk to the Pyrenees in order to determine the figure of the Earth.

Challis, James. 1803–1862. English astronomer; Professor of Astronomy at Cambridge from 1836. He accomplished much useful work, but, unfortunately, is remembered as the man who failed to discover Neptune before the success by Galle and D'Arrest at Berlin.

Charlier, Carl Vilhelm Ludwig. 1862–1934. Swedish cosmologist; Professor of Astronomy at Lund from 1897. He accomplished outstanding work with regard to the distribution of stars in our Galaxy.

Christie, William Henry Mahoney. 1845–1922. English astronomer (Astronomer Royal from 1881 to 1910). He modernized Greenwich Observatory, and fully maintained its great reputation achieved under Airy.

Clairaut, Alexis Claude. 1713–1765. French mathematical genius, who published his first important paper at the age of twelve. He studied the motion of the Moon, and worked out the perihelion passage of Halley's Comet in 1759 to within a month of the actual date.

Clavius, Christopher Klau. 1537–1612. German Jesuit mathematical teacher, who laid down the calendar reform of 1582 at the request of Pope Gregory.

Nicolaus Copernicus by Lohrmann (Royal Society of London)

Copernicus, Nicolaus. 1473–1543. The Latinized name of Mikołaj Kopernik, born in Toruń in Poland. He entered the Church, and became Canon of Frombork. He had a varied career, including medicine and also the defence of his country against the Teutonic Knights, but is remembered for his great book *De Revolutionbus Orbium Cœlestium,* finally published during the last days of his life (he had previously withheld it because he was well aware of Church opposition). It was this book which revived the heliocentric theory according to which the Earth moves round the Sun, and sparked off the 'Copernican revolution' which came to its end with the work of Newton more than a century later.

Curtis, Heber Doust. 1872–1942. American astronomer, who worked at Lick, Allegheny and Michigan Observatories. He was an outstanding spectroscopist, and in 1920 took part in the 'Great Debate' with Shapley about the size of the Galaxy and the status of the resolvable nebulæ; Curtis was wrong about the size of the Galaxy, but correct in maintaining that the spiral nebulæ were independent galaxies. He also played a major rôle in the establishment of the McMath-Hulbert Observatory, renowned for its solar research.

D'Arrest, Heinrich Ludwig. 1822–1875. German astronomer, born in Berlin. While Assistant at the Berlin Observatory, he joined Galle in the successful search for Neptune. From 1857 he worked at Copenhagen Observatory. He specialized in comet and asteroid work, and also published improved positions for about 2000 nebulæ.

Darwin, George Howard. 1845–1912. Son of Charles Darwin. From 1883 he was Professor of Astronomy at Cambridge, and drew up his famous though now rejected tidal theory of the origin of the planets. He was knighted in 1906.

Dawes, William Rutter. 1799–1868. English clergyman, and a keen-eyed amateur observer who specialized in observations of the Sun, planets and double stars. He discovered Saturn's Crêpe Ring independently of Bond.

Delambre, Jean-Baptiste Joseph. 1749–1822. French astronomer, best remembered for his work on the history of the science but also a skilled computer of planetary tables.

De la Rue, Warren. 1815–1889. English astronomer, born in Guernsey. He was a pioneer of astronomical photography; in 1852 he obtained the first good photographs of the Moon, and in 1857 of the Sun. His photographs of the total solar eclipse of 1860 finally proved that the prominences are solar rather than lunar.

Delaunay, Charles. 1816–1872. French astronomer, who specialized in studies of the Moon's motion. He became Director of the Paris Observatory in 1870, but was drowned in a boating accident two years later.

Democritus. *c.* 460–360 BC. Greek philosopher, born at Abdera in Thrace. He adopted Leucippus' atomic theory, and was the first to claim that the Milky Way is made up of stars.

Descartes, René. 1596–1650. French astronomer, author of the theory that matter originates as vortices in an all-pervading ether. He also made great improvements in optics. His books were published in Holland, but he died in Sweden.

De Sitter, Willem. 1872–1934. Dutch astronomer and cosmologist, born in Friesland; he went to the Cape, and from 1908 was Professor of Astronomy at Leiden. He studied the motions of Jupiter's satellites and also the rotation of the Sun, but is best remembered for his pioneer work in relativity theory. The 'De Sitter universe', finite but unbounded, was calculated to be 2000 million light-years in radius and to contain 80 000 million galaxies.

Deslandres, Henri Alexander. 1853–1948. French astronomer (originally an Army officer); from 1907 Director of the Meudon Observatory, and from 1927 Director of the Paris Observatory also. He was a pioneer spectroscopist, and developed the spectroheliograph independently of Hale.

Dollond, John. 1706–1761. English optician, who re-invented the achromatic lens in 1758 and thus improved refractors beyond all recognition.

Donati, Giovanni Battista. 1826–1873. Italian astronomer who discovered the great comet of 1858, and was the first to obtain the spectrum of a comet (Tempel's of 1864). From 1859 he was Director of the observatory at Florence, and in 1872 was largely responsible for the creation of the now-celebrated observatory at Arcetri.

Dreyer, John Louis Emil. 1852–1926. Danish astronomer, born in Copenhagen, who went to Ireland as astronomer to Lord Rosse at Birr Castle and became Director of the Armagh Observatory in 1882. He was a great astronomical historian,

Sir Frank Watson Dyson at the opening of an observatory at Bedford College London (then a women's college) in 1930 (Radio Times Hulton Picture Library)

but is best remembered for his *New General Catalogue* of Clusters and Nebulæ (the N.G.C) still regarded as a standard work. In 1916 he retired from Armagh and went to Oxford, where he lived for the rest of his life.

Dyson, Frank Watson. 1868–1939. English astronomer (Astronomer Royal, 1910–1933). A great administrator as well as an energetic observer of eclipses; he also carried out important work in the field of astrophysics and stellar motions.

Eddington, Arthur Stanley. 1882–1945. English astronomer, born at Kendal. After working at Cambridge and at Greenwich, he was appointed Professor of Astronomy at Cambridge in 1913. He was a pioneer of the theory of the evolution and constitution of the stars, and an outstanding relativist; in 1919 he confirmed Einstein's prediction of the displacement of star positions near the eclipsed Sun. He was knighted in 1930. In addition to his outstanding work, Eddington was one of the best of all writers of popular scientific books, and was a splendid broadcaster.

Einstein, Albert. 1879–1955. German Jew, whose name will be remembered as long as Newton's; in 1905 he laid down the Special Theory of Relativity, and from 1915 to 1917 he developed the General Theory. In 1933 he left Germany, fearing persecution of the Jews, and settled in America.

Elger, Thomas Gwyn. 1838–1897. English amateur astronomer; he was first Director of the Lunar

Albert Einstein, photographed about 1920 (Radio Times Hulton Picture Library)

Section of the British Astronomical Association, and in 1895 published an excellent outline map of the Moon.

Encke, Johann Franz. 1791–1865. German astronomer; from 1825, Director of the Berlin Observatory. He was responsible for compiling the star maps which enabled Galle and D'Arrest to locate Neptune. In 1818 he computed the orbit of a faint comet, and successfully predicted its return; this was Encke's Comet, which has the shortest period of any known comet (3·3 years).

Eratosthenes. *c.* 276–196 BC. Greek philosopher, born in Cyrene; he became Librarian at Alexandria, and made a remarkably accurate measurement of the circumference of the Earth.

Eudoxus. *c.* 408–355 BC. Greek astronomer, born at Cnidus. He went to Athens, and attended lectures by Plato. Finally he settled in Sicily. He developed the theory of concentric spheres – the first truly scientific attempt to explain the movements of the celestial bodies.

Euler, Leonhard, 1707–1783. Swiss mathematician, born in Basle. He was a brilliant mathematician, who pioneered studies of the lunar theory, the movements of planets and comets, and the tides. He lost his sight in 1766, but this did not stop him from working; he undertook the complicated calculations mentally!

Fabricius, David. 1564–1617. Dutch minister and amateur astronomer, who observed Mira Ceti in 1596 (though without recognizing it

as a variable) and made pioneer telescopic observations, notably of the Sun. In 1617 he announced from the pulpit that he knew the identity of a member of his congregation who had stolen one of his geese – and he was presumably correct, since he was assassinated before he could divulge the name of the culprit!

Fabricius, Johann. 1587–1616. Son of David Fabricius, and also a pioneer observer of the Sun by telescopic means; he discovered sunspots independently of Galileo and Scheiner. He died a year before his father.

Fallows, Fearon. 1789–1831. English astronomer, born in Cumberland, who went to South Africa in 1821 as the first Director of the Cape Observatory. He managed to establish the observatory, working under almost incredible difficulties, but the primitive living conditions undermined his health. The reduction of his Cape observations was undertaken by Airy.

Fauth, Philipp Johann Heinrich. 1867–1943. German astronomer, who compiled a large map of the Moon. Unfortunately he believed in the absurd theory that the Moon is ice-covered, and this influenced all his work.

Ferguson, James. 1710–1776. Scottish popularizer of astronomy, who began life as a shepherd-boy but whose books gained great influence. He was also one of the first to suggest an evolutionary origin of the Solar System.

Flammarion, Camille. 1842–1925. French astronomer, renowned both for his observations of Mars

John Flamsteed, first Astronomer Royal, stone bust by G. Elliott at the Royal Observatory, Herstmonceux (photo, Patrick Moore)

and for his popular books. He set up his own observatory at Juvisy, and founded the Société Astronomique de France.

Flamsteed, John. 1646–1720. English astronomer (Astronomer Royal, 1675–1720, though at first the title was 'unofficial'). His main work was the compilation of a new star catalogue, the final version of which was published posthumously. Flamsteed was also Rector of Burstow in Surrey.

Fleming, Wilhelmina. 1857–1911. Scottish woman astronomer, who emigrated to America and worked at Harvard College Observatory, where she was in charge of the famous Draper star-catalogue. She discovered 10 novæ and 222 variable stars.

Fontana, Francisco. 1585–1656. Italian amateur (a lawyer by profession); one of the earliest telescopic observers. He left sketches of Mars and Venus, though the 'markings' which he recorded were certainly illusory.

Fowler, Alfred. 1868–1940. English astronomer, whose spectroscopic work in connection with the Sun, stars and comets was of great importance.

Franklin-Adams, John. 1843–1912. English businessman who took up astronomy as a hobby at the age of 47, and compiled a photographic chart of the stars which is still regarded as a standard work.

Fraunhofer, Joseph von. 1787–1826. Outstanding German optical worker, orphaned in early childhood and rescued from poverty by the Elector of Bavaria. He joined the Physical and Optical Institute of Munich, and was Director from 1823. He invented the diffraction grating, constructed the best lenses in the world, and studied the dark lines in the solar spectrum (the 'Fraunhofer Lines'). He made the Dorpat refractor for Struve (the first telescope to be clock-driven) and also the Königsberg heliometer. His comparatively early death was a tragedy for science.

Galilei, Galileo. 1564–1642. The first great telescopic observer – and also the true founder of experimental mechanics. He worked successively at Pisa, Padua and Florence. The story of his remarkable telescopic discoveries (including the satellites of Jupiter, the phases of Venus and the gibbous aspect of Mars, the starry nature of the Milky Way and many more), and of how his defence of the Copernican theory brought him into conflict with the Church, is one of the most famous in scientific history. He was condemned by the Inquisition in 1633, and was kept a virtual prisoner in his villa at Arcetri; in his last years he also lost his sight.

Galileo Galilei, copy of a portrait by Justus Sustermans (Royal Society of London)

Galle, Johann Gottfried. 1812–1910. German astronomer, best remembered as being the first (with D'Arrest) to locate Neptune in 1846. He discovered three comets, and in 1872, while director of the Breslau Observatory, was the first to use an asteroid for measuring solar parallax.

Gassendi, Pierre. 1592–1655. French mathematician and astronomer. In 1631 he made the first of all observations of a transit of Mercury.

Gauss, Karl Friedrich. 1777–1855. German mathematical genius. In 1801 he calculated the orbit of the first asteroid, Ceres, from a few observations, and enabled Olbers to recover it in the following year. He invented the 'method of least squares', known to every mathematician.

Gill, David. 1843–1914. Scottish astronomer. In 1877 he used observations of Mars to redetermine the solar parallax, and in 1879 went to South Africa as HM Astronomer at the Cape. It was his photograph of the comet of 1882 which showed him the importance of mapping the sky photographically – since his plate showed many stars as well as the comet. He was also deeply involved in cataloguing the southern stars. He was knighted in 1900.

Goldschmidt, Hermann. 1802–1866. German astronomer, who settled in Paris. Using small telescopes poked through his attic window, he discovered 14 asteroids between 1852 and 1861.

Goodacre, Walter. 1856–1938. English amateur astronomer, who published an excellent map of the Moon in 1910.

Goodricke, John. 1764–1786. Born of English parents in Holland. He was a deaf-mute, but with a brilliant brain. It was he who found that Algol is an eclipsing binary rather than true variable, and he also discovered the fluctuations of the intrinsic variable δ Cephei.

Gould, Benjamin Apthorp. 1824–1896. American astronomer, who founded the *Astrophysical Journal*. From Cordoba Observatory, in Argentina, he compiled the *Uranimetria Argentina*, the first major catalogue of the southern stars.

Green, Charles. 1735-1771. English astronomer who went with Captain Cook to study the 1769 transit of Venus. He died on the return voyage.

Gregory, James. 1638–1675. Scottish mathematician. In 1663 he described the principle of the reflecting telescope, but never actually made one.

Grimaldi, Francesco Maria. 1618–1663. Italian Jesuit, who made observations of the Moon which were used in the lunar map compiled by his friend Riccioli. Grimaldi discovered the refraction of light.

Gruithuisen, Franz von Paula. 1774–1852. German astronomer; from 1826 Professor of Astronomy at Munich. He was an assiduous observer of the Moon and planets, but his vivid imagination tended to discredit his work; at one stage he even reported the discovery of artificial structures on the Moon. He also proposed the impact theory of lunar crater formation.

Gum, Colin. 1924–1960. Australian astronomer, who carried out work of vital importance in the surveying of southern radio sources. The famous 'Gum Nebula' in Vela/Puppis is named after him. He was killed in a skiing accident at Zermatt in Switzerland.

Hadley, John. 1682–1743. English astronomer; friend of Bradley. He made the first really good reflecting telescope (6 in. aperture) in 1723, and 1731 constructed his 'reflecting quadrant', which replaced the astrolabe and the cross-staff in navigation.

Hale, George Ellery. 1868–1938. American astronomer. A pioneer solar observer, who invented the spectroheliograph and discovered the magnetic fields of sunspots. In 1897 he became Director of the Yerkes Observatory, and transferred to Mount Wilson in 1905. He master-minded the building of the Mount Wilson 60-in. and 100-in. reflectors as well as the Yerkes refractor, and was mainly responsible for the building of the Palomar 200-in. reflector, which was unfortunately not completed in his lifetime.

Hall, Asaph. 1829–1907. American astronomer, noted for his planetary work. At Washington, in 1877, he discovered the two satellites of Mars. From 1896 he was Professor of Astronomy at Harvard.

Edmond Halley, a portrait by Thomas Murray (Royal Society of London)

Halley, Edmond. 1656–1742. English astronomer (Astronomer Royal, 1720–1742). Though best known for his prediction of the return of the great comet which now bears his name, Halley accomplished much other valuable work; he catalogued the southern stars from St. Helena, studied star-clusters and nebulæ, and discovered the proper motions of some of the bright stars. Even more importantly, he was responsible for the writing of Newton's *Principia*, and personally financed its publication.

Harding, Karl Ludwig. 1765–1834. German astronomer, who was at first assistant to Schröter and then was appointed Professor of Astronomy at Göttingen. In 1804 he discovered the third asteroid, Juno.

Harriot, Thomas. 1560–1621. English scholar, once tutor to Sir Walter Raleigh. There is no doubt that he compiled the first telescopic map of the Moon, and completed it some months before Galileo began his work.

Harrison, John. 1693–1776. English clockmaker, who invented the marine chronometer which revolutionized navigation. Several of his original chronometers are now on display at the National Maritime Museum in London.

Hartmann, Johannes Franz. 1865–1936. German astronomer, Director of the Göttingen Observatory from 1909 to 1921, when he went to Argentina to superintend the national observatory there. His important work was connected with stellar and nebular radial velocities, in the course of which he discovered interstellar absorption lines in the spectrum of δ Orionis.

Hay, William Thompson. 1888–1949. 'Will Hay' deserves mention here as probably the only skilled amateur astronomer who was by profession a stage and screen comedian! In 1933 he discovered the famous white spot on Saturn – the most prominent ever seen on that planet.

Heis, Eduard. 1806–1877. German astronomer; Professor at Munster from 1852. He was a leading authority on the Zodiacal Light, meteors and variable stars, and he published a valuable star catalogue. He was renowned for his keen eyesight, and is said to have counted 19 naked-eye stars in the Pleiades.

Hencke, Karl Ludwig. 1793–1866. German amateur astronomer – postmaster at Driessen. In 1845, after fifteen years' search, he discovered the fifth asteroid, Astræa.

Henderson, Thomas, 1798–1844. Scottish astronomer, who spent a brief period (1832–3) as HM Astronomer at the Cape. While there, he made the measurements which enabled him to measure the parallax of α Centauri. In 1834 he became the first Astronomer Royal for Scotland.

Herschel, Friedrich Wilhelm' (always known as William Herschel). 1738–1822. Probably the greatest observer of all time. He was born in Hanover, but spent most of his life in England. He was the best telescope maker of his day, and in 1781 became famous by his discovery of the planet Uranus. He made innumerable discoveries of double stars, clusters and nebulæ; he found that many doubles are physically-associated or binary systems, and he was the first to give a reasonable idea of the shape of the Galaxy. He was knighted in 1816, and received every honour that the scientific world could bestow. George III appointed him King's Astronomer (not Astronomer Royal).

Herschel, Caroline. 1750–1848. William Herschel's sister and constant assistant in his astronomical work. She discovered eight comets.

Herschel, John Frederick William. 1792–1871. William Herschel's son. He graduated from Cambridge in 1813, and from 1832 to 1838 took a large telescope to the Cape to make the first really systematic observation of the southern heavens. He discovered 3347 double stars and 525 nebulæ, and may be said to have completed his father's pioneering work.

Hertzsprung, Ejnar. 1873–1967. Danish astronomer, who worked successively at Frederiksberg, Copenhagen, Göttingen, Mount Wilson and Leiden (Director of the Leiden Observatory from 1935). In 1905 he discovered the giant and dwarf subdivisions of late-type stars, and this led on to the compilation of H-R or Hertzsprung-Russell Diagrams, which are of fundamental importance in astronomy.

Hevelius. 1611–1687. The Latinized name' of Johannes Hewelcke of Danzig (now Gdańsk). From his private observatory he drew up a catalogue of 1500 stars, and observed planets, the Moon and comets, using the unwieldy long-focus,

Sir William Herschel, from a drawing by J. Russell

Sir John Herschel, photographed by Julia Margaret Cameron in 1867 (Radio Times Hulton Picture Library)

small-aperture refractors of his day. His observatory was burned down in 1679, but he promptly constructed another. His original map of the Moon has been lost; tradition says that the copper engraving was melted down and made into a teapot after his death.

Hind, John Russell. 1823–1895. English astronomer, who discovered 11 asteroids, the 1848 nova in Ophiuchus, and his 'variable nebula' round T Tauri. He also computed many cometary orbits, and from 1853 was superintendent of the *Nautical Almanac*.

Hipparchus. Fl. 140 BC. Great Greek astronomer, who lived in Rhodes. He drew up a star catalogue, later augmented by Ptolemy. Among his many discoveries was that of precession; he also constructed trigonometric tables. Unfortunately all his original works have been lost.

Hooke, Robert. 1653–1703. English scientific genius, contemporary with (though no friend of!) Newton. He built various astronomical instruments, and made some useful observations, including sketches of lunar craters.

Horrocks, Jeremiah. 1619–1641. English astronomer who, with his friend Crabtree, was the first to observe a transit of Venus (1639). He also worked on lunar theory. His early death was a great tragedy for science.

Hubble, Edwin Powell. 1889–1953. American astronomer, who served in the Army during the first world war and was also a boxing champion. In 1923, using the Mount Wilson 100-in. reflector, he discovered short-period variables in the Andromeda Spiral, and proved the Spiral to be an independent galaxy. He also established the velocity/distance relationship known as Hubble's Law.

Huggins, William. 1824–1910. Pioneer English spectroscopist, who had his private observatory at Tulse Hill, near London. He was a pioneer of stellar spectroscopy; he established that the irresolvable nebulæ are gaseous; he was the first to determine stellar radial motions by means of the Doppler shifts in their spectral lines, and he carried out important solar and planetary work. He was knighted in 1897.

Humason, Milton La Salle. 1891–1972. Born in Minnesota. He was mainly self-taught, but joined the staff of Mount Wilson Observatory in 1920, and from then on worked closely with Hubble, studying the forms, spectra, radial motions and nature of the galaxies; he also photographed the spectra of supernovæ in external systems. In 1919 he carried out a photographic search for a trans-Neptunian planet at the request of W. H. Pickering, who had made independent calculations similar to Lowell's. Humason took several plates, but failed to locate the planet. When the plates were re-examined years later, after Pluto had been discovered at Flagstaff, it was found that Humason had recorded the planet twice – but once the image was masked by a

star, and on the other occasion it fell on a flaw in the plate!

Huygens, Christiaan. 1629–1695. Dutch astronomer; probably the best telescopic observer of his time. He discovered Saturn's brightest satellite (Titan) in 1655, and was the first to realize that the curious appearance of the planet was due to a system of rings. He was also the first to see markings on Mars. His activities extended into many fields of science; in particular, he invented the pendulum clock.

Innes, Robert Thorburn Ayton. 1861–1933. Scottish astronomer, who emigrated first to Australia (becoming a wine merchant) and then went to South Africa, as director of the observatory at Johannesburg. He specialized in double star work, discovering more than 1500 new pairs; he also discovered Proxima Centauri, the nearest star beyond the Sun.

Janssen, Pierre Jules César. 1824–1907. French astronomer, who specialized in solar work (in 1870 he escaped from the besieged city of Paris by balloon to study a total eclipse). Independently of Lockyer, he discovered the means of observing the Sun's chromosphere and prominences without waiting for an eclipse. From 1876 he was Director of the Meudon Observatory, and in 1904 published an elaborate solar atlas, containing more than 8000 photographs. The square at the entrance to the Meudon Observatory is still called the Place Janssen, and his statue is to be seen there.

Jansky, Karl Guthe. 1905–1949. American radio engineer, of Czech descent. He joined the Bell Telephone Laboratories, and was using an improvised aerial to investigate problems of static when he detected radio waves which he subsequently showed to come from the Milky Way. This was, in fact, the beginning of radio astronomy; but for various reasons Jansky paid little attention to it after 1937, and virtually abandoned the problem.

Jeans, James Hopwood. 1877–1946. English astronomer. He elaborated the plausible but now rejected theory of the tidal origin of the planets, but his major work was in connection with stellar constitution, in which he made notable advances. He was also an expert writer of popular scientific books,

and was famous as a lecturer and broadcaster.

Jones, Harold Spencer. 1890–1960. English astronomer (Astronomer Royal, 1933–1955). A Cambridge graduate, who was HM Astronomer at the Cape from 1923 until his appointment to Greenwich. From the Cape he carried out much important work, mainly in connection with star catalogues and stellar radial velocities. During his régime as Astronomer Royal he redetermined the solar parallax by means of the world-wide observations of Eros, and published several excellent popular books as well as technical papers. He played a major rôle in the removal of the main equipment from Greenwich to the new site at Herstmonceux, in Sussex, and himself transferred to Herstmonceux in 1948, though it was not until 1958 that the move was completed. He was knighted in 1943.

Kant, Immanuel. 1724–1804. German philosopher, remembered astronomically for proposing a theory of the origin of the Solar System which had some points of resemblance to Laplace's later Nebular Hypothesis.

Kapteyn, Jacobus Cornelius. 1851–1922. Dutch astronomer and cosmologist. His most celebrated discovery was that of 'star-streaming'.

Johannes Kepler

Kepler, Johannes. 1571–1630. German astronomer, born in Württemberg. He was the last assistant to Tycho Brahe, and after Tycho's death used the mass of observations to establish his three famous Laws of Planetary Motion. He observed the 1604 supernova, and also several comets, as well as making improvements to the refracting telescope, but his main achievements were theoretical. He ranks with Copernicus and Galileo as one of the main figures in the story of the'Copernican revolution'.

Kirch, Gottfried. 1639–1710. German astronomer; Director of the Berlin Observatory from 1705. He was one of the earliest of systematic observers of comets, star-clusters and variable stars. In 1686 he discovered the variability of χ Cygni.

Kirchhoff, Gustav Robert. 1824–1887. Professor of physics at Heidelberg. One of the greatest of German physicists, who explained the dark lines in the Sun's spectrum. His great map of the solar spectrum was published from Berlin in 1860.

Kirkwood, Daniel. 1814–1895. American astronomer; an authority on asteroids and meteors. He drew attention to gaps in the asteroid belt, known today as the Kirkwood Gaps; they are due to the gravitational influence of Jupiter.

Kuiper, Gerard P. 1905–1973. American astronomer, who made notable advances in planetary and lunar work and was deeply involved with the programmes of sending probes beyond the Earth. The first crater to be identified on Mercury from Mariner 10 was named in his honour.

Kulik, Leonid. 1883–1942. Russian scientist, who was also trained as a forester and who achieved fame because of his work in meteorite research. In particular, he led several expeditions to study the Tunguska object of 1908. He died in a German prison camp in 1942 after having been captured by the invading forces.

Lacaille, Nicolas Louis de. 1713–1762. French astronomer, who went to the Cape to draw up the first good southern-star catalogue.

Lagrange, Joseph Louis de. 1736–1813. French mathematical genius, and author of the classic *Mécanique Analytique*. He wrote numerous astronomical papers, dealing, among other topics, with the Moon's libration and the stability of the Solar System.

Laplace, Pierre Simon. 1749–1827. French mathematician who made great advances in dynamical astronomy. In 1796 he wrote *Système du Monde*, in which he outlined his Nebular Hypothesis of the origin of the planets. In its original form this has now been discarded, but modern theories have many points of resemblance to it.

Lassell, William. 1799–1880. English astronomer, who would have taken part in the hunt for Neptune but for a sprained ankle. He discovered Triton, Neptune's larger satellite, and (independently of Bond) Hyperion, the seventh satellite of Saturn, as well as two satellites of Uranus (Ariel and Umbriel). He set up a 24-in. reflector in Malta, and with it discovered six hundred nebulæ.

Leavitt, Henrietta Swan. 1868–1921. American woman astronomer, best remembered for her observations of Cepheids in the Small Magellanic Cloud (1912), based on photographs taken in South America; these led on to the discovery of the vital period-luminosity law for Cepheids. Miss Leavitt also discovered four novæ, several asteroids, and over 2400 variable stars.

Lemaître, Georges. 1894–1966. French priest, who was a leading mathematician; from 1927, Professor at Louvain University. His most important paper, leading to what is now called the 'Big Bang' theory of the universe, appeared in 1927, but did not become well-known until publicized by Eddington three years later. During the first world war Lemaître served in the Belgian Army, and won the Croix du Guerre.

Le Monnier, Pierre Charles. 1715–1799. French astronomer who was concerned in star cataloging. He observed the planet Uranus several times, but did not check his observations, and missed the chance of a classic discovery. It was said that he never failed to quarrel with anyone whom he met!

Le Verrier, Urbain Jean Joseph. 1811–1877. French astronomer, whose calculations led in 1846 to the discovery of Neptune. He was an authority on meteors, and in 1867 computed the orbit of the Leonids. He also developed solar and planetary theory, and believed in the existence of a planet (Vulcan) closer to the Sun than Mercury – now known to be a myth. He was forced to resign the Directorship of the Paris Observatory in 1870 because of his irritability, but was reinstated on the death by drowning of his successor, Delaunay.

Lexell, Anders John. 1740–1784. Finnish astronomer, born at Abö. He became Professor of Mathematics at St Petersburg. He discovered the periodical comet of 1770 (now lost), and was one of the first to prove that the object discovered by Herschel in 1781 was a planet rather than a comet.

Lindsay, Eric Mervyn. 1907–1974. Irish astronomer, Director of the Armagh Observatory from 1936 until his death. His main work was in connection with the Magellanic Clouds and with quasars. He had close connections with the Boyden Observatory in South Africa (where he had previously been assistant astronomer) and forged close links between it, Harvard, Dunsink (Dublin) and Armagh. He was an excellent lecturer on popular astronomy, and founded the Armagh Planetarium in 1966.

Lockyer, Joseph Norman. 1836–1920. English astronomer, and an independent discoverer of the method of studying the solar chromosphere and prominences at time of non-eclipse. He was knighted in 1897. He founded the Norman Lockyer Observatory at Sidmouth in Devon, which still exists even though no astronomical work is now carried on there.

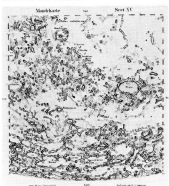

Part of Lohrmann's map of the Moon (photo, John R. Freeman & Co.)

Lohrmann, Wilhelm Gotthelf. 1796–1840. German land surveyor, who began an elaborate lunar map but was unable to complete it owing to ill-health. The map was completed forty years later by Julius Schmidt.

Lomonosov, Mikhail. 1711–1765. Russian astronomer; he was also termed 'the founder of Russian literature'. His father was a fisherman. In 1735 he went to the University of St. Petersburg, and then on to Marburg in Germany to study chemistry. On his return to Russia in 1741 he insulted some of his colleagues at the St Petersburg Academy and was imprisoned for several months, during which time he wrote two of his most famous poems. However, he later became Professor of Chemistry at St Petersburg, and in 1746 became a Secretary of State. He drew up the first accurate map of the Russian Empire, described a 'solar furnace', and investigated electrical phenomena. he also studied auroræ. In 1761 he observed the transit of Venus, and rightly concluded that Venus has a considerable atmosphere. His most important contribution was his championship of the Copernican theory and of Newton's theories, neither of which had really taken root in Russia before Lomonosov's work.

Lowell, Percival. 1855–1916. American astronomer, who founded the Lowell Observatory at Flagstaff, Arizona, in 1894. He paid great attention to Mars, and believed the 'canals' to be artificial waterways. His calculations led to the discovery of the planet Pluto, though the planet was not actually found until 1930 – by Clyde Tombaugh, at the Lowell Observatory. Lowell himself was a great astronomer who did much for science, and it is regrettable that he is today remembered mainly because of his erroneous theories about the Martian canals.

Lyot, Bernard. 1897–1953. Great French astronomer; Director of the Meudon Observatory. He made many advances in instrumental techniques, and invented the coronagraph, which enables the inner corona to be studied at times of non-eclipse. He died suddenly while taking part in an eclipse expedition to Africa.

Maclear, Thomas. 1794–1879. Irish astronomer, who in 1833 succeeded Henderson as HM Ast-

ronomer at the Cape. He made an accurate measurement of an arc of meridian as well as verifying Henderson's parallax of α Centauri; he also studied comets and nebulæ. He was knighted in 1860.

Mädler, Johann Heinrich von. 1794–1874. German astronomer, who was the main observer in the great lunar map by himself and Beer, published in 1837–8 – a map which remained the standard for several decades. In 1840 he left his Berlin home to become Director of the Dorpat Observatory in Estonia. He erroneously believed that η Tauri (Alcyone) was the star lying at the centre of the Galaxy. He retired in 1865 and spent his last years in Hanover.

Maraldi, Giacomo Filippo. 1665–1729. Italian astronomer; nephew of G. D. Cassini. He was renowned for his observations of the planets, particularly Mars, and assisted his uncle at the Paris Observatory.

Maskelyne, Nevil. 1732–1811. English astronomer (Astronomer Royal, 1765–1811). Educated at Cambridge; he then went to St Helena, at the suggestion of Bradley, to observe the transit of Venus, and decided to make a serious study of navigation. During his régime as Astronomer Royal he founded the *Nautical Almanac.*

Méchain, Pierre François Andre. 1744–1805. French astronomer, who discovered eight comets between 1781 and 1799.

Menzel, Donald H. 1901-1976. American astronomer, celebrated for his research into problems of the Sun and planets as well as in stellar studies. He was also an excellent lecturer, and a skilled writer of popular books.

Messier, Charles. 1730–1817. French astronomer, interested mainly in comets. Though he discovered 13 comets, he is remembered chiefly because of his catalogue of star-clusters and nebulæ, published in 1781.

Michell, John. 1725–1793. English clergyman, and an amateur astronomer who made the first suggestion that many double stars may be physically associated or binary systems.

Milne, Edward Arthur. 1896–1950. English astronomer, who graduated from Cambridge and then went successively to Manchester

and Oxford. He made important contributions to astrophysics, and developed his theory of 'kinematic relativity', which was for a time regarded as an alternative to general Einsteinian relativity.

Minkowski, Rudolf. 1895–1976. German astronomer, who went to Mount Wilson in 1935 and remained there. He was one of the leading authorities on novæ and planetary nebulæ, and after the war became a pioneer in the new science of radio astronomy. His studies of rapidly-moving gases in radio galaxies led to the rejection of the 'colliding galaxies' theory.

Montanari, Geminiano. 1633–1687. Italian astronomer, who worked at Bologna and then at Padua. In 1669 he discovered the variability of Algol.

Nevill, Edmund Neison. 1851–1940. English astronomer, who published an important book and map concerning the Moon in 1876; he wrote under the name of Neison. From 1882 to 1910 he was Director of the Natal Observatory at Durban in South Africa, returning to England when the Observatory was closed.

Newcomb, Simon. 1835–1909. American astronomer, for some years head of the American Nautical Almanac office. His chief work was in mathematical astronomy, to which he made valuable contributions. He is also remembered as the man who proved to his own satisfaction that no heavier-than-air machine could ever fly!

Newton, Isaac. 1643–1727. Probably the greatest of all astronomers. To list all his contributions here would be pointless; suffice to say that his *Principia*, published in 1687, has been described as the greatest mental effort ever made by one man. In addition to his scientific work, he sat briefly in Parliament, and served as Master of the Mint. He was knighted in 1705, and on his death was buried in Westminster Abbey.

Olbers, Heinrich Wilhelm Matthias. 1758–1840. German doctor, who was a skilled amateur astronomer and established his private observatory in Bremen. He discovered two of the first four asteroids (Pallas and Vesta) and rediscovered the first (Ceres); he carried out important work in connection with

Sir Isaac Newton, a study by
J. Vanderbank for a portrait now owned
by the Royal Society (Royal Society of
London)

cometary orbits, and discovered a periodical comet which has a period of 69·5 years, and last returned in 1956. Olbers also wrote about his celebrated paradox: 'Why is it dark at night'?

Perrine, Charles Dillon. 1867–1951. American astronomer, who discovered two of Jupiter's satellites as well as nine comets. He worked at the Lick Observatory until 1909, when he became Director of the Cordoba Observatory in Argentina, where he constructed a 30-in. reflector and made many observations of southern galaxies. He also planned a major star catalogue, but was politically unpopular, and after a narrow escape from assassination he retired (1936).

Peters, Christian Heinrich Friedrich. 1813–1890. Danish astronomer, who emigrated to America in 1848. He discovered 48 asteroids.

Piazzi, Giuseppe. 1746–1826. Italian astronomer who became Director of the Palermo Observatory in Sicily. During the compilation of a star catalogue he discovered the first asteroid, Ceres (on 1 January 1801, the first day of the new century).

Pickering, Edward Charles. 1846–1919. American astronomer; for 43 years, from 1876, Director of the Harvard College Observatory. He concentrated upon photometry, variable stars, and above all stellar spectra; in the famous *Draper Catalogue*, the stars were classified according to their spectra. During

his régime the Harvard Observatory was modernized, and a southern outstation was set up at Arequipa in Peru.

Pickering, William Henry. 1858–1938. Brother of E. C. Pickering, who worked with him at Harvard and also served for a while as astronomer-in-charge of the Arequipa outstation. In 1898 he discovered Saturn's ninth satellite, Phœbe. He made extensive studies of the Moon and Mars, mainly from the Harvard station in Jamaica which was set up in 1800. Independently of Lowell, he calculated the position of the planet Pluto.

Plutarch. *c.* 46–120. Greek biographer, mentioned here because of his authorship of *De Facie in Orbe Lunæ* – On the Face in the Orb of the Moon – in which he claims that the Moon is a world of mountains and valleys.

Pond, John. 1767–1836. English astronomer (Astronomer Royal, 1811–1835). Though an excellent and painstaking astronomer, Pond was handicapped by ill-health during the latter part of his régime at Greenwich, and was eventually asked to resign. He tried unsuccessfully to obtain star-distances by the parallax method.

Pons, Jean Louis. 1761–1831. French astronomer, whose first post at an observatory (Marseilles) was that of caretaker! He was self-taught, and concentrated on hunting for comets; he found 36 in all, and ended his career as Director of the Museum Observatory in Florence.

Proctor, Richard Anthony. 1837–1888. English astronomer, who was an excellent cosmologist but is best known for his many popular books. In 1881 he emigrated to America, and remained there for the rest of his life. Proctor paid considerable attention to the planets, and constructed a map of Mars.

Ptolemy (Claudius Ptolemæus). *c.* 120–180. The 'Prince of Astronomers', who lived and worked in Alexandria. Nothing is known about his life, but his great work has come down to us through its Arab translation (the *Almagest*). Ptolemy's star catalogue was based on that of Hipparchus but with many contributions of his own; he also brought the geocentric system to its highest state of perfection, so that it is always known as the Ptolemaic theory. He

constructed a reasonable map of the Mediterranean world, and even showed Britain, though it is true that he joined Scotland on to England in a back-to-front position.

Purbach, Georg von. 1423–1461. Austrian astronomer, who became a professor at Vienna in 1450. He founded a new school of astronomy, compiled tables, and began to write an *Epitome of Astronomy* based on Ptolemy's *Almagest*. After Purbach's premature death, the book was completed by his friend and pupil Regiomontanus.

Pythagoras. *c.* 440–500 BC. The great Greek geometer, mentioned here because he was one of the very first to maintain that the Earth is spherical rather than flat. He seems also to have studied the movements of the planets.

Ramsden, Jesse. 1735–1800. English maker of astronomical instruments. His meridian circles were the first to be lit through the hollow axis.

Rayet, Georges Antoine. 1839–1906. French astronomer; in 1867, with Wolf, drew attention to the Wolf-Rayet stars, which have bright lines in their spectra. He went from Paris to Bordeaux, and became Director of the Observatory there.

Redman, Richard Oliver. 1905–1975. English astronomer, who graduated from Cambridge. He made extensive studies of the Sun, stellar velocities, galactic rotation and the photometry of galaxies. In 1937 he went to the Radcliffe Observatory in Pretoria, and designed the spectrograph for the 74-in. reflector. In 1947 he returned to Cambridge as Professor of Astrophysics and Director of the Observatories. Many programmes were carried through, and Redman also devoted much time and energy in the planning and construction of the 153-in. Anglo-Australian telescope at Siding Spring.

Regiomontanus. 1436–1476. The Latinized name of Johann Müller, Purbach's pupil. He completed the *Epitome of Astronomy*, and at Nürnberg set up a printing press, publishing the first printed astronomical ephemerides. He died in Rome, where he had been invited to help in reforming the calendar.

Rhæticus, Georg Joachim. 1514–1576. German astronomer, who became Professor of Astronomy at Wittenberg in 1536. He was an early convert to the Copernican system; he visited Copernicus at Frombork, and persuaded him to send his great book for publication.

Riccioli, Giovanni Battista. 1598–1671. Italian Jesuit astronomer, who taught at Padua and Bologna. He was a pioneer telescopic observer, and drew up a lunar map, inaugurating the system of nomenclature which is still in use. Oddly enough, he never accepted the Copernican system!

Robinson, Romney. 1792–1882. Irish astronomer, who was Director of the Armagh Observatory from 1823 to his death. He published the Armagh catalogue of over 5000 stars, and made many other contributions; he also invented the cup anemometer. It is on record that when the railway company planned to build a line close to Armagh, Robinson managed to have it diverted, since he maintained that the trains would shake his telescopes!

Rømer, Ole. 1644–1710. Danish astronomer. In 1675 he used the eclipse times of Jupiter's satellites to make an accurate measurement of the velocity of light. In 1681 he became Director of the Copenhagen Observatory. Among his numerous inventions are the transit instrument and the meridian circle.

Rosse, third Earl of. 1800–1867. Irish amateur astronomer, who in 1845 completed the building of a 72-in. reflector and erected it at his home at Birr Castle. The 72-in., with its metal mirror, was much the largest telescope ever built up to that time, and Lord Rosse used it well; his greatest discovery was that many of the galaxies are spiral in form.

Rosse, fourth Earl of. 1840–1908. The fourth Earl continued his father's work, and was also the first man to measure the tiny quantity of heat coming from the Moon. After his death the telescope was not used again, though the tube, between the stone walls which served as an observatory, may still be seen at Birr Castle, and a museum has been erected on the site.

Rowland, Henry Augustus. 1848–1901. American scientist; Professor of Physics in Baltimore from 1876. His great map of the solar spectrum was published in 1895–7; it showed 20 000 absorption lines.

Russell, Henry Norris. 1877–1957. American astronomer; Director of Princeton Observatory from 1908. He devoted much of his energy to studies of stellar constitution and evolution, and independently of Hertzsprung he discovered the giant and dwarf subdivisions of stars of late spectral type. This led on to the H-R or Hertzsprung-Russell Diagram, in which luminosity (or the equivalent) is plotted against spectral type.

Rutherfurd, Lewis Morris. 1816–1892. American barrister, who gave up his profession to devote himself to astronomy. He was a pioneer in astronomical photography, and his pictures of the Moon were outstanding; his ruled solar gratings for solar spectra were the best of their time.

Scheiner, Christoph. 1575–1650. German Jesuit, who was for some time a professor of mathematics in Rome. He discovered sunspots independently of his contemporaries, and wrote a book, *Rosa Ursina,* which contains solar drawings and observations for the years 1611–1625. He was unfriendly toward Galileo, and indeed played a rather discreditable part in the events leading up to Galileo's trial and condemnation.

Giovanni Virginio Schiaparelli (Radio Times Hulton Picture Library)

Schiaparelli, Giovanni Virginio. 1835–1910. Italian astronomer, who

graduated from Turin and became Director of the Brera Observatory in Milan. He discovered the connection between meteors and comets, but his most famous work was in connection with the planets. It was he who first drew attention to the 'canal network' on Mars, in 1877.

Schlesinger, Frank. 1871–1943. American astronomer, born in New York. His main work was in connection with stellar parallaxes. He was Director successively of the Yale and Allegheny Observatories, and was responsible for the Yale 'southern station' in Johannesburg. His major works, *General Catalogue of Parallaxes* and its supplement, dealt with more than 2000 stars. He also pioneered the determination of star positions by using wide-angle cameras.

Schmidt, Julius (actually Johann Friedrich Julius). 1825–1884. German astronomer, who became Director of the Athens Observatory in 1858 and spent most of his life in Greece. He concentrated upon lunar work, producing an elaborate map (based on Lohrmann's early work) and making great improvements in selenography. It was he who drew attention to the alleged change in the lunar crater Linné, in 1866. Schmidt also discovered the outburst of the recurrent nova T Coronæ, in 1866.

Schönfeld, Eduard. 1828–1891. German astronomer, who collaborated with Argelander in preparing the *Bonn Durchmusterung* and later extended it to the southern hemisphere.

Schröter, Johann Hieronymus. 1745–1816. Chief magistrate of Lilienthal, near Bremen. He set up a private observatory, and made outstanding observations of the Moon and planets. Unfortunately many of his notebooks were lost in 1813, with the destruction of his observatory by the invading French troops.

Schwabe, Heinrich. 1789–1875. German apothecary, who became a noted amateur astronomer concentrating on the Sun. His great discovery was that of the 11-year sunspot cycle.

Schwarzschild, Karl. 1873–1916. German astronomer, who worked successively at Vienna, Göttingen and (as Observatory director) Potsdam. His early work dealt with photometry, but he was also a pioneer of theoretical astrophysics. Military service during the first war broke his health and led to his premature death.

Secchi, Angelo. 1818–1878. Italian Jesuit astronomer; one of the great pioneers of stellar spectroscopy, classifying the stars into four types (a system superseded later by that of Harvard). He was also an authority in solar work, and his planetary observations were equally outstanding.

Seyfert, Carl. 1911–1960. American astronomer, who concentrated upon studies of galaxies. In 1942 he drew attention to those galaxies with very condensed nuclei, now always termed Seyfert galaxies.

Shapley, Harlow. 1885–1972. Great American astronomer, who began his main work at Princeton under H. N. Russell. In 1914 he advanced the pulsation theory of Cepheid variables, and was soon able to use the variables in globular clusters to give the first accurate picture of the shape and size of the Galaxy. In 1921 he became Director of the Harvard College Observatory. In later years he concentrated upon studies of galaxies and upon the international aspect of astronomy. He was also an excellent lecturer, and a skilled writer of popular books.

Sosigenes. Greek astronomer, who flourished about 46 BC. He was entrusted by Julius Cæsar with the reform of the calendar. Nothing is known about his life.

Smyth, William Henry. 1788–1865. English naval officer, rising to the rank of Admiral, who in 1830 established a private observatory at Bedford and made numerous observations. He is best remembered for his famous book, *Cycle of Celestial Objects*.

Smyth, Piazzi (actually Charles Piazzi). 1819–1900. Son of Admiral Smyth. He served as Astronomer Royal for Scotland between 1844 until his death. He was a skilled astronomer who carried out much valuable work, including spectroscopic examinations of the Zodiacal Light, but also an eccentric who wrote a large and totally valueless volume about the significance of the Great Pyramid!

South, James. 1785–1867. English amateur astronomer, who founded a private observatory in Southwark and

collaborated with John Herschel in studies of double stars. In 1822 he observed an occultation of a star by Mars, and the virtually instantaneous disappearance convinced him that the Martian atmosphere must be extremely tenuous.

Spörer, Friedrich Wilhelm Gustav. 1822–1895. German astronomer, who joined the staff at Potsdam Observatory. He concentrated upon the Sun, and discovered the variation in latitude of spot zones over the course of a solar cycle (Spörer's Law).

Struve, Friedrich Georg Wilhelm. 1793–1864. German astronomer, born in Altona. He went to Dorpat in Estonia (then, as now, controlled by Russia) and in 1818 became Director of the Observatory. Using the 9-in. Fraunhofer refractor – the first telescope to be clock-driven – he began his classic work on double stars. In 1839 he went to Pulkova, to become director of the new observatory set up by Tsar Nicholas. Here he continued his double-star work, and his *Mensuræ Micrometricæ* gives details of over 3000 pairs. Struve also measured the parallax of Vega; his value was announced in 1840.

Struve, Otto (Wilhelm). 1819–1905. Son of F. G. W. Struve; he was born at Dorpat, and became assistant to his father, accompanying him to Pulkova. He continued his father's work, and also became a leading authority on double stars. He succeeded to the directorship of Pulkova Observatory in 1861, retiring in 1889 and returning to Germany.

Struve, Karl Hermann. 1854–1920. Son of Otto Struve; born at Pulkova, later becoming assistant to his father. His main work was in connection with planetary satellites. In 1895 he went to Königsberg, and in 1904 became Director of the Berlin Observatory, which was reorganized by him and transferred to Babelsberg during his term of office.

Struve, Gustav Wilhelm Ludwig. 1858–1920. Son of Otto Struve, and brother of Karl. He too was born at Pulkova and acted as assistant to his father. He went to Dorpat in 1886, and from 1894 was Director of the Kharkov Observatory. He was concerned mainly with statistical astronomy and with the motion of the Sun.

Struve, Otto. 1897–1963. Son of Gustav; often known as Otto Struve II. He was born in Kharkov, and fought during the first world war; he then joined the White Army under Wrangel and Denikin, and after their defeat reached Constantinople, where he worked as a labourer. Finally his plight became known, and he was offered a post at the Yerkes Observatory, where he arrived in 1921. He spent the rest of his life in America; in 1932 he became Director at Yerkes, after which he founded the McDonald Observatory in Texas and was its director from 1939 to 1947, when he became Chairman of the Department of Astronomy at Chicago. In 1959 he began a new career as the first Director of the National Radio Astronomy Observatory, but ill-health forced his resignation in 1962. He was a brilliant astrophysicist, dealing mainly with spectroscopic binaries, stellar rotation and interstellar matter; he was also an author of popular books. He is (so far!) the last of the famous Struve astronomers. It is notable that all four were in succession awarded the Gold Medal of the Royal Astronomical Society – a sequence unique in astronomical history.

Swift, Lewis. 1820–1913. American astronomer who specialized in hunting for comets and nebulæ. He found 13 comets (including the Great Comet of 1862) and 900 nebulæ.

Tempel, Ernst Wilhelm. 1821–1889. German astronomer, who became Director of the Arcetri Observatory. In 1859 he discovered the nebula in the Pleiades; he also discovered six asteroids and several comets, including the comet of 1865–6 which is associated with the Leonid meteors.

Thales. c. 624–547 BC. The first of the great Greek philosophers. He believed the Earth to be flat, and floating in an ocean, but he was a pioneer mathematician and observer, and successfully predicted the eclipse of 585 BC. which stopped the war between the Lydians and the Medes.

Timocharis. Fl. c. 280 BC. Greek astronomer, who made some accurate measurements of star positions; one of these (of Spica) enabled

Hipparchus, 150 years later, to demonstrate the precession of the equinoxes.

Turner, Herbert Hall. 1861–1930. English astronomer who played an important rôle in the organization and preparation of the International Astrographic Chart. In 1903 he discovered Nova Geminorum.

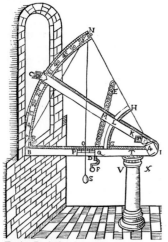

Tycho Brahe's observations are renowned for their accuracy but were made before the invention of the telescope. This drawing shows his quadrant with plumb-line NS and spirit-level PQ

Tycho Brahe. 1546–1601. The great Danish observer – probably the best of pre-telescopic times. He studied the supernova of 1572, and from 1576 to 1596 worked at his observatory at Hven, an island in the Baltic, making amazingly accurate measurements of star positions and the movements of the planets, particularly Mars. His observatory – Uraniborg – became a scientific centre, but Tycho was haughty and tactless (during his student days he had part of his nose sliced off in a duel, and made himself a replacement out of gold, silver and wax!), and after quarrels with the Danish Court he left Hven and went to Prague as Imperial Mathematician to the Holy Roman Emperor, Rudolph II. Here he was joined by Kepler, who acted as his assistant. When Tycho died, Kepler came into possession of the Hven observations, and used them to prove that the Earth moves round the Sun –

something which Tycho himself could never accept.

Van Maanen, Adriaan. 1884–1947. Dutch astronomer, who emigrated to America and joined the Mount Wilson staff in 1912. He specialized in stellar parallaxes and proper motions, and accomplished much valuable work, though his alleged detection of movements in the spiral arms of galaxies later proved to be erroneous. He also discovered the white dwarf still known as Van Maanen's Star.

Vogel, Hermann Carl. 1842–1907. German astronomer, born and educated in Leipzig. He went to Potsdam in 1874, and concentrated upon stellar spectroscopy, pioneering research into spectroscopic binaries. In 1883 he published the first catalogue of stellar spectra.

Walther, Bernard. 1430–1504. (Often spelled 'Walter'.) German amateur astronomer, who lived in Nürnberg; he financed Regiomontanus' equipment, and carried on the work when Regiomontanus died. He was a very accurate observer, whose measurements of star and planetary positions were of great value to later astronomers.

Wargentin, Pehr Vilhelm. 1717–1783. Swedish astronomer, and Director of the Stockholm Observatory. His best work was in the preparation of extremely accurate tables of Jupiter's satellites.

Webb, Thomas William, 1806–1885. Vicar of Hardwicke in Herefordshire. He was an excellent observer, but is best remembered for his book *Celestial Objects for Common Telescopes*, which remains a classic.

Wilkins, Hugh Percy. 1896–1960. Welsh amateur astronomer (by profession a Civil Servant) who concentrated upon lunar observation, and produced a 300-in. map of the Moon. He was for many years Director of the Lunar Section of the British Astronomical Association.

Wolf, Maximilian Franz Joseph Cornelius. 1863–1932. (Better known as Max Wolf.) German astronomer, who was born and lived in Heidelberg. He studied comets, and discovered his periodical comet in 1884; he was the first to hunt for asteroids photographically, and discovered well over a thousand, and he also carried out research into dark nebulæ.

Wright, Thomas. 1711–1785. Born near Durham, and trained as a clockmaker, though he afterwards taught mathematics. He is remembered for his book published in 1750, in which he suggested that the Galaxy is disk-shaped. He also believed Saturn's rings to be composed of small particles.

Xenophanes. *c.* 570–478 BC. Greek philosopher, born in Colophon. His astronomical theories sound strange today; an infinitely thick flat Earth, a new Sun each day, and celestial bodies which – apart from the Moon – were made of fire!

Zach, Franz Xavier von. 1754–1832. Hungarian baron, who became renowned as an amateur astronomer. He published tables of the Sun and Moon, and was one of the chief organizers of the 'Celestial Police' who banded together to hunt for the supposed planet between Mars and Jupiter. He became Director of the Seeberg Observatory at Gotha, and did much for international co-operation among astronomers.

Zöllner, Johann Carl Friedrich. 1834–1882. German astronomer, born in Leipzig, becoming Professor of Astronomy there in 1874. He carried out pioneer spectroscopic work, observing the forms of solar prominences; he invented the polarising photometer, and was the first to suggest that the spectral types of stars prepresent an evolutionary sequence.

Zwicky, Fritz. 1898–1974. Swiss astronomer; he was born in Bulgaria, but remained a Swiss citizen throughout his life. He graduated from Zürich, and in 1925 went to the California Institute of Technology, where he remained permanently, becoming Professor of Astrophysics from 1942 until his retirement in 1968. He became famous for his studies of galaxies and intergalactic matter; he predicted the existence of neutron stars (1934) and even black holes. He discovered many supernovæ in external galaxies, and master-minded a catalogue of compact galaxies. He was also active in the development of astronomical instrumentation, and was a pioneer worker with Schmidt telescopes. He received the Gold Medal of the Royal Astronomical Society in 1973.